Stieltjes Differential and Integral Inequalities

SERIES ON CONCRETE AND APPLICABLE MATHEMATICS
ISSN: 1793-1142

Series Editor: Professor George A. Anastassiou
Department of Mathematical Sciences
University of Memphis
Memphis, TN 38152, USA

---

*Published**

Vol. 29 *Stieltjes Differential and Integral Inequalities*
by Svetlin G. Georgiev & Sanket Tikare

Vol. 28 *Neural Networks as Positive Linear Operators*
by George A. Anastassiou

Vol. 27 *New Operators and Their Applications in Geometric Function Theory*
by Alina Alb Lupaş

Vol. 26 *Rainbow and Properly Colored Subgraphs*
by Jie Hu & Xueliang Li

Vol. 25 *Advanced Topics on Semilinear Evolution Equations*
by Mouffak Benchohra, Gaston M. N'Guérékata & Abdelkrim Salim

Vol. 24 *Trigonometric and Hyperbolic Generated Approximation Theory*
by George A. Anastassiou

Vol. 23 *Introduction to Matrix Theory: with Applications in Economics and Engineering*
Second Edition
by Ferenc Szidarovszky, Sandor Molnar & Mark Molnar

Vol. 22 *Bloch-Type Periodic Functions: Theory and Applications to Evolution Equations*
by Yong-Kui Chang, Gaston M. N'Guérékata & Rodrigo Ponce

Vol. 21 *Functional Equations and Inequalities: Solutions and Stability Results*
by John Michael Rassias, E. Thandapani, K. Ravi & B. V. Senthil Kumar

*To view the complete list of the published volumes in the series, please visit:
https://www.worldscientific.com/series/scam

Series on Concrete and Applicable Mathematics – Vol. 29

Stieltjes Differential and Integral Inequalities

Svetlin G. Georgiev
Sorbonne University, France

Sanket Tikare
Ramniranjan Jhunjhunwala College, India

World Scientific

NEW JERSEY • LONDON • SINGAPORE • BEIJING • SHANGHAI • HONG KONG • TAIPEI • CHENNAI • TOKYO

Published by

World Scientific Publishing Co. Pte. Ltd.
5 Toh Tuck Link, Singapore 596224
USA office: 27 Warren Street, Suite 401-402, Hackensack, NJ 07601
UK office: 57 Shelton Street, Covent Garden, London WC2H 9HE

Library of Congress Control Number: 2026015739

British Library Cataloguing-in-Publication Data
A catalogue record for this book is available from the British Library.

Series on Concrete and Applicable Mathematics — Vol. 29
STIELTJES DIFFERENTIAL AND INTEGRAL INEQUALITIES

ISBN 978-981-98-2329-1 (hardcover)
ISBN 978-981-98-2330-7 (ebook for institutions)
ISBN 978-981-98-2331-4 (ebook for individuals)

For any available supplementary material, please visit
https://www.worldscientific.com/worldscibooks/10.1142/14585#t=suppl

Desk Editors: Nambirajan Karuppiah/Lai Fun Kwong

Typeset by Stallion Press
Email: enquiries@stallionpress.com

Preface

The subject of this monograph is the fusion of two theories, mathematical inequalities and Stieltjes calculus, which are presented comprehensively and systematically. Inequalities have proven to be essential tools for the advancement of various branches of mathematics. Numerous types of inequalities involving derivatives, integrals, or both are of considerable significance. Due to their utility and importance, inequalities have attracted substantial scholarly interest. The scope of nonlinear problems continues to grow, necessitating the development of innovative approaches to analyze them. On the other hand, the Stieltjes calculus is the branch of mathematics in which we study the notion of classical calculus with respect to a function, which offers a new unification and extension of the continuous and discrete calculus. More specifically, the Stieltjes derivative is a modification of the usual derivative through a nondecreasing left-continuous function. This change in the definition allows us to study several differential problems under a unique framework. In an effort to expand the applicability of Stieltjes calculus, this book aims to explore the concept of inequality across a broad spectrum of topics. The primary aim is to extend a variety of fundamental inequalities based on the Stieltjes calculus approach. This book offers a broad, comprehensive, and systematic approach to the foundations and developments of various inequalities in the context of Stieltjes. It begins with the basics of Stieltjes calculus and then explores various types of inequalities. The book is written in a self-contained manner, comprises six pedagogically organized chapters, each divided into multiple sections covering a wide range of Stieltjes inequalities. Each chapter contains a good percentage of new material and concludes with a section with practical problems. The content is presented in a straightforward manner and is aimed at senior undergraduate and postgraduate students in engineering and science programs.

In Chapter 1, we present the essential preliminaries and basic concepts of Stieltjes calculus, which are necessary for understanding the rest of the book. Chapter 2 focuses on first-order Stieltjes differential inequalities. In this chapter, we use these inequalities to discuss the existence and uniqueness of solutions, as well as maximal and minimal solutions for first-order Stieltjes differential equations. Chapter 3 establishes some basic inequalities for Stieltjes calculus, including the Stieltjes–Young inequalities, Stieltjes–Jensen inequalities, Stieltjes–Hölder inequalities, Stieltjes–Minkowski inequalities, Stieltjes–Steffensen inequalities, and Stieltjes–Hermite–Hadamard inequalities. Chapter 4 covers Wirtinger-type inequalities for Stieltjes calculus and their extensions. Chapter 5 discusses Opial-type inequalities for Stieltjes calculus and their applications for establishing the existence of solutions. Finally, in Chapter 6, we present various Lyapunov-type inequalities for Stieltjes calculus related to Stieltjes differential equations.

The book is primarily designed for researchers seeking to understand a new mathematical theory, Stieltjes inequalities, and apply them to the study of the qualitative properties of differential equations. Here, the differential equations can include ordinary differential equations, difference equations, quantum equations, and generalized ordinary differential equations. Graduate students at various levels will find this text a valuable resource for both introductory and advanced material.

The authors wish to thank their families and friends for their constant support and encouragement. They are also grateful to the Executive Editor, *Rochelle Kronzek Miller*, and the Desk Editor, *Lai Fun Kwong*, for carefully managing the publication process of this book. Excellent typesetting and formatting of the manuscript by Nambirajan Karuppaiah is highly appreciated by the authors. Sanket is grateful to his wife, *Shaila*, and daughter, *Manasvini*, for their patience and tolerance during the writing of this book. As always, Sanket has dedicated this book to his Guru, Late Shri *Manohar Hari Risbud* (1927–2009).

Svetlin G. Georgiev
Paris, France
Sanket Tikare
Mumbai, India

About the Authors

Svetlin G. Georgiev is a mathematician who works on various aspects of mathematics. Currently, he focuses on ordinary and partial differential equations, differential geometry, dynamic geometry on time scales, integral equations on time scales, theory of distributions and harmonic analysis.

Sanket Tikare is an Assistant Professor in the Department of Mathematics, at Ramniranjan Jhunjhunwala College, which is an empowered autonomous college affiliated with the University of Mumbai, India. He specializes in applied analysis. His research areas of interest include differential equations, difference equations, dynamic equations on time scales, quantum calculus and equations. He is the co-author of four monographs and has also edited three books.

Contents

Chapter 1

Elements of Stieltjes Differential Calculus*

This chapter is devoted to a brief exposition of the Stieltjes differential calculus that provides a framework for the study of Stieltjes inequalities. A detailed discussion of the Stieltjes differential calculus is beyond the scope of this book; for this reason, the authors confine to outlining a minimal set of properties needed for proceeding further. Throughout this chapter, we assume f to be a real-valued function defined on $[a, b]$ and suppose that $[a, b] \subset \mathbb{R}$.

1.1 Functions of Bounded Variation

Definition 1.1. A function f defined on $[a, b]$ is said to be increasing (decreasing) provided $f(t) \leq f(s)$ ($f(t) \geq f(s)$) for $t < s$, $s, t \in [a, b]$. If $f(t) < f(s)$ ($f(t) > f(s)$) for $t < s$, $s, t \in [a, b]$, then f is said to be strictly increasing (strictly decreasing). Increasing functions and decreasing functions are called monotonic or monotone (strictly monotonic). If f decreases, then $-f$ increases. Thus, we can consider only increasing functions in many problems involving monotonic functions. Monotonic functions will always be considered finite.

Let f be an increasing function on $[a, b]$ and let $t_0 \in [a, b]$. Take $\{t_n\}_{n \in \mathbb{N}}$, so that $t_n > t_0$, $n \in \mathbb{N}$, and $\lim_{n\to\infty} t_n = t_0$. Then, $t_n > t_m$ for $n < m$, $n, m \in \mathbb{N}$, and

$$f(t_1) \geq f(t_n) \geq f(t_m) \geq f(t_0), \quad n < m, \quad n, m \in \mathbb{N}.$$

*This chapter is reproduced from Georgiev, S.G., Tikare, S. (2024). Stieltjes differential calculus with applications, Vol. 12, *Trends in Abstract and Applied Analysis*, World Scientific Publishing, Singapore.

Thus, $\{f(t_n)\}_{n\in\mathbb{N}}$ is a decreasing bounded sequence. Then, $\lim_{n\to\infty} f(t_n)$ exists and it is finite. This limit is nothing but $\inf_{t\in(t_0,b]} f(t)$, and it does not depend on the choice of the sequence $\{t_n\}_{n\in\mathbb{N}}$. It is denoted by $f(t_0+)$. The symbol $f(t_0-)$ is defined similarly. We have $f(t_0-) \le f(t_0) \le f(t_0+)$ and $f(a) \le f(a+)$, $f(b-) \le f(b)$. From here, we conclude that f is continuous at t_0 if and only if

$$f(t_0-) = f(t_0) = f(t_0+).$$

Definition 1.2. The numbers $f(t_0) - f(t_0-)$ and $f(t_0+) - f(t_0)$ are said to be the left and right saltus, respectively, of f at t_0. The number $f(t_0+) - f(t_0-)$ is said to be the saltus of f at t_0.

Theorem 1.1. *Let f be an increasing function on $[a,b]$ and $t_j \in (a,b)$, $j \in \{1,\dots,n\}$. Then,*

$$(f(a+) - f(a)) + \sum_{j=1}^{n}(f(t_j+) - f(t_j-)) + (f(b) - f(b-)) \le f(b) - f(a). \tag{1.1}$$

Corollary 1.1. *An increasing function f defined on $[a,b]$ can have only a finite number of points of discontinuity at which its saltus is greater than a given positive number σ.*

Theorem 1.2. *The set of points of discontinuity of an increasing function f defined on $[a,b]$ is at most denumerable. Moreover, if $x_1, x_2, \dots$ are all of the interior points of discontinuity, then*

$$(f(a+) - f(a)) + \sum_{j=1}^{\infty}(f(x_j+) - f(x_j-)) + (f(b) - f(b-)) \le f(b) - f(a). \tag{1.2}$$

Definition 1.3. Let f be an increasing function on $[a,b]$. Define the saltus function $s\colon [a,b] \to \mathbb{R}$ as

$$s(t) = \begin{cases} 0, & t = a, \\ (f(a+) - f(a)) + \sum_{t_j<t}(f(t_j+) - f(t_j-)) & \\ \quad + (f(t) - f(t-)), & t \in (a,b]. \end{cases}$$

Theorem 1.3. *Let f be an increasing function defined on $[a,b]$. Then, the function $\phi = f - s$ is an increasing and continuous function on $[a,b]$.*

Let A be a given abstract set and let f be a given function defined on A. Then, we denote

$$f(E) = \{f(t) \colon t \in E\}.$$

Thus, function f induces a mapping of the family of all subsets E of A into the family of all subsets of the image set $f(A)$. In other words, $f(E)$ consists of those elements s for which the equation $f(t) = s$ has at least one solution in the set E.

Remark 1.1. If $E_1 \subset E_2 \subset A$, then $f(E_1) \subset f(E_2)$.

Remark 1.2. Let $E_j \subset A$, $j \in \mathbb{N}$. Then, $f\left(\cup_{j=1}^{\infty} E_j\right) = \cup_{j=1}^{\infty} f(E_j)$.

Example 1.1. Let $A = [a, b]$ and $f \colon [a, b] \to \mathbb{R}$ be a continuous and increasing function. Then, $f(A) = [f(a), f(b)]$.

Remark 1.3. Let $E_j \subset A$, $j \in \mathbb{N}$, and f be defined on A. Then,

$$f\left(\cap_{j=1}^{\infty} E_j\right) = \cap_{j=1}^{\infty} f(E_j).$$

Remark 1.4. Let $E_1, E_2 \subset A$ with $E_1 \cap E_2 = \emptyset$ and f be defined on A. Then, $f(E_1) \cap f(E_2) = \emptyset$.

Definition 1.4. The number λ is said to be a derived number of f at x_0 provided there exists a sequence $\{h_n\}_{n \in \mathbb{N}}$, $h_n \neq 0$, $n \in \mathbb{N}$, such that $h_n \to 0$ as $n \to \infty$ and

$$\lambda = \lim_{n \to \infty} \frac{f(t_0 + h_n) - f(t_0)}{h_n}.$$

Obviously, the derived number is denoted by $Df(t_0)$.

Example 1.2. Let $f'(t_0)$ exists. Then, $Df(t_0) = f'(t_0)$.

Example 1.3. Consider the Dirichlet function

$$f(t) = \begin{cases} 1 & \text{for } t \in \mathbb{Q}, \\ 0 & \text{for } t \in \mathbb{R} \backslash \mathbb{Q}. \end{cases}$$

Let $h, t_0 \in \mathbb{R}$ be arbitrarily chosen. We have the following cases:

(1) Let $t_0 \in \mathbb{Q}$. Then,

$$\frac{f(t_0 + h) - f(t_0)}{h} = \begin{cases} 0 & \text{for } h \in \mathbb{Q}, \\ -\frac{1}{h} & \text{for } h \in \mathbb{R} \backslash \mathbb{Q}. \end{cases}$$

(2) Let $t_0 \in \mathbb{R}\backslash\mathbb{Q}$. Then,

$$\frac{f(t_0+h)-f(t_0)}{h} = \begin{cases} 0 & \text{for } t_0+h, \quad h \in \mathbb{R}\backslash\mathbb{Q}, \\ \frac{1}{h} & \text{for } t_0+h \in \mathbb{Q}. \end{cases}$$

Thus, the Dirichlet function has three derived numbers: $0, \infty$ and $-\infty$.

Theorem 1.4. *Let function f be defined on $[a,b]$. Then, its derived numbers exist at every point in $[a,b]$.*

Theorem 1.5. *A function f defined on $[a,b]$ has derivative $f'(t_0)$ at $t_0 \in [a,b]$ if and only if all derived numbers of f at t_0 are equal.*

Theorem 1.6. *Let f be a monotonically increasing function defined on $[a,b]$. If, at every point t of $E \subset [a,b]$, there exists at least one derived number $Df(t)$ such that $Df(t) \leq p$ for some $p > 0$, then $m^*(f(E)) \leq pm^*(E)$, where $m^*(A)$ denotes the outer measure of A.*

Theorem 1.7. *Let f be a strictly increasing function defined on $[a,b]$. If, at every point t of $E \subset [a,b]$, there exists at least one derived number $Df(t)$ such that $Df(t) \geq q$ for some $q \geq 0$, then $m^*(f(E)) \geq qm^*(E)$.*

Corollary 1.2. *The set of points of which at least one derived number of an increasing function defined on $[a,b]$ is infinite is of measure zero.*

Theorem 1.8. *Let f be an increasing function defined on $[a,b]$ and let p,q be two numbers such that $p < q$ and $E_{p,q} = \{t \in [a,b] : p < t < q\}$. If, at every point t of the set $E_{p,q} \subset [a,b]$, there exist two derived numbers $D_1 f(t)$ and $D_2 f(t)$ such that $D_1 f(t) < p < q < D_2 f(t)$, then $m(E_{p,q}) = 0$.*

Theorem 1.9. *An increasing function f defined on the interval $[a,b]$ has a finite derivative at almost all points of $[a,b]$.*

Theorem 1.10. *Let f be an increasing function defined on $[a,b]$. Then, its derivative f' is measurable and*

$$\int_a^b f'(t)dt \leq f(b) - f(a),$$

i.e., f' is summable.

Theorem 1.11. *For every subset E of measure zero on the closed interval $[a,b]$, there exists a continuous increasing function σ such that $\sigma'(t) = \infty$, $t \in E$.*

Suppose that f is a function defined on the bounded closed interval $[a, b]$ and P is a partition of $[a, b]$ which subdivides $[a, b]$ into parts by means of the points

$$P \colon a = t_0 < t_1 < \cdots < t_n = b, \quad n \in \mathbb{N}. \tag{1.3}$$

Consider the sum

$$\bigvee(f, P) = \sum_{j=0}^{n-1} |f(t_{j+1}) - f(t_j)|.$$

Definition 1.5. The least upper bound of the set of all possible sums $\bigvee(f, P)$ is called the total variation of f on $[a, b]$. It is denoted by $\bigvee_a^b(f)$. The function f is said to be a function of bounded variation on $[a, b]$ provided $\bigvee_a^b(f) < \infty$. We also state that f has finite variation on $[a, b]$.

Note 1.1. A function of bounded variation is also known as a function of finite variation.

Remark 1.5. A function of bounded variation need not be continuous. For example, $f(t) = \lfloor t \rfloor$, where $\lfloor t \rfloor$ denotes the greatest integer not greater than x, a function of bounded variation on $[0, 2]$ but not continuous. Also, a continuous function may not be of bounded variation, see example in Remark 1.6.

Theorem 1.12. *A monotonic function f on $[a, b]$ is of bounded variation on $[a, b]$.*

Definition 1.6. A finite function f defined on $[a, b]$ is said to satisfy the Lipschitz condition with a constant $K > 0$ provided

$$|f(t) - f(s)| \leq K|t - s| \quad \text{for } s, t \in [a, b].$$

Theorem 1.13. *Let the function f be defined on $[a, b]$ and satisfy the Lipschitz condition with a constant K on $[a, b]$. Then, f is of bounded variation on $[a, b]$.*

Theorem 1.14. *A function f of bounded variation on $[a, b]$ is bounded on $[a, b]$.*

Remark 1.6. A bounded function need not be of bounded variation. For example,

$$f(t) = \begin{cases} t \sin\left(\frac{1}{t}\right) & \text{for } 0 < t \leq 1, \\ 0 & \text{for } t = 0. \end{cases}$$

Then, f is bounded on $[0,1]$, but it is not of bounded variation on $[0,1]$. To see this, fix $n \in \mathbb{N}$ and let $P\colon 0 = t_0 < t_1 < \cdots < t_n = 1$. Note that $f(t_k) = (-1)^k t_k$, so

$$\begin{aligned} |f(t_{k+1}) - f(t_k)| &= t_{k+1} - t_k \\ &\geq 2t_{k+1} \\ &\geq \frac{4}{3\pi} \frac{1}{k+1}. \end{aligned}$$

Consequently,

$$\bigvee(f,P) \geq \frac{4}{3\pi} \sum_{k=0}^{n-1} \frac{1}{k+1} \to \infty \quad \text{as } n \to \infty.$$

Theorem 1.15. *The sum, difference, and multiplication of two functions f and g of bounded variation on $[a,b]$ are functions of bounded variation on $[a,b]$.*

Theorem 1.16. *Let f and g be functions of bounded variation on $[a,b]$ and $g \geq \sigma > 0$ on $[a,b]$. Then, $\frac{f}{g}$ is a function of bounded variation on $[a,b]$.*

Theorem 1.17. *Let f be a function of bounded variation on $[a,b]$ and $c \in (a,b)$. Then,*

$$\bigvee_a^b(f) = \bigvee_a^c(f) + \bigvee_c^b(f). \tag{1.4}$$

Theorem 1.18. *A function f defined and finite on $[a,b]$ is a function of bounded variation on $[a,b]$ if and only if it is representable as the difference of two increasing functions.*

Corollary 1.3. *If a function f is of bounded variation on $[a,b]$, then at almost every point of $[a,b]$, the derivative f' exists and is finite. Furthermore, f' is a summable function.*

Corollary 1.4. *The set of points of discontinuity of a function f of bounded variation on $[a,b]$ is at most denumerable. At every point x_0 of discontinuity, both $f(t_0+)$ and $f(t_0-)$ exist.*

Theorem 1.19. *A function of bounded variation can be written as the sum of its saltus function and a continuous function of bounded variation.*

Example 1.4. Consider the Dirichlet function f defined in Example 1.3 on $[a, b]$, where $a, b \in \mathbb{Q}$. Take $n \in \mathbb{N}$ arbitrarily and

$$a = t_0 < t_1 \in \mathbb{R}\backslash\mathbb{Q} < t_2 \in \mathbb{Q} < \cdots < t_{n-1} \in \mathbb{R}\backslash\mathbb{Q} < t_n = b \in \mathbb{Q}.$$

Then,

$$\begin{aligned}\bigvee_a^b(f) &\geq |f(t_1) - f(t_0)| + |f(t_2) - f(t_1)| + \cdots + |f(t_n) - f(t_{n-1})| \\ &= |0 - 1| + |1 - 0| + \cdots + |1 - 0| \\ &= 1 + \cdots + 1 \\ &= n.\end{aligned}$$

Thus, f is not a function of bounded variation on $[a, b]$.

Example 1.5. Consider the function

$$f(x) = \begin{cases} \frac{1}{1-t} & \text{for } t \neq 1, \\ 0 & \text{for } t = 1 \end{cases}$$

on $[0, 1]$. For $t \neq 1$, we have

$$f'(t) = \frac{1}{(1-t)^2} > 0.$$

Thus, f is an increasing function on $(0, 1)$ and is of bounded variation on any closed subinterval of $(0, 1)$. Since f has a vertical asymptote at $x = 1$, we can make the sum

$$\sum_{j=1}^{n} |f(t_j) - f(t_{j-1})|$$

as large as we like by choosing partition points close to 1. Thus, f is not a function of bounded variation on $[0, 1]$.

Example 1.6. Consider the function

$$f(t) = \begin{cases} \sqrt[3]{t} \sin\left(\frac{\pi}{t}\right) & \text{for } t \neq 0, \\ 0 & \text{for } t = 0 \end{cases}$$

on $[0,1]$. Take $n \in \mathbb{N}$ even and

$$t_0 = 0, \quad t_n = 1, \quad t_{n-(2k+1)} = \frac{1}{k+3}, \quad t_{n-2k} = \frac{2}{2k+3}, \qquad k \in \mathbb{N}.$$

Then,

$$\begin{aligned}
\bigvee_0^1(f) &\geq \sum_{j=1}^{n} |f(t_j) - f(t_{j-1})| \\
&\geq \sum_{j=1}^{\frac{n}{2}} |f(t_{2j+1}) - f(t_{2j})| \\
&= \sum_{k=1}^{n} |f(t_{n-(2k+1)}) - f(t_{n-2k})| \\
&= \sum_{k=1}^{n} \left| f\left(\frac{1}{k+3}\right) - f\left(\frac{2}{2k+3}\right) \right| \\
&= \sum_{k=1}^{n} \sqrt[3]{\frac{2}{2k+3}}.
\end{aligned}$$

Since the series

$$\sum_{k=1}^{\infty} \sqrt[3]{\frac{2}{2k+3}}$$

is divergent, we conclude that f is not of bounded variation on $[0,1]$.

Note 1.2. The function

$$f(t) = \begin{cases} t\cos\left(\frac{\pi}{2t}\right) & \text{for } t \in (0,1], \\ 0 & \text{for } t = 0 \end{cases}$$

is not a function of bounded variation on $[0,1]$.

Theorem 1.20 (Helley's Selection Principle). *Let $H = \{f\}$ be an infinite family of functions defined on $[a,b]$ such that $|f(t)| \leq K$, $t \in [a,b]$, for any $f \in H$ and for some $K > 0$. Then, for any denumerable set $E \subset [a,b]$,* *there exists a sequence* $\{f_n\}_{n\in\mathbb{N}} \subset H$ *that converges at any point of the set* E.

Theorem 1.21. *Let* $F = \{f\}$ *be an infinite family of increasing functions defined on* $[a, b]$ *such that* $|f(t)| \le K$, $t \in [a, b]$, *for any* $f \in F$ *and for some* $K > 0$. *Then, there is a sequence* $\{f_n\}_{n\in\mathbb{N}} \subset F$ *of increasing functions that converges, at every point of the interval* $[a, b]$, *to an increasing function* ϕ *on* $[a, b]$.

Theorem 1.22 (The Helley First Theorem). *Let* $F = \{f\}$ *be an infinite family of functions defined on* $[a, b]$ *such that*

$$|f(t)| \le K, \quad \bigvee_a^b(f) \le K, \quad t \in [a, b],$$

for any $f \in F$ *and for some* $K > 0$. *Then, there exists a sequence* $\{f_n\}_{n\in\mathbb{N}} \subset F$ *that converges, at every point of* $[a, b]$, *to a function* ϕ *of bounded variation on* $[a, b]$.

Theorem 1.23. *Let* f *be of bounded variation on* $[a, b]$. *If* f *is continuous at* $t_0 \in [a, b]$, *then the function*

$$g(t) = \bigvee_a^t(f), \quad t \in [a, b],$$

is continuous at t_0.

Corollary 1.5. *A continuous function* f *of bounded variation on* $[a, b]$ *can be written as the difference of two continuous increasing functions.*

Let $\mathscr{C}([a, b])$ denote the set of continuous functions defined on $[a, b]$. For any partition P, given by (1.3), denote

$$\mu(P) = \max_{k\in\{0,\dots,n-1\}}(t_{k+1} - t_k).$$

Let $f \in \mathscr{C}([a, b])$ and $\omega_k(f, P)$, $k \in \{0, \dots, n-1\}$, denote the oscillation of f on $[t_k, t_{k+1}]$, $k \in \{0, \dots, n-1\}$, i.e., $\omega_k(f, P) = \sup\{|f(x) - f(y)|: x, y \in [t_k, t_{k+1}]\}$, and

$$\Omega(f, P) = \sum_{k=0}^{n-1} \omega_k(f, P).$$

Theorem 1.24. *Let* $f \in \mathscr{C}([a,b])$. *Then,*

$$\lim_{\mu(P)\to 0} V(f,P) = \bigvee_a^b (f) \quad \textit{and} \quad \lim_{\mu(P)\to 0} \Omega(f,P) = \bigvee_a^b (f).$$

Definition 1.7. Let $f \in \mathscr{C}([a,b])$. Then, there exist

$$m = \min_{t\in[a,b]} f(t) \quad \text{and} \quad M = \max_{t\in[a,b]} f(t).$$

For $s \in [m, M]$, the Banach indicatrix of s is defined to be the number that is equal to the number of roots of the equation $f(t) = s$, $t \in [a,b]$. It is denoted by $N(s)$. Note that the Banach indicatrix is an integer-valued function.

Theorem 1.25. *The Banach indicatrix is measurable and*

$$\bigvee_m^M (N) = \bigvee_a^b (f).$$

Corollary 1.6. *A continuous function f is of bounded variation if and only if its Banach indicatrix N is summable.*

1.2 The Stieltjes Integral

Let f and g be two functions defined and finite on the closed interval $[a,b]$. Consider the partition P, given by (1.3), and choose tag points $\xi_k \in [t_k, t_{k+1}]$, $k \in \{0,\ldots,n-1\}$. Form the sum

$$\sigma(P,f,g) = \sum_{k=0}^{n-1} f(\xi_k)(g(t_{k+1}) - g(t_k)).$$

Definition 1.8. If there exists a finite limit I such that

$$I = \lim_{\mu(P)\to 0} \sigma(P,f,g),$$

then this finite limit is said to be the Stieltjes integral of f with respect to g and it is denoted by

$$\int_a^b f(t)dg(t).$$

The exact meaning of this definition is as follows: The number I is said to be the Stieltjes integral of f with respect to g provided for any given $\varepsilon > 0$, there exists a $\delta = \delta(\varepsilon) > 0$ such that for any partition P for which $\mu(P) < \delta$, the inequality

$$|\sigma(P, f, g) - I| < \varepsilon$$

holds for any choice of the tag points ξ_k, $k \in \{0, \ldots, n-1\}$.

Remark 1.7. Note that the Riemann integral is a special case of the Stieltjes integral, obtained by setting $g(t) = t$, $t \in [a, b]$.

Example 1.7. Consider the functions

$$f(t) = \begin{cases} 0 & \text{for } t \in [-1, 0], \\ 1 & \text{for } t \in (0, 1], \end{cases}$$

$$g(t) = \begin{cases} 0 & \text{for } t \in [-1, 0), \\ 1 & \text{for } t \in [0, 1]. \end{cases}$$

We compute

$$I_1 = \int_{-1}^{0} f(t)dg(t), \quad I_2 = \int_{0}^{1} f(t)dg(t), \quad \text{and} \quad I_3 = \int_{-1}^{1} 2f(t)dg(t).$$

Consider the partitions P_1 of $[-1, 0]$ and P_2 of $[0, 1]$, given by

$$P_1\colon -1 = t_0 < t_1 < \cdots < t_m = 0 \quad \text{and} \quad P_2\colon 0 = s_0 < s_1 < \cdots < s_n = 1,$$

respectively. Take $\xi_k \in [t_k, t_{k+1}]$, $k \in \{0, \ldots, m-1\}$, and $\eta_k \in [s_k, s_{k+1}]$, $k \in \{0, \ldots, n-1\}$. Then,

$$\sigma_1(P_1, f, g) = \sum_{k=0}^{m-1} f(\xi_k)(g(t_{k+1}) - g(t_k)) = 0$$

and

$$\sigma_2(P_2, f, g) = \sum_{k=0}^{n-1} f(\eta_k)(g(s_{k+1}) - g(s_k)) = 0,$$

whereupon $I_1 = 0$ and $I_2 = 0$. Now, let

$$P_3\colon\ -1 = u_0 < u_1 < \cdots < u_i < 0 < u_{i+1} < \cdots < u_p = 1$$

be a partition of $[-1, 1]$ and $\zeta_k \in [u_k, u_{k+1}]$, $k \in \{0, \ldots, p-1\}$. Then,

$$\begin{aligned}\sigma(P_3, f, g) &= \sum_{k=0}^{p-1} f(\zeta_k)(g(u_{k+1}) - g(u_k)) \\ &= f(\zeta_i)(g(u_{i+1}) - g(u_i)) \\ &= f(\zeta_i) \\ &= \begin{cases} 0 & \text{if } \zeta_i \in [-1, 0] \\ 1 & \text{if } \zeta_i \in (0, 1]. \end{cases}\end{aligned}$$

Thus, I_3 does not exist.

We list the following obvious properties of the Stieltjes integral.

Theorem 1.26. *Let the functions f, f_1, f_2, g, g_1, g_2 be defined and finite on $[a, b]$.*

(1) *Suppose that the Stieltjes integrals*

$$\int_a^b f_1(t)dg(t) \quad \text{and} \quad \int_a^b f_2(t)dg(t)$$

exist. Then,

$$\int_a^b (f_1(t) + f_2(t))dg(t) = \int_a^b f_1(t)dg(t) + \int_a^b f_2(t)dg(t).$$

(2) *Suppose that the Stieltjes integrals*

$$\int_a^b f(t)dg_1(t) \quad \text{and} \quad \int_a^b f(t)dg_2(t)$$

exist. Then,

$$\int_a^b f(t)d(g_1 + g_2)(t) = \int_a^b f(t)dg_1(t) + \int_a^b f(t)dg_2(t).$$

(3) *Suppose that the Stieltjes integral*

$$\int_a^b f(t)dg(t)$$

exist. Then,

$$\int_a^b kf(t)d(lg)(t) = kl\int_a^b f(t)dg(t)$$

for any $k, l \in \mathbb{R}$.

(4) *Suppose that* $a < c < b$ *and the Stieltjes integrals*

$$\int_a^c f(t)dg(t), \quad \int_c^b f(t)dg(t), \quad \textit{and} \quad \int_a^b f(t)dg(t)$$

exist. Then,

$$\int_a^b f(t)dg(t) = \int_a^c f(t)dg(t) + \int_c^b f(t)dg(t). \tag{1.5}$$

Remark 1.8. Let f and g be as in Example 1.7. Also, let $a = -1$, $c = 0$, $b = 1$. By Example 1.7, I_1 and I_2 exist, but I_3 does not exist. Therefore, (1.5) does not hold.

In the following, we state the formula for integration by parts for the Stieltjes integral.

Theorem 1.27. *Let f and g be defined and finite on $[a, b]$. Then, the existence of one of the Stieltjes integrals*

$$\int_a^b f(t)dg(t) \quad \textit{and} \quad \int_a^b g(t)df(t)$$

implies the existence of the other. In this case, the equality

$$\int_a^b f(t)dg(t) + \int_a^b g(t)df(t) = f(b)g(b) - f(a)g(a) \tag{1.6}$$

holds.

Theorem 1.28. *Let f and g be defined and finite on $[a, b]$. The Stieltjes integral*

$$\int_a^b f(t)dg(t)$$

exists provided $f \in \mathscr{C}([a, b])$ and g is of bounded variation on $[a, b]$.

Theorem 1.29. *Let $f \in \mathscr{C}([a,b])$ and g be Riemann integrable over $[a,b]$ such that g' exists at every point of $[a,b]$. Then,*

$$\int_a^b f(t)dg(t) = \int_a^b f(t)g'(t)dt.$$

Theorem 1.30. *Let $f \in \mathscr{C}([a,b])$ and g be a constant function on each of the intervals $(a,c_1),(c_1,c_2),\ldots,(c_m,b)$, where $a < c_1 < \cdots < c_m < b$. Then,*

$$\begin{aligned}\int_a^b f(t)dg(t) = f(a)(g(a+) - g(a)) + \sum_{k=1}^{m} f(c_k)(g(c_k+) - g(c_k-)) \\ + f(b)(g(b) - g(b-)).\end{aligned} \tag{1.7}$$

Example 1.8. Let $f(t) = t^2 - t$, $t \in [0,3]$ and

$$g(t) = \begin{cases} -1 & \text{for } t = 0, \\ 1 & \text{for } t \in (0,1), \\ -2 & \text{for } t \in (1,2), \\ 0 & \text{for } t \in (2,3), \\ 1 & \text{for } t = 3. \end{cases}$$

We shall find

$$\int_0^3 f(t)dg(t),$$

where $a = 0$, $b = 3$, $c_1 = 1$, $c_2 = 2$, and $f(0) = 0$, $f(1) = 0$, $f(2) = 2$, $f(3) = 6$. Then, applying Theorem 1.30, we get

$$\begin{aligned}\int_0^3 f(t)dg(t) &= f(0)(g(0+) - g(0)) + f(1)(g(1+) - g(1-)) \\ &\quad + f(2)(g(2+) - g(2-)) + f(3)(g(3) - g(3-)) \\ &= 2(0 - (-2)) + 6(1 - 0) \\ &= 4 + 6 \\ &= 10.\end{aligned}$$

Theorem 1.31. *Let $f \in \mathscr{C}([a,b])$ and g be of bounded variation on $[a,b]$. Then,*

$$\left| \int_a^b f(t)dg(t) \right| \leq \max_{t\in[a,b]} |f(t)| \bigvee_a^b (g).$$

Theorem 1.32. *Let $\{f_n\}_{n\in\mathbb{N}}$ be a sequence of continuous functions on $[a,b]$ that converges uniformly to the continuous function f on $[a,b]$. Also, let g be a function of finite variation on $[a,b]$. Then,*

$$\lim_{n\to\infty} \int_a^b f_n(t)dg(t) = \int_a^b f(t)dg(t).$$

Theorem 1.33 (The Helly Second Theorem). *Let $f \in \mathscr{C}([a,b])$ and $\{g_n\}_{n\in\mathbb{N}}$ be a sequence of functions finite and defined on $[a,b]$ that converges to a finite function g at every point of $[a,b]$. If*

$$\bigvee_a^b (g_n) < K, \quad n \in \mathbb{N},$$

for some positive constant K, then

$$\lim_{n\to\infty} \int_a^b f(t)dg_n(t) = \int_a^b f(t)dg(t).$$

1.3 The Stieltjes Derivative

In this section, we introduce the Stieltjes derivative and explore some of its properties. Let $I \subset \mathbb{R}$. Here, and henceforth, we assume that $g\colon I \to \mathbb{R}$ is a monotone nondecreasing function which is continuous from the left everywhere.

Definition 1.9. Let $f\colon I \to \mathbb{R}$ be a given function. The derivative with respect to g, or g-derivative, or Stieltjes derivative of f at point $t \in I$ is denoted and defined as follows:

(1)

$$f'_g(t) = \lim_{s\to t} \frac{f(s)-f(t)}{g(s)-g(t)}, \tag{1.8}$$

when g is continuous at t, or

(2)

$$f'_g(t) = \lim_{s \to t+} \frac{f(s) - f(t)}{g(s) - g(t)}, \tag{1.9}$$

when g is discontinuous at t.

We state that f is g-differentiable or Stieltjes differentiable at $t_0 \in I$ provided $f'_g(t_0)$ exists.

Remark 1.9. The limit (1.8) does not make any sense at the points of the set

$$C_g = \{t \in I \colon g \text{ is a constant on } (t - \varepsilon, t + \varepsilon) \text{ for some } \varepsilon > 0\},$$

and no matter how f is. This seeming drawback is not important because the set C_g is negligible in a sense to be precise.

Remark 1.10. If $f'(t)$ and $g'(t)$ exist and $g'(t) > 0$, then

$$f'_g(t) = \frac{f'(t)}{g'(t)}.$$

Define the set

$$D_g = \{t \in I \colon g(t+) - g(t) > 0\},$$

where $g(t+)$ stands for the right-hand limit of g at t. Note that D_g is countable because g is monotone and for each $t \in D_g$, the limit (1.9) exists if and only if $f(t+)$ exists, and, in that case, we have

$$f'_g(t) = \frac{f(t+) - f(t)}{g(t+) - g(t)}.$$

Definition 1.10. For $k \in \mathbb{N}$, define the kth g-derivative or kth Stieltjes derivative of f at $t \in I$ as

$$f_g^{(k)}(t) = \left(f_g^{(k-1)}\right)'_g(t),$$

where $f_g^0(t) = f(t)$.

Example 1.9. Let $I = \mathbb{R}$ and

$$g(t) = \begin{cases} 2 & \text{for } t \in (1, \infty), \\ t + 3 & \text{for } t \in (-\infty, 1], \end{cases}$$

and $f(x) = x^2 + x + 1$, $t \in \mathbb{R}$. We shall find $f_g^{(k)}(1)$, $k \in \mathbb{N}$. Note that g is nondecreasing on $\mathbb{R}$, it is continuous from the left everywhere, and it is discontinuous at 1. Moreover, $g(1+) = 2$ and $g(1) = 4$. Also, observe that $f \in \mathscr{C}(\mathbb{R})$. Then,

$$f_g'(1) = \frac{f(1+) - f(1)}{g(1+) - g(1)} = \frac{3-3}{2-4} = 0.$$

Hence,

$$f_g^{(k)}(1) = 0, \quad k \in \mathbb{N}, \quad k \geq 2.$$

Example 1.10. Let $I = \mathbb{R}$ and

$$g(t) = \begin{cases} t^2 & \text{for } t \in (0, \infty), \\ 7 & \text{for } t \in (-\infty, 0], \end{cases}$$

and

$$f(t) = \begin{cases} 4 & \text{for } t \in (0, \infty), \\ t+2 & \text{for } t \in (-\infty, 0]. \end{cases}$$

Note that g is nondecreasing on $\mathbb{R}$ and it is continuous from the left everywhere, and it is discontinuous at 0. Moreover, we have

$$g(0+) = 0, \quad g(0) = 7, \quad f(0+) = 4, \quad f(0) = 2.$$

Then,

$$f_g'(0) = \frac{f(0+) - f(0)}{g(0+) - g(0)} = \frac{4-2}{0-7} = -\frac{2}{7}.$$

Now, we will explore some of the properties of the Stieltjes derivative.

Theorem 1.34. *Let $t \in I \backslash D_g$ be such that $f_g'(t)$ exists. Then, $t \notin C_g$ and f is continuous from the right at t provided that $g(s) > g(t)$ for $s > t$.*

Theorem 1.35. *Let $t \in I \backslash D_g$ be such that $f_g'(t)$ exists. Then, $t \notin C_g$ and f is continuous from the left at t provided that $g(s) < g(t)$ for $s < t$.*

Theorem 1.36. *For $t \in D_g$, $f_g'(t)$ exists if and only if $f(t+)$ exists.*

Theorem 1.37. *Let f_1 and f_2 be real-valued functions defined on a neighbourhood of a point $t \in I$. Also, let $g(s) \neq g(t)$ for $s \neq t$. If f_1 and f_2 are Stieltjes differentiable at t, then for $c_1, c_2 \in \mathbb{R}$, we have*

$$(c_1 f_1 + c_2 f_2)'_g(t) = c_1 f'_{1g}(t) + c_2 f'_{2g}(t).$$

Theorem 1.38. *Let f_1 and f_2 be real-valued functions defined on a neighbourhood of a point $t \in I$. Also, let $g(s) \neq g(t)$ for $s \neq t$. If f_1 and f_2 are Stieltjes differentiable at t, then*

$$(f_1 f_2)'_g(t) = f'_{1g}(t) f_2(t+) + f_1(t+) f'_{2g}(t).$$

Theorem 1.39. *Let f_1 and f_2 be real-valued functions defined on a neighbourhood of a point $t \in I$. Also, let $g(s) \neq g(t)$ for $s \neq t$. If f_1 and f_2 are Stieltjes differentiable at t and $f_2(t)f_2(t+) \neq 0$, then*

$$\left(\frac{f_1}{f_2}\right)'_g(t) = \frac{f'_{1g}(t) f_2(t) - f_1(t) f'_{2g}(t)}{f_2(t) f_2(t+)}.$$

Example 1.11. Let $I = \mathbb{R}$ and $g(t) = 3t + \sin t$, $f(t) = (t^2+t+1)(t^2-t+1)$, $t \in \mathbb{R}$. We shall find $f'_g(t)$, $t \in \mathbb{R}$. Note that $g'(t) = 3 + \cos t > 0$, $t \in \mathbb{R}$. Therefore, g is an increasing function on $\mathbb{R}$. Moreover, $g \in \mathscr{C}(\mathbb{R})$.

Set $f(t) = f_1(t) f_2(t)$, $t \in \mathbb{R}$, where $f_1(t) = t^2+t+1$ and $f_2(t) = t^2-t+1$, $t \in \mathbb{R}$. Then, $f'_1(t) = 2t+1$ and $f'_2(t) = 2t-1$. Hence,

$$f'_{1g}(t) = \frac{f'_1(t)}{g'(t)} = \frac{2t+1}{3+\cos t} \quad \text{and} \quad f'_{2g}(t) = \frac{f'_2(t)}{g'(t)} = \frac{2t-1}{3+\cos t}.$$

Now, applying Theorem 1.38, we arrive at

$$\begin{aligned} f'_g(t) &= f'_{1g}(t) f_2(t) + f_1(t) f'_{2g}(t) \\ &= \frac{2t+1}{3+\cos t}(t^2-t+1) + (t^2+t+1)\frac{2t-1}{3+\cos t} \\ &= \frac{(2t+1)(t^2-t+1) + (t^2+t+1)(2t-1)}{3+\cos t} \\ &= \frac{2t^3 - 2t^2 + 2t + t^2 - t + 1 + 2t^3 - t^2 + 2t^2 - t + 2t - 1}{3+\cos t} \\ &= \frac{6t^3+2t}{3+\cos t}. \end{aligned}$$

Thus,

$$f'_g(t) = \frac{6t^3+2t}{3+\cos t}, \quad t \in \mathbb{R}.$$

Example 1.12. Let g be as in Example 1.11 and

$$f(t) = \frac{t^2+t+1}{t^2-t+1}, \quad t \in \mathbb{R}.$$

We will find $f'_g(t)$, $t \in \mathbb{R}$. We write

$$f(t) = \frac{f_1(t)}{f_2(t)}, \quad t \in \mathbb{R},$$

where f_1 and f_2 are as in Example 1.11. By the computations in Example 1.11 and applying Theorem 1.39, we find

$$\begin{aligned} f'_g(t) &= \frac{f'_{1g}(t)f_2(t) - f_1(t)f'_{2g}(t)}{(f_2(t))^2} \\ &= \frac{\frac{2t+1}{3+\cos t}(t^2-t+1) - (t^2+t+1)\frac{2t-1}{3+\cos t}}{(t^2-t+1)^2} \\ &= \frac{(2t+1)(t^2-t+1) - (t^2+t+1)(2t-1)}{(3+\cos t)(t^2-t+1)^2} \\ &= \frac{2t^3-2t^2+2t+t^2-t+1-2t^3+t^2-2t^2+t-2t+1}{(3+\cos t)(t^2-t+1)^2} \\ &= \frac{2(-t^2+1)}{(3+\cos t)(t^2-t+1)^2}, \quad t \in \mathbb{R}. \end{aligned}$$

Thus,

$$f'_g(t) = \frac{2(-t^2+1)}{(3+\cos t)(t^2-t+1)^2}, \quad t \in \mathbb{R}.$$

We have three versions of the chain rule for Stieltjes derivatives of the composition of functions. Note that, in these versions, the usual derivative is involved.

Theorem 1.40 (The Chain Rule I). *Let f be a real-valued function defined in a neighbourhood of $t \in I \backslash D_g$ and let h be a real-valued function defined in a neighbourhood of $f(t)$. If $h'(f(t))$ and $f'_g(t)$ both exist, then*

$$(h \circ f)'_g(t) = h'(f(t))f'_g(t) \quad \textit{for } t \in I \backslash D_g.$$

Theorem 1.41 (The Chain Rule II). *Let f be a real-valued function defined in a neighbourhood of $t \in I \backslash D_g$ and let h be a real-valued function defined in a neighbourhood of $f(t)$. If $h'_g(f(t))$, $g'(f(t))$, and $f'_g(t)$ exist, then*

$$(h \circ f)'_g(t) = h'_g(f(t))g'(f(t))f'_g(t) \quad \textit{for } t \in I \backslash D_g.$$

Theorem 1.42 (The Chain Rule III). *Let $f\colon \mathbb{R} \to \mathbb{R}$ be continuously differentiable and suppose $l\colon I \to \mathbb{R}$ is Stieltjes differentiable. Then, $f \circ l\colon I \to \mathbb{R}$ is Stieltjes differentiable, and the formula*

$$(f \circ l)'_g(t) = \left\{ \int_0^1 f'(l(t) + h\Delta g(t) l'_g(t))dh \right\} l'_g(t) \tag{1.10}$$

holds.

Proof. Note that

$$\begin{aligned} f(l(t+)) - f(l(s)) &= \int_{l(s)}^{l(t+)} f'(y)dy \\ &= (l(t+) - l(s)) \int_0^1 f'(h(l(t+)) + (1-h)l(s))dh. \end{aligned}$$

Let $\varepsilon > 0$ and $t \in I$. Set

$$\varepsilon^* = \frac{\varepsilon}{1 + 2\int_0^1 |f'(hl(t+) + (1-h)l(t))|dh}.$$

We choose $\varepsilon > 0$, so that

$$\frac{\varepsilon}{2(1 + \varepsilon^* + |l'_g(t)|)} < 1.$$

Since l is Stieltjes differentiable at t, there exists a neighbourhood U_1 of t such that

$$|l(t+) - l(s) - l'_g(t)(g(t+) - g(s))| \leq \frac{\varepsilon^*}{2}|g(t+) - g(s)| \quad \text{for all } s \in U_1.$$

Since f' is continuous on $\mathbb{R}$, using the fact that

$$|hl(t+) + (1-h)l(s) - hl(t+) - (1-h)l(t)| = (1-h)|l(t) - l(s)|$$

for $h \in (0,1)$, there exists a neighbourhood U_2 of t such that

$$|f'(hl(t+) + (1-h)l(s)) - f'(hl(t+) + (1-h)l(t))| \leq \frac{\varepsilon}{2(1 + \varepsilon^* + |l'_g(t)|)}$$

for all $s \in U_2$. Let $U = U_1 \cap U_2$ and $s \in U$. We put $\alpha = hl(t+) + (1-h)l(s)$ and $\beta = hl(t+) + (1-h)l(t)$. Then, using the choice of ε, we have

$$|f'(\alpha) - f'(\beta)| < 1, \quad \text{i.e., } |f'(\alpha)| < 1 + |f'(\beta)|.$$

Also,

$$
\begin{aligned}
&\left|(f\circ l)(t+)-(f\circ l)(s)-(g(t+)-g(s))l_g'(t)\int_0^1 f'(\beta)dh\right|\\
&=\left|(l(t+)-l(s))\int_0^1 f'(hl(t+)+(1-h)l(s))dh\right.\\
&\quad\left.-(g(t+)-g(s))l_g'(t)\int_0^1 f'(\beta)dh\right|\\
&=\left|(l(t+)-l(s)-(g(t+)-g(s))l_g'(t))\int_0^1 f'(\alpha)dh\right.\\
&\quad\left.+(g(t+)-g(s))l_g'(t)\int_0^1 (f'(\alpha)-f'(\beta))dh\right|\\
&\le |l(t+)-l(s)-(g(t+)-g(s))l_g'(t)|\int_0^1 |f'(\alpha)|dh\\
&\quad+|g(t+)-g(s)||l_g'(t)|\int_0^1 |f'(\alpha)-f'(\beta)|dh\\
&\le \frac{\varepsilon^*}{2}|g(t+)-g(s)|\int_0^1 |f'(\alpha)|dh\\
&\quad+(\varepsilon^*+|l_g'(t)|)|g(t+)-g(s)|\int_0^1 |f'(\alpha)-f'(\beta)|dh\\
&\le |g(t+)-g(s)|\frac{\varepsilon}{2\left(1+2\int_0^1|f'(\beta)|dh\right)}\int_0^1 |f'(\alpha)|dh\\
&\quad+(\varepsilon^*+|l_g'(t)|)|g(t+)-g(s)|\frac{\varepsilon}{2(\varepsilon^*+|l_g'(t)|)}\\
&\le |g(t+)-g(s)|\frac{\varepsilon}{2\left(1+2\int_0^1|f'(\beta)|dh\right)}\left(1+\int_0^1 |f'(\beta)|dh\right)\\
&\quad+\frac{\varepsilon}{2}|g(t+)-g(s)|\\
&\le \frac{\varepsilon}{2}|g(t+)-g(s)|+\frac{\varepsilon}{2}|g(t+)-g(s)|\\
&=\varepsilon|g(t+)-g(s)|.
\end{aligned}
$$

Therefore, $f\circ l$ is Stieltjes differentiable at t, and its derivative satisfies (1.10). □

Definition 1.11. Let $f\colon I \to \mathbb{R}$ be a given function. A function $F\colon I \to \mathbb{R}$ is said to be a Stieltjes primitive of f provided

$$F'_g(t) = f(t), \quad x \in I.$$

Remark 1.11. A Stieltjes primitive of a given function $f\colon I \to \mathbb{R}$ may not be unique.

Theorem 1.43. *Let $I = [a, b]$ and $f \in \mathscr{C}([a, b])$ and define*

$$F(t) = \int_a^t f(\tau)dg(\tau) \quad \textit{for } t \in [a, b].$$

Then, F is a Stieltjes primitive of f and $F'_g(t) = f(t)$ for $t \in [a, b]$.

Theorem 1.44 (The Leibniz–Newton–Stieltjes Formula). *Let $I = \mathbb{R}$. Assume that $f\colon [a, b] \to \mathbb{R}$ is Stieltjes integrable and $F \in \mathscr{C}([a, b])$ is its Stieltjes primitive. Then,*

$$\int_a^b f(t)dg(t) = F(b) - F(a).$$

Example 1.13. Let $I = \mathbb{R}$, $f(t) = t^3 - 12t$, and

$$g(t) = \begin{cases} 1 & \text{for } t \in (1, \infty), \\ t + 4 & \text{for } t \in (-\infty, 1]. \end{cases}$$

Also, let

$$F(t) = \frac{t^4}{4} - 6t^2, \quad t \in [-3, 0].$$

Then,

$$\begin{aligned} F'_g(t) &= \frac{F'(t)}{g'(t)} \\ &= t^3 - 12t, \quad t \in [-3, 0]. \end{aligned}$$

Thus, F is a Stieltjes primitive of f on $[-3, 0]$.

Now, we will introduce the Stieltjes monomials $h_k\colon I \times I \to \mathbb{R}$, $k \in \mathbb{N}_0$, as follows:

$$\begin{aligned} h_0(x, y) &= 1, \\ h_k(x, y) &= \int_y^x h_{k-1}(z, y)dg(z), \quad k \in \mathbb{N}, \quad x, y \in I. \end{aligned}$$

Note that

$$(h_k)'_{g,x}(x,y) = h_{k-1}(x,y), \quad k \in \mathbb{N}, \quad x, y \in I.$$

Example 1.14. Let $I = \mathbb{R}$ and $g(t) = t^2 + 1$, $t \in \mathbb{R}$. Then, we have the following:

$$\begin{aligned}
h_0(x,y) &= 1, \quad x, y \in \mathbb{R}, \\
h_1(x,y) &= \int_y^x h_0(z,y)dg(z) \\
&= \int_y^x dg(z) \\
&= g(x) - g(y) \\
&= x^2 + 1 - y^2 - 1 \\
&= x^2 - y^2, \quad x, y \in \mathbb{R},
\end{aligned}$$

and

$$\begin{aligned}
h_2(x,y) &= \int_y^x h_1(z,y)dg(z) \\
&= \int_y^z (z^2 - y^2)g'(z)dz \\
&= 2\int_y^x (z^2 - y^2)zdz \\
&= \int_y^x (z^2 - y^2)d(z^2 - y^2) \\
&= \frac{(z^2 - y^2)^2}{2}\bigg|_{z=y}^{z=x} \\
&= \frac{(x^2 - y^2)^2}{2}, \quad x, y \in \mathbb{R}.
\end{aligned}$$

Assume that

$$h_k(x,y) = \frac{(x^2 - y^2)^k}{k!}, \quad x, y \in \mathbb{R},$$

for some $k \in \mathbb{N}$. We will prove that

$$h_{k+1}(x,y) = \frac{(x^2 - y^2)^{k+1}}{(k+1)!}, \quad x, y \in \mathbb{R}.$$

In fact, we have

$$\begin{aligned} h_{k+1}(x,y) &= \int_y^x h_k(z,y)dg(z) \\ &= 2\int_y^x \frac{(z^2-y^2)^k}{k!} zdz \\ &= \int_y^x \frac{(z^2-y^2)^k}{k!} d(z^2-y^2) \\ &= \frac{(z^2-y^2)^{k+1}}{(k+1)!}\bigg|_{z=y}^{z=x} \\ &= \frac{(x^2-y^2)^{k+1}}{(k+1)!}, \quad x,y \in \mathbb{R}. \end{aligned}$$

The above example can be generalized as follows.

Theorem 1.45. *Let $I=[a,b]$ and g be Riemann integrable on $[a,b]$ and g' exists on $[a,b]$. Then,*

$$h_k(x,y) = \frac{(g(x)-g(y))^k}{k!}$$

for $x, y \in I$ and any $k \in \mathbb{N}$.

Theorem 1.46. *Let $s_0 \in I$. For $k, m \in \mathbb{N}_0$, we have*

$$h_{k+m+1}(x,s_0) = \int_{s_0}^x h_k(x,y)h_m(y,s_0)dg(y), \quad x \in I.$$

Theorem 1.47. *Let $s_0 \in I$. For $k \in \mathbb{N}_0$, we have*

$$h_k(x,s_0) \geq \frac{(x-s_0)^k}{k!}, \quad k \in \mathbb{N}, \quad x \geq s_0, \quad x \in I. \tag{1.11}$$

Now, we shall give the Stieltjes version of the Taylor formula.

Theorem 1.48 (The Stieltjes–Taylor Formula). *Let $n \in \mathbb{N}$. Suppose that $f\colon I \to \mathbb{R}$ and $f_g^{(k)}(t_0)$ exist for $k \in \{0,1,\ldots,n\}$ and for some $t_0 \in I$. Then, there exists a constant $c \in I$, so that*

$$f(t) = f(t_0) + f_g'(t_0)h_1(t,t_0) + \cdots + f_g^{(n)}(t_0)h_n(t,t_0) + f_g^{(n+1)}(c)h_{n+1}(t,t_0), \quad t \in I. \tag{1.12}$$

Definition 1.12. The polynomial (1.12) is said to be the Stieltjes–Taylor polynomial of f at t_0.

1.4 Stieltjes Regressive Functions

Definition 1.13. A function $f\colon I \to \mathbb{R}$ is said to be Stieltjes regressive provided

$$1 + \Delta g(t) f(t) \neq 0, \quad t \in I,$$

where $\Delta g(t) = g(t+) - g(t-)$, $t \in I$. The set of all Stieltjes regressive functions on I is denoted by $\mathscr{R}_g(I)$ or $\mathscr{R}_g$.

Definition 1.14. In $\mathscr{R}_g$, define the Stieltjes circle plus $\oplus_g$ as

$$(f \oplus_g h)(t) = f(t) + h(t) + \Delta g(t) f(t) h(t), \quad t \in I.$$

Example 1.15. Let $I = [0, \infty)$, $f(t) = 1 + t$,

$$g(t) = \begin{cases} 1 & \text{for } t \in [0, 2], \\ 3 & \text{for } t \in (2, \infty), \end{cases}$$

and $h(t) = t^2$, $t \in I$. We shall find $(f \oplus_g h)(2)$. We have

$$f(2) = 1 + 2 = 3, \quad \text{and} \quad h(2) = 2^2 = 4,$$
$$\Delta g(2) = g(2+) - g(2) = 3 - 1 = 2.$$

Hence,

$$\begin{aligned} (f \oplus_g h)(2) &= f(2) + h(2) + \Delta g(2) f(2) h(2) \\ &= 3 + 4 + 2 \cdot 3 \cdot 4 \\ &= 7 + 24 \\ &= 31. \end{aligned}$$

Theorem 1.49. *The set $\mathscr{R}_g$ under the operation $\oplus_g$ is an Abelian group.*

Definition 1.15. The group $(\mathscr{R}_g, \oplus_g)$ is said to be a Stieltjes regressive group.

Definition 1.16. For $f \in \mathscr{R}_g$, define the Stieltjes circle minus as follows:

$$(\ominus_g f)(t) = -\frac{f(t)}{1 + \Delta g(t) f(t)}, \quad t \in I.$$

From Theorem 1.49, it follows that $\ominus_g f \in \mathscr{R}_g$ provided $f \in \mathscr{R}_g$.

Example 1.16. Let I, g, and f be as in Example 1.15. Then,

$$\begin{aligned}(\ominus_g f)(2) &= -\frac{f(t)}{1+\Delta g(t)f(t)}\\ &= -\frac{1+2}{1+2(1+2)}\\ &= -\frac{3}{7}\end{aligned}$$

and

$$\begin{aligned}(\ominus_g g)(2) &= -\frac{g(2)}{1+\Delta g(2)g(2)}\\ &= -\frac{1}{1+2\cdot 1}\\ &= -\frac{1}{3}.\end{aligned}$$

Definition 1.17. For $f, h \in \mathscr{R}_g$, define $f \ominus_g h = f \oplus_g (\ominus_g h)$.

By definition, we get

$$\begin{aligned}(f \ominus_g h)(t) &= f(t) + (\ominus_g h)(t) + \Delta g(t)f(t)(\ominus_g h)(t)\\ &= f(t) - \frac{h(t)}{1+\Delta g(t)h(t)} - \frac{\Delta g(t)f(t)h(t)}{1+\Delta g(t)h(t)}\\ &= \frac{f(t)+\Delta g(t)f(t)h(t) - h(t) - \Delta g(t)f(t)h(t)}{1+\Delta g(t)h(t)}\\ &= \frac{f(t)-h(t)}{1+\Delta g(t)h(t)}.\end{aligned}$$

Thus,

$$(f \ominus_g h)(t) = \frac{f(t)-h(t)}{1+\Delta g(t)h(t)}, \quad t \in I.$$

Example 1.17. Let I, g, f, and h be as in Example 1.15. Then,

$$\begin{aligned}(f \ominus_g h)(2) &= \frac{f(2)-h(2)}{1+\Delta g(2)h(2)}\\ &= \frac{3-4}{1+2\cdot 4}\\ &= -\frac{1}{9}.\end{aligned}$$

Theorem 1.50. *Let $f, h \in \mathscr{R}_g$. Then, we have the following:*

(1) $\ominus_g(\ominus_g f) = f$.

(2) $f \ominus_g f = 0$.

(3) $f \ominus_g h \in \mathscr{R}_g$.

(4) $\ominus_g(f \ominus_g h) = h \ominus_g f$.

(5) $(\ominus_g(f \oplus_g h)) = (\ominus_g f) \oplus_g (\ominus_g h)$.

Definition 1.18. For $f \in \mathscr{R}_g$, define the Stieltjes circle square as

$$f^{②_g} = (-f)(\ominus_g f).$$

For $f \in \mathscr{R}_g$, we have

$$\begin{aligned} f^{②_g}(t) &= (-f(t))(\ominus_g f)(t) \\ &= -f(t)\frac{-f(t)}{1+\Delta g(t)f(t)} \\ &= \frac{(f(t))^2}{1+\Delta g(t)f(t)}, \quad t \in I. \end{aligned}$$

Hence,

$$\frac{(f(t))^2}{f^{②_g}(t)} = 1 + \Delta g(t)f(t), \quad t \in I.$$

Example 1.18. Let I, g, f, and h be as in Example 1.15. Then,

$$\begin{aligned} f^{②_g}(2) &= (-f(t))(\ominus_g f)(t) \\ &= \frac{(f(2))^2}{1+\Delta g(2)f(2)} \\ &= \frac{3^2}{1+2\cdot 3} \\ &= \frac{9}{7} \end{aligned}$$

and

$$\begin{aligned} g^{\textcircled{2}_g}(2) &= (-g(2))(\ominus_g g)(2) \\ &= \frac{(g(2))^2}{1+\Delta g(2)g(2)} \\ &= \frac{1^2}{1+2\cdot 1} \\ &= \frac{1}{3}. \end{aligned}$$

Theorem 1.51. *Suppose that $f \in \mathscr{R}_g$. Then, we have the following:*

(1) $(\ominus_g f)^{\textcircled{2}_g} = f^{\textcircled{2}_g}$.

(2) $(f + (\ominus_g f))(t) = \Delta g(t) f^{\textcircled{2}_g}(t), \quad t \in I$.

(3) $f \oplus_g f^{\textcircled{2}_g} = f + f^2$.

1.5 Stieltjes Elementary Functions

1.5.1 *Stieltjes Exponential Function*

Suppose that $f\colon I \to \mathbb{C}$ is a continuous and Stieltjes regressive function on I and $t_0, T \in I$, $t_0 < T$. Set

$$\widetilde{f}(t) = \begin{cases} f(t) & \text{for } t \in [t_0, T]\backslash D_g, \\ \frac{\log(1+f(t)\Delta g(t))}{\Delta g(t)} & \text{for } t \in [t_0, T] \cap D_g. \end{cases}$$

Definition 1.19. Define the Stieltjes exponential function as follows:

$$e_{f,g}(t, t_0) = \exp\left[\int_{t_0}^{t} \widetilde{f}(s)dg(s)\right], \quad t \in [t_0, T],$$

where $f \in \mathscr{R}_g$.

Example 1.19. Let $I = [0, \infty)$, $f(t) = t^2 + t$, $t \in I$, and

$$g(t) = \begin{cases} -1 + t & \text{for } t \in [0, 2], \\ 4 & \text{for } t \in (2, 8]. \end{cases}$$

Then,

$$\widetilde{f}(t) = \begin{cases} f(t) & \text{for } t \in [0,2), \\ \frac{\log(1+f(2)\Delta g(2))}{\Delta g(2)} & \text{for } t = 2, \\ 4 & \text{for } t \in (2,8], \end{cases}$$

$$= \begin{cases} t^2 + t & \text{for } t \in [0,2), \\ \frac{\log(1+6\cdot 3)}{3} & \text{for } t = 2, \\ t^2 + t & \text{for } t \in (2,8], \end{cases}$$

$$= \begin{cases} t^2 + t & \text{for } t \in [0,2), \\ \frac{\log(19)}{3} & \text{for } t = 2, \\ t^2 + t & \text{for } t \in (2,8]. \end{cases}$$

Therefore,

$$e_{f,g}(t,0) = \exp\left[\int_0^t \widetilde{f}(s)dg(s)\right]$$

$$= \begin{cases} \exp\left[\int_0^t \widetilde{f}(s)g'(s)ds\right] & \text{for } t \in [0,2), \\ \exp\left[\int_0^t \widetilde{f}(s)g'(s)ds + \widetilde{f}(2)(g(2+) - g(2))\right] & \text{for } t \in [2,8], \end{cases}$$

$$= \begin{cases} \exp\left[\int_0^t (s^2+s)ds\right] & \text{for } t \in [0,2), \\ \exp\left[\int_0^t (s^2+s)ds + \frac{\log(19)}{3}\cdot 3\right] & \text{for } t \in (2,8], \end{cases}$$

$$= \begin{cases} \exp\left[\left(\frac{s^3}{3} + \frac{s^2}{2}\right)\Big|_{s=0}^{s=t}\right] & \text{for } t \in [0,2), \\ \exp\left[\left(\frac{s^3}{3} + \frac{s^2}{2}\right)\Big|_{s=0}^{s=t} + \log(19)\right] & \text{for } t \in (2,8], \end{cases}$$

$$= \begin{cases} e^{\frac{t^3}{3}+\frac{t^2}{2}} & \text{for } t \in [0,2), \\ 19e^{\frac{t^3}{3}+\frac{t^2}{2}} & \text{for } t \in [2,8]. \end{cases}$$

Example 1.20. Let $I = [0, \infty)$, $f(t) = 2 + tT$, $t \in I$, and

$$g(t) = \begin{cases} -2 & \text{for } t = 0, \\ -1 & \text{for } t \in (0,1), \\ 0 & \text{for } t = 1, \\ 1 & \text{for } t \in (1,2), \\ 2 & \text{for } t = 2, \\ 3 & \text{for } t \in (2,3), \\ 5 & \text{for } t \in [3,\infty). \end{cases}$$

We shall find $e_{f,g}(3,0)$. We have

$$\Delta g(0) = 1, \quad \Delta g(1) = 1, \quad \Delta g(2) = 1, \quad \Delta g(3) = 0,$$
$$f(0) = 2, \quad f(1) = 3, \quad f(2) = 4, \quad f(3) = 5,$$

and

$$\widetilde{f}(0) = \frac{\log(1 + f(0)\Delta g(0))}{\Delta g(0)} = \frac{\log(1 + 2 \cdot 1)}{1} = \log 3,$$
$$\widetilde{f}(1) = \frac{\log(1 + f(1)\Delta g(1))}{\Delta g(1)} = \frac{\log(1 + 3 \cdot 1)}{1} = \log 4 = 2\log 2,$$
$$\widetilde{f}(2) = \frac{\log(1 + f(2)\Delta g(2))}{\Delta g(2)} = \frac{\log(1 + 4 \cdot 1)}{1} = \log 5,$$
$$\widetilde{f}(3) = f(3) = 5.$$

Therefore,

$$\begin{aligned} e_{f,g}(3,0) &= \exp\left[\int_0^3 \widetilde{f}(s)dg(s)\right] \\ &= \exp[\widetilde{f}(0)\Delta g(0) + \widetilde{f}(1)\Delta g(1) + \widetilde{f}(2)\Delta g(2) + \widetilde{f}(3)\Delta g(3)] \\ &= \exp[\log 3 + 2\log 2 + \log 5 + 0] \\ &= \exp[\log 3 + \log 4 + \log 5] \\ &= \exp[\log 60] \\ &= 60. \end{aligned}$$

In the following, we shall list some of the properties of the Stieltjes exponential function.

Theorem 1.52. *Suppose $f, h\colon I \to \mathbb{C}$ are continuous functions and Stieltjes regressive and $t_0 \in I$. Then, we have the following:*

(1) $e_{f,g}(t_0, t_0) = 1$.

(2) $(e_{f,g})'_g(t, t_0) = f(t)e_{f,g}(t, t_0), \quad t \in I$.

(3) $\frac{1}{e_{f,g}(t,t_0)} = e_{\ominus_g f,g}(t, t_0), \quad t \in I$.

(4) $e_{f,g}(t, t_0)e_{h,g}(t, t_0) = e_{f\oplus_g h}(t, t_0), \quad t \in I$.

(5) $\frac{e_{f,g}(t,t_0)}{e_{h,g}(t,t_0)} = e_{f\ominus_g h,g}(t, t_0), \quad t \in I$.

1.5.2 *Stieltjes Trigonometric Functions*

Let $I \subseteq \mathbb{R}$ and $f\colon I \to \mathbb{R}$ be the Stieltjes regressive function.

Definition 1.20. Define Stieltjes trigonometric functions as follows:

$$\sin_{f,g}(t, t_0) = \frac{e_{if,g}(t, t_0) - e_{-if,g}(t, t_0)}{2i},$$

$$\cos_{f,g}(t, t_0) = \frac{e_{if,g}(t, t_0) + e_{-if,g}(t, t_0)}{2}, \quad t \in I.$$

Some basic properties of the Stieltjes trigonometric functions are listed in the following theorem.

Theorem 1.53. *The above Stieltjes trigonometric functions have the following properties:*

(1) $(\sin_{f,g})'_g(t, t_0) = f(t)\cos_{f,g}(t, t_0), \quad t \in I$.

(2) $(\cos_{f,g})'_g(t, t_0) = -f(t)\sin_{f,g}(t, t_0), \quad t \in I$.

(3) $(\cos_{f,g}(t, t_0))^2 + (\sin_{f,g}(t, t_0))^2 = e_{if,g}(t, t_0)e_{-if,g}(t, t_0), \quad t \in I$.

(4) $\cos_{f,g}(t, t_0) + i\sin_{f,g}(t, t_0) = e_{if,g}(t, t_0), \quad t \in I$.

(5) $\cos_{f,g}(t, t_0) - i\sin_{f,g}(t, t_0) = e_{-if,g}(t, t_0), \quad t \in I$.

1.5.3 *Stieltjes Hyperbolic Functions*

Let $I \subseteq \mathbb{R}$ and $f\colon I \to \mathbb{R}$ be the Stieltjes regressive function.

Definition 1.21. Define Stieltjes hyperbolic functions as follows:

$$\sinh_{f,g}(t,t_0) = \frac{e_{f,g}(t,t_0) - e_{-f,g}(t,t_0)}{2},$$
$$\cosh_{f,g}(t,t_0) = \frac{e_{f,g}(t,t_0) + e_{-f,g}(t,t_0)}{2}, \quad t \in I.$$

Now, some basic properties of the Stieltjes hyperbolic functions are deduced in the following theorem.

Theorem 1.54. *The above Stieltjes hyperbolic functions have the following properties:*

(1) $(\sinh_{f,g})'_g(t,t_0) = f(t)\cosh_{f,g}(t,t_0)$, $t \in I$.
(2) $(\cosh_{f,g})'_g(t,t_0) = f(t)\sinh_{f,g}(t,t_0)$, $t \in I$.
(3) $(\cosh_{f,g}(t,t_0))^2 - (\sinh_{f,g}(t,gt_0))^2 = e_{f,g}(t,t_0)e_{-f,g}(t,t_0)$, $t \in I$.
(4) $\cosh_{f,g}(t,t_0) + \sinh_{f,g}(t,t_0) = e_{f,g}(t,t_0)$, $t \in I$.
(5) $\cosh_{f,g}(t,t_0) - \sinh_{f,g}(t,t_0) = e_{-f,g}(t,t_0)$, $t \in I$.

1.6 Advanced Practical Problems

Problem 1.1. Find $\bigvee_a^b(f)$, where

(1) $f(t) = t^2$, $a = 0$, $b = 1$.
(2) $f(t) = t^4 + t$, $a = 0$, $b = 2$.
(3) $f(t) = \cos t$, $a = 0$, $b = \frac{\pi}{2}$.
(4) $f(t) = \sin t$, $a = 0$, $b = \frac{\pi}{2}$.
(5) $f(t) = \frac{1}{t}$, $a = 1$, $b = 2$.

Answer 1.1.

(1) 1.
(2) 18.
(3) 1.
(4) 1.
(5) $\frac{1}{2}$.

Problem 1.2. Let f be of bounded variation and differentiable on $[a, b]$ such that f' is integrable in the Riemann sense on $[a, b]$. Then, prove that

$$\bigvee_a^b(f) = \int_a^b |f'(t)|dt.$$

Problem 1.3. Using Problem 1.2, solve Problem 1.1.

Problem 1.4. Let $E \subset [0, 1]$ and χ be its characteristic function. Then, prove that χ is of finite variation if and only if E is a finite set.

Problem 1.5. Let $f(t) = t + 1$ and $g(t) = t^3$, $t \in [0, 1]$. Then, using the definition for Stieltjes integral, compute $\int_0^1 f(t)dg(t)$.

Answer 1.2. $\frac{7}{4}$.

Problem 1.6. Let $f(t) = t^4 - t$ and $g(t) = t^3$, $t \in [0, 2]$. Then, using Theorem 1.29, compute the integral $\int_0^2 f(t)dg(t)$.

Answer 1.3. $\frac{300}{7}$.

Problem 1.7. Let $f(t) = 3t - 2$, $t \in [0, 2]$, and

$$g(t) = \begin{cases} 1 & \text{for } t = 0, \\ -1 & \text{for } t \in \left(0, \frac{1}{2}\right), \\ -2 & \text{for } t \in \left(\frac{1}{2}, 1\right), \\ -3 & \text{for } t \in (1, 2), \\ 4 & \text{for } t = 2. \end{cases}$$

Then, using Theorem 1.30, compute the integral $\int_0^2 f(t)dg(t)$.

Answer 1.4. $\frac{61}{2}$.

Problem 1.8. Let $f \in \mathscr{C}([a, b])$ and g be defined and finite on $[a, b]$ such that g' is integrable in the Riemann sense on $[a, b]$. Then, prove that

$$\left| \int_a^b f(t)dg(t) \right| \le \max_{t\in[a,b]} |f(t)| \int_a^b |g'(t)|dt.$$

Problem 1.9. Let $f \in \mathscr{C}([a,b])$ and g be monotone on $[a,b]$. Then, prove that

$$\left| \int_a^b f(t)dg(t) \right| \le \max_{t\in[a,b]} |f(t)||g(b)-g(a)|.$$

Problem 1.10. Let $f \in \mathscr{C}([a,b])$ and g be defined and finite on $[a,b]$ such that g' is summable in the Riemann sense on $[a,b]$ except the points $a < c_1 < \cdots < c_l < b$. Then, prove that

$$\int_a^b f(t)dg(t) = \int_a^b f(t)g'(t)dt + f(a)(g(a+)-g(a)) + f(b)(g(b)-g(b-))$$
$$+ \sum_{m=1}^{l} f(c_m)(g(c_m+)-g(c_m-)).$$

Problem 1.11. Let $I = \mathbb{R}$ and $g(t) = (1+t)^3$, $t \in \mathbb{R}$. Find $f_g'(t)$, $t \in \mathbb{R}$, $t \neq -1$, where

(1) $f(t) = \frac{1}{t^2+2t+2}$, $t \in \mathbb{R}$.

(2) $f(t) = e^{t^2}$, $t \in \mathbb{R}$.

(3) $f(t) = 3(\cos t)^2 + 2(\cos t)^3$, $t \in \mathbb{R}$.

(4) $f(t) = \cos(\sin t) + \sin(\cos t)$, $t \in \mathbb{R}$.

(5) $f(t) = \cos(t^3) + (\cos t)^3$, $t \in \mathbb{R}$.

Answer 1.5.

(1) $-\frac{2}{3(1+t)(t^2+2t+2)^2}$, $t \in \mathbb{R}$, $t \neq -1$.

(2) $\frac{2te^t}{3(1+t)^2}$, $t \in \mathbb{R}$, $t \neq -1$.

(3) $\frac{2t\cos(t^2)+\sin(2t)}{3(1+t)^2}$, $t \in \mathbb{R}$, $t \neq -1$.

(4) $\frac{-3\sin(2t)+6(\cos t)^2 \sin t}{3(1+t)^2}$, $t \in \mathbb{R}$, $t \neq -1$.

(5) $\frac{-\cos t \sin(\sin t) - \sin t \cos(\cos t)}{3(1+t)^2}$, $t \in \mathbb{R}$, $t \neq -1$.

(6) $\frac{-3t^2 \sin(t^3) - 3(\cos t)^2 \sin t}{3(1+t)^2}$, $t \in \mathbb{R}$, $t \neq -1$.

Problem 1.12. Let $I = \mathbb{R}$ and

$$g(t) = \begin{cases} t^2+1 & \text{for } t \in (0,\infty), \\ -5 & \text{for } t \in (-\infty, 0], \end{cases}$$

and

$$f(t) = \begin{cases} -3 & \text{for } t \in (0, \infty), \\ t^2 + t & \text{for } t \in (-\infty, 0]. \end{cases}$$

Find $f'_g(0)$.

Answer 1.6. $-\frac{1}{2}$.

Problem 1.13. Let $I = [1, \infty)$, $f(t) = t^3 - t + 1$, and $g(t) = 1 + t^2$, $t \in I$. Find $f_g^{(3)}(t)$, $t \in \mathbb{R}$.

Answer 1.7. $\frac{-3(t^2+10)}{8t^5}$, $t \in I$

Problem 1.14. Let $I = \mathbb{R}$, $f_1(t) = -1 + 2t - t^3$, $f_2(t) = 1 - t + t^2$, and $g(t) = 1 + 4t^5$, $t \in I$. Verify Theorems 1.38 and 1.39.

Problem 1.15. Let $I = [0, \infty)$ and define $f(t) = 3 + 2t$, $t \in I$, and

$$g(t) = \begin{cases} -1 & \text{for } t = 0, \\ 3 & \text{for } t \in (0, 1), \\ 5 & \text{for } t = 1, \\ 9 & \text{for } t \in (1, 2), \\ 15 & \text{for } t = 2, \\ 16 & \text{for } t \in (2, 3), \\ 17 & \text{for } t \in [3, \infty). \end{cases}$$

Then, find $e_{f,g}(3, 0)$.

Answer 1.8. 2184.

Problem 1.16. Let $I = [0, \infty)$ and define $f(t) = 1 - 2t$, $g(t) = 3t + 1$, $h(t) = 2 + 3t$, $t \in I$. Then, find

(1) $(f \oplus_g h)(t)$, $t \in I$.
(2) $(\ominus_g f)(t)$, $t \in I$.
(3) $(\ominus_g h)(t)$, $t \in I$.
(4) $(f \ominus_g h)(t)$, $t \in I$.
(5) $(h \ominus_g f)(t)$, $t \in I$.

Problem 1.17. Let $I = [0, 4]$ and define $f(t) = 2t$ and $h(t) = t + 2$, $t \in I$. Then, find $e_{f,g}(t, 0)$ and $e_{f \oplus_g h, g}(t, 0)$, $t \in I$.

Problem 1.18. Let $I = [0, \infty)$ and define $f(t) = t + 7$, $t \in I$, and

$$g(t) = \begin{cases} -3 & \text{for } t = 0, \\ -2 & \text{for } t \in (0, 1), \\ -1 & \text{for } t = 1, \\ 0 & \text{for } t \in (1, 2), \\ 1 & \text{for } t = 2, \\ 3 & \text{for } t \in (2, 3), \\ 7 & \text{for } t \in [3, \infty). \end{cases}$$

Then, find $\sinh_{f,g}(t, 0)$ and $\cosh_{f,g}(t, 0)$, $t \in I$.

Problem 1.19. Let $I = [0, \infty)$ and define $f(t) = t^2 - 1$, $t \in I$, and

$$g(t) = \begin{cases} 2 & \text{for } t = 0, \\ 4 & \text{for } t \in (0, 1), \\ 6 & \text{for } t = 1, \\ 8 & \text{for } t \in (1, 2), \\ 10 & \text{for } t = 2, \\ 12 & \text{for } t \in (2, 3), \\ 14 & \text{for } t \in [3, \infty). \end{cases}$$

Then, find $\cosh_{f,g}(t, 0) - 3 \sinh_{f,g}(t, 0)$, $t \in I$.

Chapter 2

First-Order Stieltjes Differential Inequalities

In this chapter, we derive some Gronwall-type Stieltjes differential inequalities. Then, we present some criteria for comparison of solutions of Stieltjes differential inequalities. As applications of these inequalities, we investigate a class of Stieltjes integro-differential equations, and some criteria for the existence of solutions, uniqueness of solutions, and the existence of minimal and maximal solutions are presented.

Suppose that $I \subset \mathbb{R}$ and $g\colon I \to \mathbb{R}$ is a monotone nondecreasing function that is continuous from the left everywhere and $t_0 \in I$ is its fixed point. With $\mathscr{C}_g^1(I)$, we denote the set of all Stieltjes differentiable functions on I whose first-order Stieltjes derivative is continuous on I. With $\mathscr{C}_g^k(I)$, we denote the set of all functions whose kth Stieltjes derivative exists and is continuous on I. For convenience, we write $e_{p,g}(t)$ for $e_{p,g}(t, t_0)$.

2.1 Stieltjes–Gronwall-Type Inequalities

Theorem 2.1 (The Stieltjes–Gronwall-Type Inequality I). *Suppose that $y \in \mathscr{C}_g^1(I)$ and $f \in \mathscr{R}_g(I)$, $p \in \mathscr{R}_g^+(I)$. If*

$$y'_g(t) \le p(t)y(t) + f(t), \quad t \in I, \tag{2.1}$$

then

$$y(t) \le e_{p,g}(t) \left(y(t_0) + \int_{t_0}^{t} e_{\ominus_g p,g}(\tau) f(\tau) dg(\tau) \right), \quad t \ge t_0, \quad t \in I.$$

Proof. Keeping in mind the inequality (2.1), we find $y'_g(t)-p(t)y(t) \leq f(t)$, $t \in I$. Hence,

$$\begin{aligned}
(ye_{\ominus_g p,g})'_g(t) &= y'_g(t)e_{\ominus_g p,g}(t+) + (\ominus_g p)(t)y(t)e_{\ominus_g p,g}(t) \\
&= y'_g(t)(1+(\Delta g(t))(\ominus_g p)(t))e_{\ominus_g p,g}(t) + (\ominus_g p)(t)y(t)e_{\ominus_g p,g}(t) \\
&= y'_g(t)\left(1-\frac{(\Delta g(t))p(t)}{1+(\Delta g(t))p(t)}\right)e_{\ominus_g p,g}(t) \\
&\quad -\frac{p(t)y(t)}{1+(\Delta g(t))p(t)}e_{\ominus_g p,g}(t) \\
&= \frac{y'_g(t)}{1+(\Delta g(t))p(t)}e_{\ominus_g p,g}(t) - \frac{p(t)y(t)}{1+(\Delta g(t))p(t)}e_{\ominus_g p,g}(t) \\
&= (y'_g(t)-p(t)y(t))\frac{1}{1+(\Delta g(t))p(t)}e_{\ominus_g p,g}(t) \\
&= (y'_g(t)-p(t)y(t))e_{\ominus_g p,g}(t) \\
&\leq f(t)e_{\ominus_g p,g}(t), \quad t \in I.
\end{aligned}$$

Integrating the last inequality in the sense of Stieltjes, from t_0 to t, $t \geq t_0$, we arrive at

$$(ye_{\ominus_g p,g})(s)\Big|_{s=t_0}^{s=t} \leq \int_{t_0}^{t} f(\tau)e_{\ominus_g p,g}(\tau)dg(\tau), \quad t \geq t_0, \quad t \in I,$$

$$y(t)e_{\ominus_g p,g}(t) - y(t_0) \leq \int_{t_0}^{t} f(\tau)e_{\ominus_g p,g}(\tau)dg(\tau), \quad t \geq t_0, \quad t \in I,$$

$$y(t)e_{\ominus_g p,g}(t) \leq y(t_0) + \int_{t_0}^{t} f(\tau)e_{\ominus_g p,g}(\tau)dg(\tau), \quad t \geq t_0, \quad t \in I,$$

or

$$y(t) \leq e_{p,g}(t)\left(y(t_0) + \int_{t_0}^{t} f(\tau)e_{\ominus_g p,g}(\tau)dg(\tau)\right), \quad t \geq t_0, \quad t \in I.$$

This completes the proof. □

Theorem 2.2 (The Stieltjes–Gronwall-Type Inequality II). *Suppose that $y \in \mathscr{C}_g^1(I)$ and $f \in \mathscr{R}_g(I)$, $p \in \mathscr{R}_g^+(I)$. If*

$$y'_g(t) \leq -p(t)y(t+) + f(t), \quad t \in I, \tag{2.2}$$

then

$$y(t) \leq e_{\ominus_g p,g}(t)\left(y(t_0) + \int_{t_0}^{t} e_{p,g}(\tau)f(\tau)dg(\tau)\right), \quad t \geq t_0, \quad t \in I.$$

Proof. Keeping in mind the inequality (2.2), we find $y_g'(t) + p(t)y(t+) \leq f(t)$, $t \geq t_0$. Then,

$$\begin{aligned}(ye_{p,g})_g'(t) &= y_g'(t)e_{p,g}(t) + y(t+)p(t)e_{p,g}(t)\\ &= \left(y_g'(t) + y(t+)p(t)\right)e_{p,g}(t)\\ &\leq f(t)e_{p,g}(t), \quad t \in I.\end{aligned}$$

Now, integrating the last inequality in the sense of Stieltjes, from t_0 to t, $t \geq t_0$, we obtain

$$(ye_{p,g})(s)\Big|_{s=t_0}^{s=t} \leq \int_{t_0}^{t} f(\tau)e_{p,g}(\tau)dg(\tau), \quad t \geq t_0, \quad t \in I,$$

$$y(t)e_{p,g}(t) - y(t_0) \leq \int_{t_0}^{t} f(\tau)e_{p,g}(\tau)dg(\tau), \quad t \geq t_0, \quad t \in I,$$

$$y(t)e_{p,g}(t) \leq y(t_0) + \int_{t_0}^{t} f(\tau)e_{p,g}(\tau)dg(\tau), \quad t \geq t_0, \quad t \in I,$$

or

$$y(t) \leq e_{\ominus_g p,g}(t)\left(y(t_0) + \int_{t_0}^{t} f(\tau)e_{p,g}(\tau)dg(\tau)\right), \quad t \geq t_0, \quad t \in I.$$

This completes the proof. □

Remark 2.1. Suppose that all conditions of Theorem 2.2 hold. Since

$$y(t+) = y(t) + (\Delta g(t))y_g'(t), \quad t \in I,$$

the inequality (2.2) takes the form

$$\begin{aligned}y_g'(t) &\leq p(t)y(t) - p(t)(\Delta g(t))y_g'(t) + f(t), \quad t \in I,\\ (1 + p(t)(\Delta g(t)))y_g'(t) &\leq p(t)y(t) + f(t), \quad t \in I,\\ y_g'(t) &\leq \frac{p(t)}{1 + p(t)(\Delta g(t))}y(t) + \frac{f(t)}{1 + p(t)(\Delta g(t))}, \quad t \in I,\end{aligned}$$

or

$$y'_g(t) \le -(\ominus_g p)(t)y(t) + \frac{f(t)}{1+p(t)(\Delta g(t))}, \quad t \in I.$$

Now, we apply Theorem 2.2 and get

$$\begin{aligned} y(t) &\le e_{p,g}(t)\left(y(t_0) + \int_{t_0}^{t} \frac{f(\tau)}{1+p(\tau)\Delta g(\tau)} e_{p,g}(\tau)dg(\tau)\right) \\ &\le e_{p,g}(t)\left(y(t_0) + \int_{t_0}^{t} \frac{f(\tau)}{1+p(\tau)\Delta g(\tau)} e_{p,g}(\tau+)dg(\tau)\right) \\ &= e_{p,g}(t)\left(y(t_0) + \int_{t_0}^{t} f(\tau)e_{\ominus_g p,g}(\tau)dg(\tau)\right), \quad t \ge t_0, \quad t \in I. \end{aligned}$$

Thus, we get the conclusion of Theorem 2.1.

Remark 2.2. Assume that all conditions of Theorem 2.1 hold. Since

$$y(t+) = y(t) + (\Delta g(t))y'_g(t), \quad t \in I,$$

the inequality (2.2) takes the form

$$\begin{aligned} y'_g(t) &\le -p(t)y(t) - p(t)(\Delta g(t))y'_g(t) + f(t), \quad t \in I, \\ (1+p(t)(\Delta g(t)))y'_g(t) &\le -p(t)y(t) + f(t), \quad t \in I, \\ y'_g(t) &\le -\frac{p(t)}{1+p(t)(\Delta g(t))}y(t) + \frac{f(t)}{1+p(t)(\Delta g(t))}, \quad t \in I, \end{aligned}$$

or

$$y'_g(t) \le (\ominus_g p)(t)y(t) + \frac{f(t)}{1+p(t)(\Delta g(t))}, \quad t \in I.$$

Now, we apply Theorem 2.1 and obtain

$$\begin{aligned} y(t) &\le e_{\ominus_g p,g}(t)\left(y(t_0) + \int_{t_0}^{t} \frac{e_{p,g}(\tau)}{1+p(\tau)(\Delta g(\tau))} f(\tau)dg(\tau)\right) \\ &\le e_{\ominus_g p,g}(t)\left(y(t_0) + \int_{t_0}^{t} \frac{e_{p,g}(\tau+)}{1+p(\tau)(\Delta g(\tau))} f(\tau)dg(\tau)\right) \\ &= e_{\ominus_g p,g}(t)\left(y(t_0) + \int_{t_0}^{t} e_{p,g}(\tau)f(\tau)dg(\tau)\right), \quad t \ge t_0, \quad t \in I. \end{aligned}$$

Thus, we find the conclusion of Theorem 2.2.

As above, one can prove the following results.

Theorem 2.3 (The Stieltjes–Gronwall-Type Inequality III). *Suppose that $y \in \mathscr{C}_g^1(I)$ and $f \in \mathscr{R}_g(I)$, $p \in \mathscr{R}_g^+(I)$. If*

$$y'_g(t) \geq p(t)y(t) + f(t), \quad t \in I,$$

then

$$y(t) \geq e_{p,g}(t)\left(y(t_0) + \int_{t_0}^{t} e_{\ominus_g p,g}(\tau)f(\tau)dg(\tau)\right), \quad t \geq t_0, \quad t \in I.$$

Theorem 2.4 (The Stieltjes–Gronwall-Type Inequality IV). *Suppose that $y \in \mathscr{C}_g^1(I)$ and $f \in \mathscr{R}_g(I)$, $p \in \mathscr{R}_g^+(I)$. If*

$$y'_g(t) \geq -p(t)y(t) + f(t), \quad t \in I,$$

then

$$y(t) \geq e_{\ominus_g p,g}(t)\left(y(t_0) + \int_{t_0}^{t} e_{p,g}(\tau)f(\tau)dg(\tau)\right), \quad t \geq t_0, \quad t \in I.$$

Now, we will consider applications of Theorems 2.3 and 2.4.

Theorem 2.5. *Let $y \in \mathscr{C}_g^1(I)$, $f, p \in \mathscr{R}_g(I)$ and $y > 0$, $f > 0$, $p > 0$ on I. If*

$$y'_g(t) \leq -y(t+)(p(t) + f(t)y(t)), \quad t \in I,$$

then

$$y(t) \leq \frac{1}{e_{p,g}(t)\left(\frac{1}{y(t_0)} + \int_{t_0}^{t} e_{\ominus_g p,g}(\tau)f(\tau)dg(\tau)\right)}, \quad t \in I, \quad t \geq t_0.$$

Proof. From the given condition, we obtain

$$y'_g(t) \leq -y(t)y(t+)\left(\frac{p(t)}{y(t)} + f(t)\right), \quad t \in I,$$

whereupon

$$\frac{y'_g(t)}{y(t)y(t+)} \leq -\frac{p(t)}{y(t)} - f(t), \quad t \in I.$$

We set

$$z(t) = \frac{1}{y(t)}, \quad t \in I,$$

and using the fact that

$$z_g'(t) = -\frac{y_g'(t)}{y(t)y(t+)}, \quad t \in I,$$

we arrive at the inequality

$$-z_g'(t) \le -p(t)z(t) - f(t), \quad t \in I,$$

or

$$z_g'(t) \ge p(t)z(t) + f(t), \quad t \in I.$$

Now, applying Theorem 2.3, we find

$$z(t) \ge e_{p,g}(t)\left(z(t_0) + \int_{t_0}^t e_{\ominus_g p,g}(\tau)f(\tau)dg(\tau)\right), \quad t \ge t_0, \quad t \in I,$$

from where we obtain

$$y(t) \le \frac{1}{e_{p,g}(t)\left(\frac{1}{y(t_0)} + \int_{t_0}^t e_{\ominus_g p,g}(\tau)f(\tau)dg(\tau)\right)}, \quad t \in I, \quad t \ge t_0.$$

This completes the proof. □

Theorem 2.6. *Let $y \in \mathscr{C}_g^1(I)$, $f, p \in \mathscr{R}_g(I)$ and $y > 0$, $f > 0$, $p > 0$ on I. If*

$$y_g'(t) \le -y(t+)(p(t) + f(t)y(t)), \quad t \in I,$$

then

$$y(t) \le \frac{1}{e_{\ominus_g p,g}(t)\left(\frac{1}{y(t_0)} + \int_{t_0}^t e_{p,g}(\tau)f(\tau)dg(\tau)\right)}, \quad t \in I, \quad t \ge t_0.$$

Proof. Note that

$$y_g'(t) \le -y(t)y(t+)\left(\frac{p(t)}{y(t)} + f(t)\right), \quad t \in I,$$

whereupon

$$\frac{y_g'(t)}{y(t)y(t+)} \le -\frac{p(t)}{y(t)} - f(t), \quad t \in I.$$

Let

$$z(t) = \frac{1}{y(t)}, \quad t \in I.$$

Then, using the fact that

$$z_g'(t) = -\frac{y_g'(t)}{y(t)y(t+)}, \quad t \in I,$$

we get the inequality

$$-z_g'(t) \leq -p(t)z(t) - f(t), \quad t \in I,$$

or

$$z_g'(t) \geq p(t)z(t) + f(t), \quad t \in I.$$

From this and Theorem 2.4, it follows that

$$z(t) \geq e_{\ominus_g p,g}(t) \left(z(t_0) + \int_{t_0}^{t} e_{p,g}(\tau) f(\tau) dg(\tau) \right), \quad t \geq t_0, \quad t \in I.$$

Hence,

$$y(t) \leq \frac{1}{e_{\ominus_g p,g}(t) \left(\frac{1}{y(t_0)} + \int_{t_0}^{t} e_{p,g}(\tau) f(\tau) dg(\tau) \right)}, \quad t \in I, \quad t \geq t_0.$$

This completes the proof. □

Problem 2.1. Prove Theorem 2.5 using Theorem 2.6 and the identity

$$y(t) = y(t+) - (\Delta g(t)) y_g'(t), \quad t \in I.$$

Problem 2.2. Prove Theorem 2.6 using Theorem 2.5 and the identity

$$y(t+) = y(t) + (\Delta g(t)) y_g'(t), \quad t \in I.$$

As we have proved Theorems 2.5 and 2.6, we can derive the following results.

Theorem 2.7. *Let $y \in \mathscr{C}_g^1(I)$, $f, p \in \mathscr{R}_g(I)$ and $y > 0$, $f > 0$, $p > 0$ on I. If*

$$y_g'(t) \geq -y(t+)(p(t) + f(t)y(t)), \quad t \in I,$$

then

$$y(t) \geq \frac{1}{e_{p,g}(t) \left(\frac{1}{y(t_0)} + \int_{t_0}^{t} e_{\ominus_g p,g}(\tau) f(\tau) dg(\tau) \right)}, \quad t \in I, \quad t \geq t_0.$$

Theorem 2.8. *Let $y \in \mathscr{C}_g^1(I)$, $f, p \in \mathscr{R}_g(I)$ and $y > 0$, $f > 0$, $p > 0$ on I. If*

$$y_g'(t) \geq -y(t+)(p(t) + f(t)y(t)), \quad t \in I,$$

then

$$y(t) \geq \frac{1}{e_{\ominus_g p,g}(t)\left(\frac{1}{y(t_0)} + \int_{t_0}^{t} e_{p,g}(\tau)f(\tau)dg(\tau)\right)}, \quad t \in I, \quad t \geq t_0.$$

2.2 Stieltjes Differential Inequalities for First-Order Stieltjes Differential Equations

In this section, we present some fundamental Stieltjes differential inequalities for an implicit perturbation of a nonlinear first-order Stieltjes differential equation. We start with the following result.

Theorem 2.9. *Let $x \mapsto f(t,x)$ be increasing on $\mathbb{R}$ for $t \in I$ and $x \mapsto h(t,x)$ be continuous on $\mathbb{R}$ for $t \in I$. Suppose that there are functions $y, z \in \mathscr{C}_g'(I)$ such that*

$$\begin{aligned} (f(t,y(t)))'_{g,t} &< h(t,y(t)), \\ (f(t,z(t)))'_{g,t} &\geq h(t,z(t)), \quad t \in I. \end{aligned} \tag{2.3}$$

Then, the inequality $y(t_0) < z(t_0)$ implies the inequality

$$y(t) < z(t), \quad t \in I. \tag{2.4}$$

Proof. Suppose that the inequality (2.4) fails. Define the set

$$Z^* = \{t \in I \colon y(t) \geq z(t)\}.$$

Let $t_1 = \inf Z^*$. Without any restrictions, assume that

$$y(t_1) = z(t_1) \quad \text{and} \quad y(t) < z(t) \quad \text{for } t < t_1, \quad t \in I.$$

Define the functions $Y(t) = f(t,y(t))$ and $Z(t) = f(t,z(t))$, $t \in I$. Then,

$$Y(t_1) = Z(t_1) \quad \text{and} \quad Y(t) < Z(t) \quad \text{for } t < t_1, \quad t \in I.$$

Hence,

$$\lim_{s \to t_1} \frac{Y(s) - Y(t_1)}{g(s) - g(t_1)} \geq \lim_{s \to t_1} \frac{Z(s) - Z(t_1)}{g(s) - g(t_1)}$$

if g is continuous at t_1, and

$$\lim_{s \to t_1} \frac{Y(s) - Y(t_1)}{g(s) - g(t_1)} \geq \lim_{s \to t_1} \frac{Z(s) - Z(t_1)}{g(s) - g(t_1)}$$

if g is discontinuous at t_1. Therefore, $Y_g'(t_1) \geq Z_g'(t_1)$. From this and (2.3), we arrive at $h(t_1, y(t_1)) > h(t_1, z(t_1))$. This is a contradiction because $y(t_1) = z(t_1)$. Thus, $Z^* = \emptyset$ and the inequality (2.4) holds for any $t \in I$. This completes the proof. □

As above, one can prove the following result.

Theorem 2.10. *Let $x \mapsto f(t, x)$ be increasing on $\mathbb{R}$ for $t \in I$ and $x \mapsto h(t, x)$ be continuous on $\mathbb{R}$ for $t \in I$. Suppose that there are functions $y, z \in \mathscr{C}_g'(I)$ such that*

$$\begin{aligned} &(f(t, y(t)))_{g,t}' \leq h(t, y(t)), \\ &(f(t, z(t)))_{g,t}' > h(t, z(t)), \quad t \in I. \end{aligned} \tag{2.5}$$

Then, the inequality $y(t_0) < z(t_0)$ implies the inequality

$$y(t) < z(t), \quad t \in I.$$

Theorem 2.11. *Let $x \mapsto f(t, x)$ be increasing on $\mathbb{R}$ for $t \in I$ and $x \mapsto h(t, x)$ be continuous on $\mathbb{R}$ for $t \in I$. Let $u \in \mathscr{C}_g'(I)$ satisfy*

$$(f(t, u(t)))_g' < h(t, u(t)), \quad t \in I, \tag{2.6}$$

and the function $v \in \mathscr{C}_g^1(I)$ satisfy

$$(f(t, v(t)))_g' > h(t, v(t)), \quad t \in I. \tag{2.7}$$

If w is any solution to the Stieltjes differential equation

$$(f(t, w(t)))_{g,t}' = h(t, w(t)), \quad t \in I, \tag{2.8}$$

and $u(t_0) < w(t_0) < v(t_0)$, then $u(t) < w(t) < v(t)$, $t \in I$.

Proof. Since $u(t_0) < w(t_0)$ and (2.6), (2.7) hold, it follows that all conditions of Theorem 2.9 hold. Therefore, $u(t) < w(t)$, $t \in I$. Also, by the inequality $w(t_0) < v(t_0)$ and (2.7), (2.8), we conclude that all conditions of Theorem 2.9 hold, and then $w(t) < v(t)$, $t \in I$. This completes the proof. □

Theorem 2.12. *Suppose that $x \mapsto f(t, x)$ is increasing on $\mathbb{R}$ for any $t \in I$ and $k_1, k_2 \in \mathscr{C}(I)$ are such that*

$$k_1(t, x) < k_2(t, x), \quad (t, x) \in I \times \mathbb{R}. \tag{2.9}$$

Let u_1 and u_2 be solutions of the following Stieltjes differential equations:

$$(f(t, u(t)))'_{g,t} = k_1(t, u(t)), \quad t \in I, \tag{2.10}$$

$$(f(t, u(t)))'_{g,t} = k_2(t, u(t)), \quad t \in I, \tag{2.11}$$

respectively. If $u_1(t_0) < u_2(t_0)$, then $u_1(t) < u_2(t)$, $t \in I$.

Proof. Since u_2 is a solution to the Stieltjes differential equation (2.11), we have

$$(f(t, u_2(t)))'_{g,t} = k_2(t, u_2(t)), \quad t \in I.$$

Also, since u_1 is a solution to the Stieltjes differential equation (2.10), using (2.9), we find

$$\begin{aligned}(f(t, u_1(t)))'_{g,t} &= k_1(t, u_1(t)) \\ &< k_2(t, u_1(t)), \quad t \in I.\end{aligned}$$

Thus, all conditions of Theorem 2.9 hold, and hence $u_1(t) < u_2(t)$, $t \in I$. This completes the proof. □

Theorem 2.13. *Suppose that $x \mapsto f(t, x)$ is increasing on $\mathbb{R}$ for $t \in I$ and there exists a bounded nonnegative function $h \in \mathscr{C}'_g(I)$ such that the function*

$$x \mapsto l(t, x) = -h(t)f(t, x)$$

is nonincreasing on $\mathbb{R}$ for any $t \in I$ and $1 - h(t)(\Delta g(t)) > 0$, $t \in I$. Assume that there exist functions $y, z \in \mathscr{C}(I)$ such that

$$\begin{aligned}(f(t, y(t)))'_g &\le l(t, y(t)), \quad t \in I, \\ y(t_0) &\le x_0\end{aligned} \tag{2.12}$$

and

$$\begin{aligned}(f(t, z(t)))'_g &\ge l(t, z(t)), \quad t \in I, \\ z(t_0) &\ge x_0.\end{aligned} \tag{2.13}$$

Then,

$$y(t) \le z(t), \quad t \in I. \tag{2.14}$$

Proof. Suppose that (2.14) fails. Then, there exist $t_1, t_2 \in I$, $t_1 < t_2$, such that $y(t_1) = z(t_1)$ and $y(t) > z(t)$, $t \in (t_1, t_2)$. Define the functions

$$Y(t) = f(t, y(t)), \quad Z(t) = f(t, z(t)), \quad m(t) = Y(t) - Z(t), \quad t \in I.$$

Then,

$$\begin{aligned} Y(t_1) &= f(t_1, y(t_1)) = f(t_1, z(t_1)) = Z(t_1), \\ m(t_1) &= Y(t_1) - Z(t_1) = 0, \end{aligned}$$

and

$$\begin{aligned} m(t) &= Y(t) - Z(t) = f(t, y(t)) - f(t, z(t)) \\ &> f(t, z(t)) - f(t, z(t)) = 0, \quad t \in (t_1, t_2). \end{aligned}$$

Now, we arrive at

$$\begin{aligned} m_g'(t) &= Y_g'(t) - Z_g'(t) \\ &\leq l(t, y(t)) - l(t, z(t)) \\ &= -h(t) f(t, y(t)) + h(t) f(t, z(t)) \\ &= -h(t)(f(t, y(t)) - f(t, z(t))) \\ &= -h(t) m(t), \quad t \in (t_1, t_2). \end{aligned}$$

Applying the Stieltjes–Gronwall-type inequality, Theorem 2.1, we find

$$m(t) \leq e_{-h,g}(t) m(t_1) = 0, \quad t \in (t_1, t_2),$$

which is a contradiction. This completes the proof. □

Theorem 2.14. *Assume that $x \mapsto f(t, x)$ is increasing on $\mathbb{R}$ for $t \in I$ and $x \mapsto l(t, x)$ is continuous on $\mathbb{R}$ for $t \in I$, and there exists a number $L > 0$ such that*

$$l(t, x_1) - l(t, x_2) \leq L(f(t, x_1) - f(t, x_2)), \quad t \in I,$$

for $x_1, x_2 \in \mathbb{R}$, $x_1 \geq x_2$. Suppose that there exist functions $y, z \in \mathscr{C}_g'(I)$ such that (2.12) and (2.13) hold. Then, (2.14) holds.

Proof. Take $\varepsilon > 0$ arbitrarily. Set $f(t, z_\varepsilon) = f(t, z) + \varepsilon e_{2L}(t)$, $t \in I$, so that $f(t, z_\varepsilon) > f(t, z)$ implies $z_\varepsilon > z$. Define

$$Z_\varepsilon(t) = f(t, z_\varepsilon(t)) \quad \text{and} \quad Z(t) = f(t, z(t)), \quad t \in I.$$

Then,

$$\begin{aligned} Z'_{\varepsilon,g}(t) &= (f(t, z_\varepsilon(t)))'_g \\ &= (f(t, z(t)))'_g + 2\varepsilon L e_{2L}(t) \\ &\geq l(t, z(t)) + 2\varepsilon L e_{2L}(t), \quad t \in I. \end{aligned}$$

Now, using the fact that

$$l(t, z_\varepsilon(t)) - l(t, z(t)) \leq L(f(t, z_\varepsilon(t)) - f(t, z(t))), \quad t \in I,$$

we get

$$\begin{aligned} l(t, z(t)) &\geq l(t, z_\varepsilon(t)) - L(f(t, z_\varepsilon(t)) - f(t, z(t))) \\ &= l(t, z_\varepsilon(t)) - \varepsilon L e_{2L}(t), \quad t \in I, \end{aligned}$$

and

$$\begin{aligned} Z'_{\varepsilon,g}(t) &\geq l(t, z_\varepsilon(t)) + 2\varepsilon L e_{2L}(t) - \varepsilon L e_{2L}(t) \\ &= l(t, z_\varepsilon(t)) + \varepsilon L e_{2L}(t) \\ &> l(t, z_\varepsilon(t)), \quad t \in I, \end{aligned}$$

i.e.,

$$D_{g,t}(f(t, z_\varepsilon(t))) > l(t, z_\varepsilon(t)), \quad t \in I.$$

Note that $z_\varepsilon(t_0) > z(t_0) > y(t_0)$. Now, applying Theorem 2.10, we get

$$y(t) < z_\varepsilon(t), \quad t \in I.$$

Taking the limit as $\varepsilon \to 0$, we find $y(t) \leq z(t)$, $t \in I$. This completes the proof. $\square$

2.3 First-Order Stieltjes Integro-Differential Inequalities

In this section, we present some important Stieltjes integro-differential inequalities. Our first result in this section is as follows.

Theorem 2.15. *Let $x \mapsto f(t,x)$ be continuous on $\mathbb{R}$ for $t \in I$ and $x \mapsto l(t,x)$ be continuous and increasing on $\mathbb{R}$ for $t \in I$. Suppose that there are functions $y, z \in \mathscr{C}_g^1(I)$ such that*

$$\begin{aligned}(y(t) - f(t, y(t)))'_{g,t} &\leq \int_{t_0}^{t} l(s, y(s))dg(s),\\ (z(t) - f(t, z(t)))'_{g,t} &\geq \int_{t_0}^{t} l(s, z(s))dg(s), \quad t \in I.\end{aligned} \tag{2.15}$$

Then, the inequality $y(t_0) < z(t_0)$ implies the inequality

$$y(t) < z(t), \quad t \in I. \tag{2.16}$$

Proof. Assume that the inequality (2.16) fails. Define the set

$$Z = \{t \in I \colon y(t) \geq z(t)\}.$$

Then, $Z \neq \emptyset$ and let $t_1 = \inf Z$. Without any restrictions, assume that

$$y(t_1) = z(t_1) \quad \text{and} \quad y(t) < z(t) \quad \text{for } t < t_1, \quad t \in I.$$

Define the functions

$$Y(t) = y(t) - f(t, y(t)) \quad \text{and} \quad Z(t) = z(t) - f(t, z(t)), \quad t \in I.$$

Then, $Y(t_1) = Z(t_1)$ and $Y(t) < Z(t)$ for $t < t_1$, $t \in I$. From here,

$$\lim_{s \to t_1} \frac{Y(s) - Y(t_1)}{g(s) - g(t_1)} \geq \lim_{s \to t_1} \frac{Z(s) - Z(t_1)}{g(s) - g(t_1)}$$

if g is continuous at t_1, and

$$\lim_{s \to t_1+} \frac{Y(s) - Y(t_1)}{g(s) - g(t_1)} \geq \lim_{s \to t_1} \frac{Z(s) - Z(t_1)}{g(s) - g(t_1)}$$

if g is discontinuous at t_1. Therefore, $Y'_g(t_1) \geq Z'_g(t_1)$. Hence, applying (2.15), we obtain

$$\begin{aligned}\int_{t_0}^{t_1} g(s, y(s))dg(s) &\geq Y'(t_1)\\ &\geq Z'(t_1)\\ &> \int_{t_0}^{t_1} g(s, z(s))dg(s).\end{aligned}$$

On the other hand, since $x \mapsto g(t, x)$ is increasing on $\mathbb{R}$ for $t \in I$, we have

$$g(t, y(t)) \leq g(t, z(t)), \quad t \in [t_0, t_1),$$

whereupon

$$\int_{t_0}^{t_1} g(s, y(s))dg(s) \leq \int_{t_0}^{t_1} g(s, z(s))dg(s),$$

which is a contradiction. This completes the proof. □

As above, we can prove the following result.

Theorem 2.16. *Let $x \mapsto f(t, x)$ be continuous on $\mathbb{R}$ for any $t \in I$ and $x \mapsto l(t, x)$ be continuous and increasing on $\mathbb{R}$ for any $t \in I$. Suppose that there are functions $y, z \in \mathscr{C}_g^1(I)$ such that*

$$\begin{aligned}(y(t) - f(t, y(t)))'_{g,t} &< \int_{t_0}^{t} l(s, y(s))dg(s),\\ (z(t) - f(t, z(t)))'_{g,t} &\geq \int_{t_0}^{t} l(s, z(s))dg(s), \quad t \in I.\end{aligned}$$

Then, the inequality $y(t_0) < z(t_0)$ implies the inequality $y(t) < z(t)$, $t \in I$.

Theorem 2.17. *Let $x \mapsto f(t, x)$ be continuous on $\mathbb{R}$ for $t \in I$ and $x \mapsto l(t, x)$ be continuous and increasing on $\mathbb{R}$ for $t \in I$. Let $u \in \mathscr{C}_g^1(I)$ satisfy*

$$(u(t) - f(t, u(t)))'_{g,t} < \int_{t_0}^{t} l(s, u(s))dg(s), \quad t \in I, \quad t \geq t_0, \tag{2.17}$$

and $v \in \mathscr{C}_g^1(I)$ satisfy

$$(v(t) - f(t, v(t)))'_{g,t} > \int_{t_0}^{t} l(s, v(s))dg(s), \quad t \in I, \quad t \geq t_0. \tag{2.18}$$

If w is any solution to the Stieltjes integro-differential equation

$$(w(t) - f(t, w(t)))'_{g,t} = \int_{t_0}^{t} l(s, w(s))dg(s), \quad t \in I, \quad t \geq t_0, \tag{2.19}$$

and $u(t_0) < w(t_0) < v(t_0)$, then $u(t) < w(t) < v(t)$, $t \in I$, $t \geq t_0$.

Proof. Since $u(t_0) < w(t_0)$ and (2.17), (2.18) hold, we find that all conditions of Theorem 2.16 hold. Therefore, $u(t) < w(t)$, $t \in I$. By the inequality $w(t_0) < v(t_0)$ and (2.18), (2.19), we conclude that all conditions of Theorem 2.16 hold, and then $w(t) < v(t)$, $t \in I$. This completes the proof. □

Theorem 2.18. *Suppose that $x \mapsto f(t, x)$ is continuous on $\mathbb{R}$ for $t \in I$ and $k_1, k_2 \in \mathscr{C}(I)$ are such that*

$$k_1(t, x) < k_2(t, x), \quad (t, x) \in I \times \mathbb{R}. \tag{2.20}$$

Let u_1 and u_2 be solutions of the Stieltjes integro-differential equations:

$$(u(t) - f(t, u(t)))'_{g,t} = \int_{t_0}^{t} k_1(s, u(s))dg(s), \quad t \in I, \quad t \geq t_0, \tag{2.21}$$

and

$$(u(t) - f(t, u(t)))'_{g,t} = \int_{t_0}^{t} k_2(s, u(s))dg(s), \quad t \in I, \quad t \geq t_0, \tag{2.22}$$

respectively. If $u_1(t_0) < u_2(t_0)$, then $u_1(t) < u_2(t)$, $t \in I$, $t \geq 0$.

Proof. Since u_2 is a solution to the Stieltjes integro-differential equation (2.22), we have

$$(u_2(t) - f(t, u_2(t)))'_{g,t} = \int_{t_0}^{t} k_2(s, u_2(s))dg(s), \quad t \in I, \quad t \geq t_0.$$

Since u_1 is a solution to the Stieltjes integro-differential equation (2.21), using (2.20), we find

$$\begin{aligned}(u_1(t) - f(t, u_1(t)))'_{g,t} &= \int_{t_0}^{t} k_1(s, u_1(s))dg(s) \\ &< \int_{t_0}^{t} k_2(s, u_1(s))dg(s), \quad t \in I, \quad t \geq t_0.\end{aligned}$$

Therefore, all conditions of Theorem 2.16 hold, and then $u_1(t) < u_2(t)$, $t \in I$. This completes the proof. □

Theorem 2.19. *Suppose that $x \mapsto x - f(t, x)$ is continuous and increasing on $\mathbb{R}$ for $t \in I$ and there exists a bounded nonnegative function $h \in \mathscr{C}(I)$ such that*

$$x \mapsto \int_{t_0}^{t} l(s, x(s))dg(s) = -h(t)(x(t) - f(t, x(t))), \quad t \in I,$$

and $1 - h(t)(\Delta g(t)) > 0$, $t \in I$. Assume that there exist functions $y, z \in \mathscr{C}_g^1(I)$ such that

$$\begin{aligned} (y(t) - f(t, y(t)))'_g &\leq \int_{t_0}^{t} l(s, y(s))dg(s), \quad t \in I, \quad t \geq t_0, \\ y(t_0) &\leq x_0 \end{aligned} \tag{2.23}$$

and

$$\begin{aligned} (z(t) - f(t, z(t)))'_g &\geq \int_{t_0}^{t} l(s, z(s))dg(s), \quad t \in I, \quad t \geq t_0, \\ z(t_0) &\geq x_0. \end{aligned} \tag{2.24}$$

Then,

$$y(t) \leq z(t), \quad t \in I. \tag{2.25}$$

Proof. Suppose that (2.25) fails. Then, there exist $t_1, t_2 \in I$, $t_1 < t_2$, such that $y(t_1) = z(t_1)$ and $y(t) > z(t)$, $t \in (t_1, t_2)$. Define the functions

$$Y(t) = y(t) - f(t, y(t)), \quad Z(t) = z(t) - f(t, z(t)), \quad m(t) = Y(t) - Z(t), \quad t \in I.$$

Then,

$$\begin{aligned} Y(t_1) = y(t_1) - f(t_1, y(t_1)) = z(t_1) - f(t_1, z(t_1)) = Z(t_1), \\ m(t_1) = Y(t_1) - Z(t_1) = 0, \end{aligned}$$

and

$$\begin{aligned} m(t) &= Y(t) - Z(t) \\ &= y(t) - f(t, y(t)) - z(t) + f(t, z(t)) \\ &> z(t) - f(t, z(t)) - z(t) - f(t, z(t)) \\ &= 0, \quad t \in (t_1, t_2). \end{aligned}$$

Now, we obtain

$$\begin{aligned} m'_g(t) &= Y'_g(t) - Z'_g(t) \\ &\leq \int_{t_0}^t l(s, y(s))dg(s) - \int_{t_0}^t l(s, z(s))dg(s) \\ &= -h(t)(y(t) - f(t, y(t))) + h(t)(z(t) - f(t, z(t))) \\ &= -h(t)(y(t) - f(t, y(t)) - z(t) + f(t, z(t))) \\ &= -h(t)m(t), \quad t \in (t_1, t_2). \end{aligned}$$

Applying the Stieltjes–Gronwall-type inequality, Theorem 2.1, we find

$$m(t) \leq e_{-h,g}(t)m(t_1) = 0, \quad t \in (t_1, t_2),$$

which is a contradiction. This completes the proof. □

Theorem 2.20. *Assume that* $x \mapsto x - f(t, x)$ *is increasing on* $\mathbb{R}$ *for* $t \in I$ *and* $x \mapsto l(t, x)$ *is continuous on* $\mathbb{R}$ *for* $t \in I$, *and there exists a number* $L > 0$ *such that*

$$\begin{aligned} &\int_{t_0}^t (l(s, x_1(s)) - l(s, x_2(s)))dg(s) \\ &\qquad \leq L(x_1(t) - f(t, x_1(t)) - x_2(t) + f(t, x_2(t))), \end{aligned}$$

for $t \in I$, $t \geq t_0$, $x_1 \geq x_2$. *Suppose that there exist functions* $y, z \in \mathscr{C}(I)$ *such that* (2.23) *and* (2.24) *hold. Then,* (2.25) *holds.*

Proof. Take $\varepsilon > 0$ arbitrarily. Set

$$z_\varepsilon(t) - f(t, z_\varepsilon(t)) = z(t) - f(t, z(t)) + \varepsilon e_{2L}(t), \quad t \in I,$$

so that $z_\varepsilon(t) - f(t, z_\varepsilon(t)) > z(t) - f(t, z(t))$ implies $z_\varepsilon(t) > z(t)$, $t \in I$. Define

$$Z_\varepsilon(t) = z_\varepsilon(t) - f(t, z_\varepsilon(t)) \quad \text{and} \quad Z(t) = z(t) - f(t, z(t)), \quad t \in I.$$

Then,

$$\begin{aligned} Z'_{\varepsilon,g}(t) &= (z_\varepsilon(t) - f(t, z_\varepsilon(t)))'_g \\ &= (z(t) - f(t, z(t)))'_g + 2\varepsilon L e_{2L}(t) \\ &\geq \int_{t_0}^t l(s, z(s))dg(s) + 2\varepsilon L e_{2L}(t), \quad t \in I. \end{aligned}$$

Next, using the fact that

$$\int_{t_0}^{t}(l(s,z_\varepsilon(s))-l(s,z(s)))dg(s)$$
$$\leq L(z_\varepsilon(t)-f(t,z_\varepsilon(t))-z(t)+f(t,z(t))),\quad t\in I,$$

we get

$$\begin{aligned}\int_{t_0}^{t} l(s,z(s))dg(s)&\\ &\geq \int_{t_0}^{t} l(s,z_\varepsilon(s))dg(s)-L(z_\varepsilon(t)-f(t,z_\varepsilon(t))-z(t)+f(t,z(t)))\\ &=\int_{t_0}^{t} l(s,z_\varepsilon(s))dg(s)-\varepsilon Le_{2L}(t),\quad t\in I,\quad t\geq t_0,\end{aligned}$$

and

$$\begin{aligned}Z'_{\varepsilon,g}(t)&\geq \int_{t_0}^{t} l(s,z_\varepsilon(s))dg(s)+2\varepsilon Le_{2L}(t)-\varepsilon Le_{2L}(t)\\ &=\int_{t_0}^{t} l(s,z_\varepsilon(s))dg(s)+\varepsilon Le_{2L}(t)\\ &>\int_{t_0}^{t} l(s,z_\varepsilon(s))dg(s),\quad t\in I,\end{aligned}$$

i.e.,

$$D_{g,t}(z_\varepsilon(t)-f(t,z_\varepsilon(t)))>\int_{t_0}^{t} l(s,z_\varepsilon(s))dg(s),\quad t\in I,\quad t\geq t_0.$$

Note that $z_\varepsilon(t_0)>z(t_0)>y(t_0)$. Now, applying Theorem 2.16, we find

$$y(t)<z_\varepsilon(t),\quad t\in I,\quad t\geq t_0.$$

Taking the limit as $\varepsilon\to 0$, we find $y(t)\leq z(t)$, $t\in I$. This completes the proof. □

2.4 Existence of Solutions for First-Order Stieltjes Integro-Differential Equations

In this section, we investigate for the existence of solutions the following Stieltjes initial value problem (SIVP):

$$\begin{aligned} (x(t) - f(t, x(t)))'_{g,t} &= \int_{t_0}^{t} l(s, x(s))dg(s), \quad t \in J, \\ x(t_0) &= x_0, \end{aligned} \tag{2.26}$$

where $J = [t_0, t_0 + a] \subset I$ for some $a > 0$, $t_0 \geq 0$, $x_0 \in \mathbb{R}$ is given, and

(A1) The function $x \mapsto x - f(t, x)$ is increasing on $\mathbb{R}$ for $t \in J$.
(A2) There exist constants L and M such that $L \leq M$ and

$$|f(t, x) - f(t, y)| \leq \frac{L|x - y|}{M + |x - y|}, \quad t \in J, \quad x, y \in \mathbb{R}.$$

(A3) There exists a continuous function h on J such that

$$|l(t, x)| \leq h(t), \quad t \in J, \quad x \in \mathbb{R}.$$

To prove our existence result, we use the following fixed point theorem due to Dhage (2004).

Theorem 2.21 (The Dhage Fixed Point Theorem I (Dhage, 2004)). *Let S be a bounded, closed, and convex subset of the Banach space E, and let $A\colon E \to E$ and $B\colon S \to E$ be two operators such that*

(1) *B is a nonlinear contraction,*
(2) *B is continuous and compact,*
(3) *$x = Ax + By$ for all $y \in S$ implies $x \in S$.*

Then, the operator $A + B$ has a fixed point in S.

Before stating our existence result, we need the following useful lemma.

Lemma 2.1. *Assume that* (A1) *holds. Then, $x \in \mathscr{C}(J)$ is a solution to the SIVP* (2.26) *if and only if it satisfies the following Stieltjes integral equation:*

$$x(t) = x_0 + f(t, x(t)) - f(t_0, x_0) + \int_{t_0}^{t} (t - s)l(s, x(s))dg(s), \quad t \in J. \tag{2.27}$$

Proof. Suppose $x \in \mathscr{C}(J)$ is a solution to the SIVP (2.26). Then, integrating the first equation of (2.26), in the sense of Stieltjes, from t_0 to t and using the second equation of (2.26), we find

$$(x(s) - f(s, x(s)))\Big|_{s=t_0}^{s=t} = \int_{t_0}^{t} \int_{t_0}^{s} l(\tau, x(\tau))dg(\tau)dg(s), \quad t \in J,$$

i.e.,

$$x(t) = x_0 + f(t, x(t)) - f(t_0, x_0) + \int_{t_0}^{t} \int_{t_0}^{s} l(\tau, x(\tau))dg(\tau)dg(s), \quad t \in J.$$

Let

$$m(t) = \int_{t_0}^{t} \int_{t_0}^{s} l(\tau, x(\tau))dg(\tau)dg(s) - \int_{t_0}^{t} (t-s)l(s, x(s))dg(s), \quad t \in J.$$

Then,

$$\begin{aligned} x_g'(t) &= \int_{t_0}^{t} g(s, x(s))dg(s) - (t-t)l(t, x(t)) - \int_{t_0}^{t} l(s, x(s))dg(s) \\ &= 0, \quad t \in J. \end{aligned}$$

Thus, l is a constant on J and $m(t) = m(t_0) = 0$, $t \in J$. Hence,

$$\int_{t_0}^{t} \int_{t_0}^{s} l(\tau, x(\tau))dg(\tau)dg(s) = \int_{t_0}^{t} (t-s)l(s, x(s))dg(s), \quad t \in J,$$

and x is a solution to the Stieltjes integral equation (2.27).

Conversely, assume that $x \in \mathscr{C}(J)$ be a solution to the Stieltjes integral equation (2.27). We put $t = t_0$ in (2.27) and find

$$\begin{aligned} x(t_0) &= x_0 + f(t_0, x(t_0)) - f(t_0, x_0) + \int_{t_0}^{t_0} l(s, x(s))dg(s) \\ &= x_0 + f(t_0, x(t_0)) - f(x_0, t_0). \end{aligned}$$

From here, $x(t_0) - f(t_0, x(t_0)) = x_0 - f(t_0, x_0)$. Applying $(A1)$, we get

$$x(t_0) = x_0.$$

By (2.27), we find

$$x(t) - f(t, x(t)) = x_0 - f(t_0, x_0) + \int_{t_0}^{t} (t - s)l(s, x(s))dg(s), \quad t \in J,$$

which we differentiate in the sense of Stieltjes and obtain the first equation of (2.26). Therefore, x is a solution to the SIVP (2.26). This completes the proof. □

Theorem 2.22. *Suppose that (A1)–(A3) hold. Then, the SIVP (2.26) has at least one solution in $\mathscr{C}(J)$.*

Proof. Let $E = \mathscr{C}(J)$ and $S = \{x \in E\colon \|x\| \le N\}$, where

$$N = |x_0 - f(t_0, x_0)| + L + F_0 + \|h\|a(t_0 + a) \quad \text{and} \quad F_0 = \sup_{t \in J} |f(t, 0)|.$$

For $x \in E$, define the operator

$$Ax(t) = f(t, x(t)), \quad t \in J.$$

We have $A\colon E \to E$. For $x \in S$, define the operator

$$Bx(t) = x_0 - f(t_0, x_0) + \int_{t_0}^{t} (t - s)l(s, x(s))dg(s), \quad t \in J.$$

Note that $B\colon S \to E$. For $x, y \in E$, we have

$$\begin{aligned} |Ax(t) - Ay(t)| &= |f(t, x(t)) - f(t, y(t))| \\ &\le \frac{L|x(t) - y(t)|}{M + |x(t) - y(t)|}, \quad t \in J, \end{aligned}$$

whereupon

$$|Ax(t) - Ay(t)|(M + |x(t) - y(t)|) \le L|x(t) - y(t)|, \quad t \in J,$$

and

$$\sup_{t \in J}(|Ax(t) - Ay(t)|(M + |x(t) - y(t)|)) \le L \sup_{t \in J} |x(t) - y(t)|, \quad t \in J,$$

or

$$\|Ax - Ay\|(M + \|x - y\|) \le L\|x - y\|,$$

and

$$\|Ax - Ay\| \le \frac{L\|x-y\|}{M + \|x-y\|}.$$

Thus, $A\colon E \to E$ is a nonlinear contraction. Now, let $\{x_n\}_{n\in\mathbb{N}}$ be a sequence of elements of S that converges to $x \in S$. Then,

$$\begin{aligned}
\lim_{n\to\infty} Bx_n(t) &= \lim_{n\to\infty} \left(x_0 - f(t_0, x_0) + \int_{t_0}^{t} (t-s) l(s, x_n(s)) dg(s) \right) \\
&= x_0 - f(t_0, x_0) + \lim_{n\to\infty} \int_{t_0}^{t} (t-s) l(s, x_n(s)) dg(s) \\
&= x_0 - f(t_0, x_0) + \int_{t_0}^{t} (t-s) \lim_{n\to\infty} l(s, x_n(s)) dg(s) \\
&= x_0 - f(t_0, x_0) + \int_{t_0}^{t} (t-s) l(s, x(s)) dg(s) \\
&= Bx(t), \quad t \in J.
\end{aligned}$$

Thus, $B\colon S \to E$ is continuous. Now, take $x \in S$ arbitrarily. Then, applying $(A2)$, we find

$$\begin{aligned}
|Bx(t)| &= \left| x_0 - f(t_0, x_0) + \int_{t_0}^{t} (t-s) l(s, x(s)) dg(s) \right| \\
&\le |x_0 - f(t_0, x_0)| + \left| \int_{t_0}^{t} (t-s) l(s, x(s)) dg(s) \right| \\
&\le |x_0 - f(t_0, x_0)| + \int_{t_0}^{t} (t-s) |l(s, x(s))| dg(s) \\
&\le |x_0 - f(t_0, x_0)| + \int_{t_0}^{t} (t-s) h(s) dg(s) \\
&\le |x_0 - f(t_0, x_0)| + a(t_0 + a)\|h\|, \quad t \in J.
\end{aligned}$$

From here,

$$\|Bx\| \le |x_0 - f(t_0, x_0)| + a(t_0 + a)\|h\|.$$

Thus, $B\colon S \to E$ is uniformly bounded. Again, let $t_1, t_2 \in J$. Then, for any $x \in S$, we have

$$\begin{aligned}
&|Bx(t_1) - Bx(t_2)| \\
&= \bigg| x_0 - f(t_0, x_0) + \int_{t_0}^{t_1} (t_1 - s) l(s, x(s)) dg(s) \\
&\quad - x_0 + f(t_0, x_0) - \int_{t_0}^{t_2} (t_2 - s) l(s, x(s)) dg(s) \bigg| \\
&= \bigg| \int_{t_0}^{t_1} (t_1 - s) l(s, x(s)) dg(s) - \int_{t_0}^{t_2} (t_2 - s) l(s, x(s)) dg(s) \bigg| \\
&= \bigg| \int_{t_0}^{t_1} (t_1 - s) l(s, x(s)) dg(s) - \int_{t_0}^{t_1} (t_2 - s) l(s, x(s)) dg(s) \\
&\quad + \int_{t_0}^{t_1} (t_2 - s) l(s, x(s)) dg(s) - \int_{t_0}^{t_2} (t_2 - s) l(s, x(s)) dg(s) \bigg| \\
&= \bigg| \int_{t_0}^{t_1} (t_1 - t_2) l(s, x(s)) dg(s) - \int_{t_1}^{t_2} (t_2 - s) l(s, x(s)) dg(s) \bigg| \\
&\leq \bigg| \int_{t_0}^{t_1} (t_1 - t_2) l(s, x(s)) dg(s) \bigg| + \bigg| \int_{t_1}^{t_2} (t_2 - s) l(s, x(s)) dg(s) \bigg| \\
&\leq \bigg| \int_{t_0}^{t_1} |t_1 - t_2| |l(s, x(s))| |dg(s)| \bigg| + \bigg| \int_{t_1}^{t_2} |t_2 - s| |l(s, x(s))| |dg(s)| \bigg| \\
&\leq \bigg| \int_{t_0}^{t_0 + a} |t_1 - t_2| h(s) dg(s) \bigg| + \bigg| \int_{t_1}^{t_2} (t_0 + a) h(s) dg(s) \bigg| \\
&\leq a |t_1 - t_2| \|h\| + (t_0 + a) |t_1 - t_2| \|h\| \\
&= (t_0 + 2a) |t_1 - t_2| \|h\|.
\end{aligned}$$

Hence, for any $\varepsilon > 0$, there is a $\delta = \delta(\varepsilon) > 0$ such that the inequality $|t_1 - t_2| < \delta$ implies $|Bx(t_1) - Bx(t_2)| < \varepsilon$. This shows that $B(s)$ is an equicontinuous family in E. Applying the Arzelá–Ascoli theorem, we conclude that $B\colon S \to E$ is compact. Now, let $x \in E$ and $y \in S$ be such that $x = Ax + By$. Then, applying $(A2)$, we find

$$\begin{aligned}
|x(t)| &= |Ax(t) + By(t)| \\
&= \left| x_0 - f(t_0, x_0) + f(t, x(t)) + \int_{t_0}^{t} (t-s) l(s, y(s)) dg(s) \right| \\
&\leq |x_0 - f(t_0, x_0)| + |f(t, x(t))| + \left| \int_{t_0}^{t} (t-s) l(s, y(s)) dg(s) \right| \\
&\leq |x_0 - f(t_0, x_0)| + |f(t, x(t)) - f(t, 0)| + |f(t, 0)| \\
&\quad + \int_{t_0}^{t} (t_0 + a) |l(s, y(s))| dg(s) \\
&\leq |x_0 - f(t_0, x_0)| + \frac{L|x(t)|}{M + |x(t)|} + F_0 + \int_{t_0}^{t} (t_0 + a) h(s) dg(s) \\
&\leq |x_0 - f(t_0, x_0)| + L + F_0 + a(t_0 + a)\|h\| \\
&= N, \quad t \in J.
\end{aligned}$$

This gives $\|x\| \leq N$. Then, $x \in S$. Thus, all conditions of Theorem 2.4 hold, and hence the operator $A + B$ has a fixed point in S that is a solution to the SIVP (2.26). This completes the proof. □

2.5 Maximal and Minimal Solutions for First-Order Stieltjes Integro-Differential Equations

In this section, we prove the existence of maximal and minimal solutions for the SIVP (2.26). We need the following definition.

Definition 2.1. A solution y of the SIVP (2.26) is said to be maximal (minimal) provided for any other solution x of the SIVP (2.26), we have

$$x(t) \leq (\geq) y(t), \quad t \in J.$$

For an arbitrary small $\varepsilon > 0$, consider the SIVP

$$\begin{aligned}
(x(t) - f(t, x(t)))'_{g,t} &= \int_{t_0}^{t} l(s, x(s)) dg(s) + \varepsilon, \quad t \in J, \\
x(t_0) &= x_0 + \varepsilon,
\end{aligned} \tag{2.28}$$

where f and g satisfy (A1)–(A3). As in the previous section, a function $x \in \mathscr{C}(J)$ is a solution to the SIVP (2.28) if and only if it satisfies the following Stieltjes integral equation:

$$x(t) = x_0 + \varepsilon - f(t_0, x_0 + \varepsilon) + f(t, x(t)) + \int_{t_0}^{t} (t-s)l(s, x(s))dg(s) + \varepsilon(t - t_0), \quad t \in J.$$

If (A1)–(A3) hold, then the SIVP (2.28) has a solution. The main result in this section is as follows.

Theorem 2.23. *Let (A1)–(A3) hold. Then, the SIVP (2.26) has a maximal and a minimal solution.*

Proof. We will prove that the SIVP (2.26) has a maximal solution. In a similar way, one can also prove that the SIVP (2.26) has a minimal solution. Let $\{\varepsilon_n\}_{n\in\mathbb{N}}$ be a decreasing sequence of positive numbers such that

$$\lim_{n\to\infty} \varepsilon_n = 0.$$

With $y(t, \varepsilon_n)$, $t \in J$, $n \in \mathbb{N}$, we denote a solution to the following SIVP:

$$(x(t) - f(t, x(t)))'_{g,t} = \int_{t_0}^{t} l(s, x(s))dg(s) + \varepsilon_n, \quad t \in J,$$
$$x(t_0) = x_0 + \varepsilon_n.$$

By Theorem 2.17, it follows that for any solution x of the SIVP (2.26), we have

$$x(t) < y(t, \varepsilon_n), \quad t \in J, \quad n \in \mathbb{N}. \tag{2.29}$$

Moreover, the sequence $\{y(t, \varepsilon_n)\}_{n\in\mathbb{N}}$ is a decreasing bounded sequence of real numbers for any fixed $t \in J$. Then, there exists the limit

$$y(t) = \lim_{n\to\infty} y(t, \varepsilon_n)$$

for any fixed $t \in J$. We will show that the sequence $\{y(t, \varepsilon_n)\}_{n\in\mathbb{N}}$ is an equi-continuous family in $\mathscr{C}(J)$. Let $t_1, t_2 \in J$ be arbitrarily chosen. Then,

$$y(t_1, \varepsilon_n) = x_0 + \varepsilon_n - f(t_0, x_0) + f(t_1, y(t_1, \varepsilon_n)) + \int_{t_0}^{t_1} (t_1 - s)l(s, x(s))dg(s) + \varepsilon_n(t_1 - t_0),$$

$$y(t_2,\varepsilon_n) = x_0 + \varepsilon_n - f(t_0,x_0) + f(t_1, y(t_2,\varepsilon_n)) + \int_{t_0}^{t_2} (t_2 - s)l(s,x(s))dg(s)$$
$$+ \varepsilon_n(t_1 - t_0),$$

$n \in \mathbb{N}$, and therefore

$$\begin{aligned}
&|y(t_1,\varepsilon_n) - y(t_2,\varepsilon_n)| \\
&\quad = \Big| x_0 + \varepsilon_n - f(t_0,x_0) + f(t_1,y(t_1,\varepsilon_n)) + \int_{t_0}^{t_1} (t_1 - s)l(s,x(s))dg(s) \\
&\qquad + \varepsilon_n(t_1 - t_0) - x_0 - \varepsilon_n + f(t_0,x_0) - f(t_1,y(t_2,\varepsilon_n)) \\
&\qquad - \int_{t_0}^{t_2} (t_2 - s)l(s,x(s))dg(s) - \varepsilon_n(t_1 - t_0)\Big| \\
&\quad = \Big| f(t_1,y(t_1,\varepsilon_n)) - f(t_2,y(t_2,\varepsilon_n)) \\
&\qquad + \int_{t_0}^{t_1} (t_1 - s)l(s,x(s))dg(s) - \int_{t_0}^{t_2} (t_2 - s)l(s,x(s))dg(s) + \varepsilon_n(t_1 - t_2)\Big| \\
&\quad \le |f(t_1,y(t_1,\varepsilon_n)) - f(t_2,y(t_2,\varepsilon_n))| \\
&\qquad + \left| \int_{t_0}^{t_1} (t_1 - s)l(s,x(s))dg(s) - \int_{t_0}^{t_2} (t_2 - s)l(s,x(s))dg(s) \right| + \varepsilon_n|t_1 - t_2| \\
&\quad = |f(t_1,y(t_1,\varepsilon_n)) - f(t_2,y(t_2,\varepsilon_n))| \\
&\qquad + \Big| \int_{t_0}^{t_1} (t_1 - s)l(s,x(s))dg(s) - \int_{t_0}^{t_1} (t_2 - s)l(s,x(s))dg(s) \\
&\qquad + \int_{t_0}^{t_1} (t_2 - s)l(s,x(s))dg(s) - \int_{t_0}^{t_2} (t_2 - s)l(s,x(s))dg(s)\Big| + \varepsilon_n|t_1 - t_2| \\
&\quad \le |f(t_1,y(t_1,\varepsilon_n)) - f(t_2,y(t_2,\varepsilon_n))| \\
&\qquad + \left| \int_{t_0}^{t_1} (t_1 - t_2)l(s,x(s))dg(s) + \int_{t_2}^{t_1} (t_2 - s)l(s,x(s))dg(s) \right| + \varepsilon_n|t_1 - t_2| \\
&\quad \le |f(t_1,y(t_1,\varepsilon_n)) - f(t_2,y(t_2,\varepsilon_n))| \\
&\qquad + \left| \int_{t_0}^{t_1} (t_1 - t_2)l(s,x(s))dg(s) \right| + \left| \int_{t_2}^{t_1} (t_2 - s)l(s,x(s))dg(s) \right| + \varepsilon_n|t_1 - t_2|
\end{aligned}$$

$$
\begin{aligned}
&\le \left|f(t_1,y(t_1,\varepsilon_n))-f(t_2,y(t_2,\varepsilon_n))\right| \\
&\quad + \left|\int_{t_0}^{t_1} |t_1-t_2||l(s,x(s))|dg(s)\right| + \left|\int_{t_2}^{t_1} (t_0+a)|l(s,x(s))|dg(s)\right| \\
&\quad + \varepsilon_n|t_1-t_2| \\
&\le \left|f(t_1,y(t_1,\varepsilon_n))-f(t_2,y(t_2,\varepsilon_n))\right| \\
&\quad + \left|\int_{t_0}^{t_1} |t_1-t_2|h(s)dg(s)\right| + \left|\int_{t_2}^{t_1} (t_0+a)h(s)dg(s)\right| + \varepsilon_n|t_1-t_2| \\
&\le \left|f(t_1,y(t_1,\varepsilon_n))-f(t_2,y(t_2,\varepsilon_n))\right| + 2(t_0+a)|t_1-t_2|\|h\| + \varepsilon_n|t_1-t_2|,
\end{aligned}
$$

i.e.,

$$
\begin{aligned}
|y(t_1,\varepsilon_n)-y(t_2,\varepsilon_n)| \le &\left|f(t_1,y(t_1,\varepsilon_n))-f(t_2,y(t_2,\varepsilon_n))\right| \\
&+ 2(t_0+a)|t_1-t_2|\|h\| + \varepsilon_n|t_1-t_2|. \qquad (2.30)
\end{aligned}
$$

Since f is continuous on $J\times[-N,N]$, then f is uniformly continuous on $J\times[-N,N]$. Then,

$$
|f(t_1,y(t_1,\varepsilon_n))-f(t_2,y(t_2,\varepsilon_n))| \to 0 \quad \text{as } t_1\to t_2.
$$

From this and (2.30), it follows that for $\varepsilon>0$, there is a $\delta=\delta(\varepsilon)>0$ such that the inequality $|t_1-t_2|<\delta$ implies $|y(t_1,\varepsilon_n)-y(t_2,\varepsilon_n)|<\varepsilon$, $n\in\mathbb{N}$. Therefore, the sequence $\{y(t,\varepsilon_n)\}_{n\in\mathbb{N}}$ is an equicontinuous family in S and for any $t\in J$, we have

$$
\begin{aligned}
y(t) &= \lim_{n\to\infty} y(t,\varepsilon_n) \\
&= \lim_{n\to\infty}\Bigg(x_0-f(t_0,x_0)+f(t,y(t,\varepsilon_n)) \\
&\quad + \int_{t_0}^{t}(t-s)l(s,y(s,\varepsilon_n))dg(s)+\varepsilon_n(t-t_0)\Bigg) \\
&= x_0-f(t_0,x_0)+f(t,y(t))+\int_{t_0}^{t}(t-s)l(s,y(s))dg(s), \quad t\in J.
\end{aligned}
$$

Hence, applying Lemma 2.1, it follows that y is a solution to the SIVP (2.26), and by the inequality (2.29), we obtain

$$
\begin{aligned}
x(t) &\le \lim_{n\to\infty} y(t,\varepsilon_n) \\
&= y(t), \quad t\in J,
\end{aligned}
$$

for any solution x of the SIVP (2.26). Consequently, y is a maximal solution to the SIVP (2.26). This completes the proof. □

2.6 Uniqueness of Solutions for First-Order Stieltjes Integro-Differential Equations

In this section, we give conditions under which the SIVP (2.26) has a unique solution. Our main result in this section is as follows.

Theorem 2.24. *Suppose that (A1)–(A3) hold. Assume that there exists a function $G\colon J \times \mathbb{R}_+ \to \mathbb{R}_+$ such that*

$$\begin{aligned}&|l(t, x_1(t)) - l(t, x_2(t))|\\ &\quad \leq G\left(s, \sup_{t_0 \leq s \leq t} |(x_1(s) - f(s, x_1(s))) - (x_2(s) - f(s, x_2(s)))|\right),\end{aligned}$$

for $t \in J$ and $x \in E$, and the SIVP

$$\begin{aligned}m'_g(t) &= \int_{t_0}^{t} G(s, m(s))dg(s), \quad t \in J,\\ m(t_0) &= 0,\end{aligned}$$

has a unique solution, the zero solution.

Proof. By Theorem 2.23, it follows that the SIVP (2.26) has a solution. Now, assume that the SIVP (2.26) has two solutions $x_1, x_2 \in E$. For them, we have

$$\begin{aligned}(x_1(t) - f(t, x_1(t)))'_{g,t} &= \int_{t_0}^{t} l(s, x_1(s))dg(s),\\ (x_2(t) - f(t, x_2(t)))'_{g,t} &= \int_{t_0}^{t} l(s, x_2(s))dg(s), \quad t \in J.\end{aligned}$$

Define the function $m(t) = (x_1(t) - f(t, x_1(t))) - (x_2(t) - f(t, x_2(t)))$, $t \in J$.

Then,

$$\begin{aligned}
|m(t)|'_g &\le |m'_g(t)| \\
&= |((x_1(t) - f(t, x_1(t))) - (x_2(t) - f(t, x_2(t))))'_{g,t}| \\
&= \left|(x_1(t) - f(t, x_1(t)))'_{g,t} - (x_2(t) - f(t, x_2(t)))'_{g,t}\right| \\
&= \left|\int_{t_0}^{t} l(s, x_1(s))dg(s) - \int_{t_0}^{t} l(s, x_2(s))dg(s)\right| \\
&= \left|\int_{t_0}^{t} (l(s, x_1(s)) - l(s, x_2(s)))dg(s)\right| \\
&\le \int_{t_0}^{t} |l(s, x_1(s)) - l(s, x_2(s))|dg(s) \\
&\le \int_{t_0}^{t} G(s, |m(s)|)dg(s), \quad t \in J.
\end{aligned}$$

Note that

$$\begin{aligned}
m(t_0) &= (x_1(t_0) - f(t_0, x_1(t_0))) - (x_2(t_0) - f(t_0, x_2(t_0))) \\
&= (x_0 - f(t_0, x_0)) - (x_0 - f(t_0, x_0)) \\
&= 0.
\end{aligned}$$

Thus, we get the following SIVP:

$$\begin{aligned}
|m(t)|'_{g,t} &\le \int_{t_0}^{t} G(s, |m(s)|)dg(s), \quad t \in J, \\
|m(t_0)| &= 0.
\end{aligned}$$

Now, applying Theorem 2.17, we get $|m(t)| = 0$, $t \in J$. From here, we obtain

$$x_1(t) - f(t, x_1(t)) = x_2(t) - f(t, x_2(t)), \quad t \in J.$$

Applying (A1), we conclude that $x_1(t) = x_2(t)$, $t \in J$. This completes the proof. □

2.7 Extremal Solutions

In this section, we give a criterion for the existence of a minimal solution and a maximal solution for the SIVP (2.26). The proof is based on the theory of cone.

Definition 2.2. A nonempty closed set K in the Banach space E is said to be a cone provided

(1) $K + K \subset K$,
(2) $\lambda K \subset K$ for any $\lambda \geq 0$,
(3) $(-K) \bigcap K = \{0\}$.

We introduce the order relation "$\leq$" in E as follows.

Definition 2.3. Let K be a cone in E and let $x, y \in E$. Then, we state that $x \leq y$ provided $y - x \in K$.

Definition 2.4. A cone K in the Banach space E is said to be normal provided there is a constant $N > 0$ such that $\|x\| \leq N\|y\|$ for $x, y \in K$ such that $x \leq y$.

Definition 2.5. For any $a, b \in E$, $a \leq b$, the order interval $[a, b]$ is a set in E given by

$$[a, b] = \{x \in E \colon a \leq x \leq b\}.$$

Definition 2.6. A mapping $T\colon [a, b] \to E$ is said to be nondecreasing or monotone increasing provided

$$Tx \leq Ty$$

for $x, y \in [a, b]$ such that $x \leq y$.

To prove our criterion for the existence of minimal and maximal solutions of the SIVP (2.26), we use the following result due to Dhage (1999).

Theorem 2.25 (The Dhage Fixed Point Theorem II (Dhage, 1999)). *Let K be a cone in the Banach space E and let $a, b \in E$ be such that $a \leq b$. Suppose that $A, B\colon [a, b] \to E$ are two nondecreasing operators such that*

(1) *A is a nonlinear contraction,*
(2) *B is completely continuous,*
(3) *$Ax + Bx \in [a, b]$ for any $x \in [a, b]$.*

If the cone K is normal, then the operator equation $Ax + Bx = x$ has the least and the greatest solution in $[a, b]$.

In $E = \mathscr{C}(J)$, define the cone $K = \{x \in E\colon x(t) \geq 0,\ t \in J\}$. Note that K is a normal cone.

Definition 2.7. A function $x \in E$ is said to be a lower solution for the SIVP (2.26) provided $x \mapsto x - f(t, x)$ is continuous and

$$(x(t) - f(t, x(t)))'_{g,t} \leq \int_{t_0}^{t} l(s, x(s))dg(s), \quad t \in J,$$
$$x(t_0) \leq x_0.$$

Definition 2.8. A function $x \in E$ is said to be an upper solution for the SIVP (2.26) provided $x \mapsto x - f(t, x)$ is continuous and

$$(x(t) - f(t, x(t)))'_{g,t} \geq \int_{t_0}^{t} l(s, x(s))dg(s), \quad t \in J,$$
$$x(t_0) \geq x_0.$$

In the following, we list some important assumptions:

(B1) The SIVP (2.26) has a lower solution y and an upper solution z on J with $y \leq z$.
(B2) The function $x \mapsto x - f(t, x)$ is increasing on $[\min_{t \in J} y(t), \max_{t \in J} z(t)]$.
(B3) The functions $x \mapsto f(t, x)$ and $x \mapsto g(t, x)$ are nondecreasing on x for any $t \in J$.
(B4) There exists a continuous function $h\colon J \to \mathbb{R}$ such that $l(t, z(t)) \leq h(t)$, $t \in J$.

Our main result in this section is as follows.

Theorem 2.26. *Suppose that* (*B*1)–(*B*4) *hold. Then, the SIVP* (2.26) *has a minimal solution and a maximal solution in* $[y, z]$ *defined on* J.

Proof. For $x \in E$, define the operators

$$Ax(t) = f(t, x(t)),$$
$$Bx(t) = x_0 - f(t_0, x_0) + \int_{t_0}^{t} (t - s)l(s, x(s))dg(s), \quad t \in J.$$

By $(B1)$, it follows that $A, B\colon [y,z] \to E$. As in the proof of Theorem 2.22, we find that A is a nonlinear contraction and B is completely continuous on $[y,z]$. Let $x_1, x_2 \in E$ be such that $x_1(t) \le x_2(t)$, $t \in J$. By $(B3)$, we find

$$\begin{aligned} f(t, x_1(t)) &\le f(t, x_2(t)), \\ l(t, x_1(t)) &\le l(t, x_2(t)), \quad t \in J. \end{aligned}$$

Hence, we obtain

$$\begin{aligned} Ax_1(t) &= f(t, x_1(t)) \\ &\le f(t, x_2(t)) \\ &= Ax_2(t), \quad t \in J, \end{aligned}$$

and

$$\begin{aligned} Bx_1(t) &= x_0 - f(t_0, x_0) + \int_{t_0}^{t} l(s, x_1(s))dg(s) \\ &\le x_0 - f(t_0, x_0) + \int_{t_0}^{t} l(s, x_2(s))dg(s) \\ &= Bx_2(t), \quad t \in J. \end{aligned}$$

Thus, the operators A and B are nondecreasing on $[y,z]$. Next, for $x \in [y,z]$, we get

$$\begin{aligned} y(t) &\le x_0 - f(t_0, x_0) + f(t, y(t)) + \int_{t_0}^{t} l(s, y(s))dg(s) \\ &\le x_0 - f(t_0, x_0) + f(t, x(t)) + \int_{t_0}^{t} l(s, x(s))dg(s) \\ &\le x_0 - f(t_0, x_0) + f(t, z(t)) + \int_{t_0}^{t} l(s, z(s))dg(s) \\ &\le z(t), \quad t \in J. \end{aligned}$$

From this, we conclude that $y(t) \le Ax(t) + Bx(t) \le z(t)$, $t \in J$, for any $x \in [y,z]$. Now, applying Theorem 2.7, we conclude that the SIVP (2.26) has a minimal solution and a maximal solution in $[y,z]$ defined on J. This completes the proof. □

2.8 Advanced Practical Problems

Problem 2.3. Let $f \in \mathscr{C}(\mathbb{R})$ and $|f(x) - f(y)| \le L|x - y|$, $x, y \in \mathbb{R}$, where $L \in (0, 1)$. Using the Stieltjes–Gronwall-type inequality, prove that the SIVP

$$\begin{aligned} x'_g &= f(x), \quad t \in I, \\ x(t_0) &= x_0, \end{aligned}$$

for some $x_0 \in \mathbb{R}$, has at least one solution.

Problem 2.4. Let $y, z \in \mathscr{C}(I)$ be such that

$$\left(y(t) + \frac{1}{2}\sin(y(t))\right)'_{g,t} < (y(t))^2, \quad t \in I,$$

$$\left(z(t) + \frac{1}{2}\sin(z(t))\right)'_{g,t} \ge (z(t))^2, \quad t \in I,$$

and $y(t_0) < z(t_0)$. Then, prove that $y(t) < z(t)$, $t \in I$, $t \ge t_0$.

Problem 2.5. Let $y, z \in \mathscr{C}(I)$ be such that

$$\left(7y(t) - \frac{1}{8}\cos(y(t))\right)'_{g,t} = (y(t))^2, \quad t \in I,$$

$$\left(7z(t) - \frac{1}{8}\cos(z(t))\right)'_{g,t} = 2(z(t))^2 + 8, \quad t \in I,$$

and $y(t_0) < z(t_0)$. Then, prove that $y(t) < z(t)$, $t \in I$, $t \ge t_0$.

Problem 2.6. Let $y, z \in \mathscr{C}(I)$ be such that

$$(3y(t) - \sin(2 + y(t)))'_{g,t} < \int_{t_0}^{t} (y(s))^2 dg(s), \quad t \in I, \quad t \ge t_0,$$

$$(3z(t) - \sin(2 + z(t)))'_{g,t} \ge \int_{t_0}^{t} (z(s))^2 dg(s), \quad t \in I, \quad t \ge t_0,$$

and $y(t_0) < z(t_0)$. Then, prove that $y(t) < z(t)$, $t \in I$, $t \ge t_0$.

Problem 2.7. Consider the SIVP

$$(12x(t) - 3\sin(2x(t)))'_{g,t} = \int_{t_0}^{t} \left((x(s))^2 + 8(x(s))^4\right) dg(s), \quad t \in I, \quad t \geq t_0,$$

$$x(t_0) = x_0$$

for some $x_0 \in \mathbb{R}$. Then, prove that

(1) the considered SIVP has at least one solution,
(2) the considered SIVP has a minimal solution and a maximal solution.

Problem 2.8. Consider the SIVP

$$\left(x(t) - \frac{1}{8(2 + \sin(x(t)))}\right)'_{g,t} = \int_{t_0}^{t} \left(1 + (x(s))^4 - \sin(x(s))\right) dg(s),$$

$$t \in I, \quad t \geq t_0,$$

$$x(t_0) = x_0$$

for some $x_0 \in \mathbb{R}$. Then, prove that

(1) the considered SIVP has at least one solution,
(2) the considered SIVP has a minimal solution and a maximal solution.

Problem 2.9. Consider the SIVP

$$(640x(t) - (\sin(2x(t)))^2 - 2(\cos(x(t)))^2)'_{g,t}$$

$$= \int_{t_0}^{t} \left((x(s))^2 + \frac{1}{1 + (x(s))^4}\right) dg(s), \quad t \in I, \quad t \geq t_0,$$

$$x(t_0) = x_0$$

for some $x_0 \in \mathbb{R}$. Then, prove that

(1) the considered SIVP has at least one solution,
(2) the considered SIVP has a minimal solution and a maximal solution.

Chapter 3

Basic Stieltjes Inequalities

This chapter deals with some basic Stieltjes inequalities used in the remainder of the book. These inequalities include Jensen-type inequalities, Young-type inequalities, Hölder-type inequalities, Minkowski-type inequalities, Steffensen-type inequalities, and Hermite–Hadamard-type inequalities.

Suppose that $I \subset \mathbb{R}$ and $g\colon I \to \mathbb{R}$ is a monotone nondecreasing function that is continuous from the left everywhere on I. Let $a, t_0 \in I$, $a \geq t_0$ and $J = [t_0, a]$.

3.1 Auxiliary Results

In this section, we derive some auxiliary results needed for the main results in the following sections.

Theorem 3.1. *Let $f \in \mathscr{C}_g^2((c,d))$ be a convex function. Then, for each $t \in (c,d)$, there exists a constant $A \in \mathbb{R}$ such that*

$$f(x) - f(t) \geq A(x-t), \quad x \in (c,d), \quad x \geq t.$$

Proof. By the Stieltjes–Taylor formula, we find

$$f(x) = f(t) + f_g'\left(g^{-1}(t)\right) + \int_t^{g^{-1}(x)} f_g''\left(g^{-1}(\tau)\right) g(x,\tau)dg(\tau),$$
$$x \in (c,d), \quad x \geq t.$$

Since f is a convex function, we have $f_g''\left(g^{-1}(x)\right) \geq 0$, $x \in (c,d)$, $x \geq t$. Therefore,

$$f(x) \geq f(t) + f_g'\left(g^{-1}(t)\right)(x-t), \quad x \in (c,d), \quad x \geq t,$$

i.e.,

$$f(x) - f(t) \geq f_g'\left(g^{-1}(t)\right)(x-t), \quad x \in (c,d), \quad x \geq t.$$

Set $A = f_g'(g^{-1}(t))$, and we get the desired result. This completes the proof. $\square$

Theorem 3.2 (The Classical Young Inequality). *Let $a, b > 0$, $p.q \in (1, \infty)$, $\frac{1}{p} + \frac{1}{q} = 1$. Then,*

$$ab \leq \frac{a^p}{p} + \frac{b^q}{q}. \tag{3.1}$$

The equality holds if and only if $a^p = b^q$.

Proof. Since the mapping $x \mapsto e^x$ is convex, we get

$$\begin{aligned} ab &= e^{\log a + \log b} \\ &= e^{\frac{1}{p}\log a^p + \frac{1}{q}\log b^q} \\ &\leq \frac{1}{p} e^{\log a^p} + \frac{1}{q} e^{\log b^q} \\ &= \frac{a^p}{p} + \frac{b^q}{q}. \end{aligned}$$

Now, we will prove that the equality holds if and only if $a^p = b^q$. When $a = 0$ or $b = 0$, the assertion is evident. Assume that $a \neq 0$ and $b \neq 0$. We set $t = ab^{-\frac{q}{p}}$ or $a = tb^{\frac{q}{p}}$. Then,

$$\begin{aligned} \frac{a^p}{p} + \frac{b^q}{q} - ab &= \frac{t^p b^q}{p} + \frac{b^q}{q} - tb^{\frac{q}{p}+1} \\ &= \frac{t^p b^q}{p} + \frac{b^q}{q} - tb^{q\left(\frac{1}{p}+\frac{1}{q}\right)} \\ &= \frac{t^p b^q}{p} + \frac{b^q}{q} - tb^q \\ &= b^q \left(\frac{t^p}{p} + \frac{1}{q} - t\right). \end{aligned}$$

Hence,

$$\frac{a^p}{p} + \frac{b^q}{q} - ab = 0,$$

equivalently,

$$b^q\left(\frac{t^p}{p}+\frac{1}{q}-t\right)=0.$$

This yields

$$\frac{t^p}{p}+\frac{1}{q}-t=0 \quad \text{if and only if} \quad t=1, \quad \text{i.e.,} \quad a^p=b^q$$

(because, if we take $f(t)=\frac{t^p}{p}+\frac{1}{q}-t$, we have $f(1)=0$, $f'(t)=t^{p-1}-1$, $f'(t)\geq 0$ for $t\geq 1$ and $f'(t)\leq 0$ for $t\in[0,1]$). This completes the proof. □

Theorem 3.3. (The Classical Young Inequality with ε). *Let $a,b,\varepsilon>0$, $p,q\in(1,\infty)$, $\frac{1}{p}+\frac{1}{q}=1$. Then,*

$$ab\leq \varepsilon a^p+C(\varepsilon)b^q, \quad \textit{where} \quad C(\varepsilon)=\frac{1}{q\,(\varepsilon p)^{\frac{q}{p}}}.$$

The equality holds if and only if $a^p=\frac{b^q}{(\varepsilon p)^q}$.

Proof. Let $a_1=(\varepsilon p)^{\frac{1}{p}}\,a$ and $b_1=b\,(\varepsilon p)^{\frac{-1}{p}}$. Applying the Young inequality (3.1) for a_1 and b_1, we get

$$\begin{aligned}ab&\leq \frac{\varepsilon p a^p}{p}+\frac{1}{(\varepsilon p)^{\frac{q}{p}}\,q}b^q\\&=\varepsilon a^p+C(\varepsilon)b^q.\end{aligned}$$

The equality holds if and only if $a_1^p=b_1^q$. This completes the proof. □

Theorem 3.4. (The Hölder Inequality). *Let $\xi_j,\eta_j\in\mathbb{R}$, $j\in\{1,\ldots,l\}$, $p,q>1$, $\frac{1}{p}+\frac{1}{q}=1$. Then,*

$$\sum_{k=1}^{l}|\xi_k\eta_k|\leq\left(\sum_{k=1}^{l}|\xi_k|^p\right)^{\frac{1}{p}}\left(\sum_{k=1}^{l}|\eta_k|^q\right)^{\frac{1}{q}}. \tag{3.2}$$

The equality holds if and only if $|\xi_j|^p\left(\sum_{k=1}^{l}|\eta_k|^q\right)=|\eta_j|^q\left(\sum_{k=1}^{l}|\xi_k|^p\right)$ for $j\in\{1,\ldots,l\}$.

Proof. If $\sum_{k=1}^{l} |\xi_k|^p = 0$ $\left(\sum_{k=1}^{l} |\eta_k|^q = 0\right)$, then $\xi_k = 0$ ($\eta_k = 0$) for $k \in \{1, \ldots, l\}$, and the assertion is evident. Assume that $\sum_{k=1}^{l} |\xi_k|^p \neq 0$ and $\sum_{k=1}^{l} |\eta_k|^q \neq 0$. We set

$$a = \frac{|\xi_j|}{\left(\sum_{k=1}^{l} |\xi_k|^p\right)^{\frac{1}{p}}} \quad \text{and} \quad b = \frac{|\eta_j|}{\left(\sum_{k=1}^{l} |\eta_k|^q\right)^{\frac{1}{q}}}$$

for some $j \in \{1, \ldots, l\}$. Applying the Young inequality (3.1), we obtain

$$\frac{|\xi_j \eta_j|}{\left(\sum_{k=1}^{l} |\xi_k|^p\right)^{\frac{1}{p}} \left(\sum_{k=1}^{l} |\eta_k|^q\right)^{\frac{1}{q}}} \leq \frac{|\xi_j|^p}{p \sum_{k=1}^{l} |\xi_k|^p} + \frac{|\eta_j|^q}{q \sum_{k=1}^{l} |\eta_k|^q},$$

whereupon

$$\begin{aligned} \sum_{j=1}^{l} \frac{|\xi_j \eta_j|}{\left(\sum_{k=1}^{l} |\xi_k|^p\right)^{\frac{1}{p}} \left(\sum_{k=1}^{l} |\eta_k|^q\right)^{\frac{1}{q}}} &\leq \sum_{j=1}^{l} \frac{|\xi_j|^p}{p \sum_{k=1}^{l} |\xi_k|^p} + \sum_{j=1}^{l} \frac{|\eta_j|^q}{q \sum_{k=1}^{l} |\eta_k|^q} \\ &= \frac{\sum_{j=1}^{l} |\xi_j|^p}{p \sum_{k=1}^{l} |\xi_k|^p} + \frac{\sum_{j=1}^{l} |\eta_j|^q}{q \sum_{k=1}^{l} |\eta_k|^q} \\ &= \frac{1}{p} + \frac{1}{q} \\ &= 1, \end{aligned}$$

i.e.,

$$\sum_{j=1}^{l} \frac{|\xi_j \eta_j|}{\left(\sum_{k=1}^{l} |\xi_k|^p\right)^{\frac{1}{p}} \left(\sum_{k=1}^{l} |\eta_k|^q\right)^{\frac{1}{q}}} \leq 1,$$

which is (3.2). This completes the proof. □

Theorem 3.5. (The Minkowski Inequality). *Let $\xi_k, \eta_k \in \mathbb{R}$, $k \in \{1, \ldots, l\}$. Then, for every $p \geq 1$, we have*

$$\left(\sum_{k=1}^{l} |\xi_k + \eta_k|^p\right)^{\frac{1}{p}} \leq \left(\sum_{k=1}^{l} |\xi_k|^p\right)^{\frac{1}{p}} + \left(\sum_{k=1}^{l} |\eta_k|^p\right)^{\frac{1}{p}}.$$

Proof. If $\sum_{k=1}^{l} |\xi_k + \eta_k|^p = 0$, then the assertion is evident. Assume that $\sum_{k=1}^{l} |\xi_k + \eta_k|^p \neq 0$. For $p = 1$, we have

$$\begin{aligned}\sum_{k=1}^{l} |\xi_k + \eta_k| &\leq \sum_{k=1}^{l} (|\xi_k| + |\eta_k|) \\ &= \sum_{k=1}^{l} |\xi_k| + \sum_{k=1}^{l} |\eta_k|.\end{aligned}$$

Now, suppose $p > 1$. We choose q to be the conjugate of p. Then, $q(p-1) = p$, and using the Hölder inequality (3.2), we obtain

$$\begin{aligned}\sum_{k=1}^{l} |\xi_k + \eta_k|^p &= \sum_{k=1}^{l} |\xi_k + \eta_k|^{p-1} |\xi_k + \eta_k| \\ &\leq \sum_{k=1}^{l} |\xi_k + \eta_k|^{p-1} |\xi_k| + \sum_{k=1}^{l} |\xi_k + \eta_k|^{p-1} |\eta_k| \\ &\leq \left(\sum_{k=1}^{l} |\xi_k + \eta_k|^{(p-1)q} \right)^{\frac{1}{q}} \left(\sum_{k=1}^{l} |\xi_k|^p \right)^{\frac{1}{p}} \\ &\quad + \left(\sum_{k=1}^{l} |\xi_k + \eta_k|^{(p-1)q} \right)^{\frac{1}{q}} \left(\sum_{k=1}^{l} |\eta_k|^p \right)^{\frac{1}{p}} \\ &= \left(\sum_{k=1}^{l} |\xi_k + \eta_k|^p \right)^{\frac{1}{q}} \left(\left(\sum_{k=1}^{l} |\xi_k|^p \right)^{\frac{1}{p}} + \left(\sum_{k=1}^{l} |\eta_k|^p \right)^{\frac{1}{p}} \right),\end{aligned}$$

whereupon

$$\left(\sum_{k=1}^{l} |\xi_k + \eta_k|^p \right)^{1-\frac{1}{q}} \leq \left(\sum_{k=1}^{l} |\xi_k|^p \right)^{\frac{1}{p}} + \left(\sum_{k=1}^{l} |\eta_k|^p \right)^{\frac{1}{p}},$$

which completes the proof. □

3.2 The Stieltjes–Young Inequality

In this section, we derive the Stieltjes analogue of the classical Young inequality.

Theorem 3.6 (The Stieltjes–Young Inequality). *Let $f \in \mathscr{C}(J)$ be a strictly increasing positive function. If $f(t_0) = t_1$ and $b \in J$ is such that $b = f^{-1}(c)$ for some $c \in [t_1, f(a)]$, then we have*

$$bc \le \int_{t_0}^{b} f(t+)dg(t) + \int_{t_1}^{c} f^{-1}(t+)dg(t).$$

Proof. Since $f \in \mathscr{C}(J)$ is a strictly increasing function on J, it follows that f^{-1} is a strictly increasing function on $[t_1, f(a)]$. Then, we have

$$\int_{t_1}^{c} f^{-1}(t+)dg(t) \ge \int_{t_1}^{c} f^{-1}(t)dg(t).$$

Next, we find

$$\begin{aligned}\int_{t_0}^{f^{-1}(c)} tf_g'(t)dg(t) &= \int_{f(t_0)}^{f(f^{-1}(c))} f^{-1}(t)dg(t)\\ &= \int_{t_1}^{c} f^{-1}(t)dg(t).\end{aligned}$$

On the other hand, using integration by parts, we arrive at

$$\begin{aligned}\int_{t_0}^{f^{-1}(c)} tf_g'(t)dg(t) &= tf(t)\Big|_{t=t_0}^{t=f^{-1}(c)} - \int_{t_0}^{f^{-1}(c)} f(t+)dg(t)\\ &= cf^{-1}(c) - t_0 f(t_0) - \int_{t_0}^{f^{-1}(c)} f(t+)dg(t)\\ &= bc - t_0 t_1 + \int_{f^{-1}(c)}^{t_0} f(t+)dg(t).\end{aligned}$$

Then,

$$\begin{aligned}\int_{t_1}^{c} f^{-1}(t+)dg(t) &\ge \int_{t_1}^{c} f^{-1}(t)dg(t)\\ &= \int_{t_0}^{f^{-1}(c)} tf_g'(t)dg(t)\\ &= bc - t_0 t_1 + \int_{f^{-1}(c)}^{t_0} f(t+)dg(t).\end{aligned}$$

Hence,

$$\int_{t_0}^{b} f(t+)dg(t) + \int_{t_1}^{c} f^{-1}(t+)dg(t)$$
$$\geq \int_{t_0}^{b} f(t+)dg(t) + bc - t_0 t_1 + \int_{f^{-1}(c)}^{t_0} f(t+)dg(t)$$
$$= bc - t_0 t_1.$$

This completes the proof. □

Problem 3.1. Let $p, q > 1$ and $\frac{1}{p} + \frac{1}{q} = 1$. Also, let $t_1 = (t_0)^{p-1}$, $b = c^{\frac{q}{p}}$ for some $c \in [t_1, a^{p-1}]$. Then, prove that

$$bc \leq \int_{t_0}^{b} (t)^{p-1} dg(t) + \int_{t_1}^{c} (t)^{\frac{q}{p}} dg(t).$$

Hint 3.7. Use Theorem 3.6 for $f(t) = t^{p-1}$, $t \in J$.

3.3 Stieltjes–Jensen Inequalities

We start with the following Stieltjes version of Jensen-type inequality.

Theorem 3.7. *Let $a, b \in \mathbb{R}$, $a \leq b$, $l \in \mathscr{C}((a,b))$ be such that*

$$l(x) \geq (g(b) - g(a)) \int_a^b l(t)dg(t), \quad x \in [a,b].$$

Also, let F be a convex function on $\mathbb{R}$. Then,

$$F\left(\frac{\int_a^b l(t)dg(t)}{g(b)-g(a)}\right) \leq \frac{1}{g(b)-g(a)} \int_a^b F(l(t))dg(t).$$

Proof. Let

$$\tau = \frac{1}{g(b)-g(a)} \int_a^b l(t)dg(t).$$

Then,

$$l(x) \geq \frac{1}{g(b)-g(a)} \int_a^b l(t)dg(t) \quad \text{for some } x \in [a,b].$$

Since F is a convex function, applying Theorem 3.1, that there exists a constant $A \in \mathbb{R}$ such that $F(l(t)) - F(\tau) \geq A(l(t) - \tau)$, $t \in [a, b]$. Hence,

$$\begin{aligned}
\int_a^b F(l(t))dg(t) - (g(b) - g(a))F\left(\frac{\int_a^b l(t)dg(t)}{g(b) - g(a)}\right) \\
&= \int_a^b F(l(t))dg(t) - (g(b) - g(a))F(\tau) \\
&= \int_a^b F(l(t))dg(t) - \int_a^b F(\tau)dg(t) \\
&= \int_a^b (F(l(t)) - F(\tau))dg(t) \\
&\geq A\int_a^b (l(t) - \tau)dg(t) \\
&= A\left(\int_a^b l(t)dg(t) - \tau(g(b) - g(a))\right) \\
&= A\left(\int_a^b l(t)dg(t) - \int_a^b l(t)dg(t)\right) \\
&= 0.
\end{aligned}$$

This completes the proof. □

Theorem 3.8. *Let $a, b \in \mathbb{R}$, $a \leq b$, $l, h \in \mathscr{C}((a, b))$ be such that*

$$\int_a^b |h(t)|dg(t) > 0$$

and

$$l(x) \geq \frac{\int_a^b |h(t)|l(t)dg(t)}{\int_a^b |h(t)|dg(t)}, \quad x \in [a, b].$$

Also, let F be a convex function on $\mathbb{R}$. Then,

$$F\left(\frac{\int_a^b |h(t)|l(t)dg(t)}{\int_a^b |h(t)|dg(t)}\right) \leq \frac{\int_a^b |h(t)|F(l(t))dg(t)}{\int_a^b |h(t)|dg(t)}.$$

Proof. Let

$$\tau = \frac{\int_a^b |h(t)|l(t)dg(t)}{\int_a^b |h(t)|dg(t)}.$$

Then, $l(x) \geq \tau$, $x \in [a, b]$. Since F is a convex function, applying Theorem 3.1, there exists a constant $A \in \mathbb{R}$ such that $F(l(t)) - F(\tau) \geq A(l(t) - \tau)$, $t \in [a, b]$. Hence,

$$\begin{aligned}
&\int_a^b |h(t)|F(l(t))dg(t) - \int_a^b |h(t)|dg(t) \left(F\left(\frac{\int_a^b |h(t)|l(t)dg(t)}{\int_a^b |h(t)|dg(t)} \right)\right) \\
&= \int_a^b |h(t)|F(l(t))dg(t) - F(\tau) \int_a^b |h(t)|dg(t) \\
&= \int_a^b |h(t)|F(l(t))dg(t) - \int_a^b |h(t)|F(\tau)dg(t) \\
&= \int_a^b |h(t)|(F(l(t)) - F(\tau))dg(t) \\
&\geq A \int_a^b |h(t)|(l(t) - \tau)dg(t) \\
&= A \left(\int_a^b |h(t)|l(t)dg(t) \right. \\
&\qquad \left. - \int_a^b |h(t)|dg(t) \frac{\int_a^b |h(t)|l(t)dg(t)}{\int_a^b |h(t)|dg(t)} \right) \\
&\geq A \left(\int_a^b |h(t)|l(t)dg(t) - \int_a^b |h(t)|l(t)dg(t) \right) \\
&= 0.
\end{aligned}$$

This completes the proof. □

Corollary 3.1. *Let $a, b \in \mathbb{R}$, $a \leq b$, $l, h \in \mathscr{C}((a, b))$ be such that $g > 0$ on $[a, b]$, and*

$$\int_a^b |h(t)|dg(t) > 0$$

and

$$l(x) \geq \frac{\int_a^b |h(t)| l(t) dg(t)}{\int_a^b |h(t)| dg(t)}, \quad x \in [a, b].$$

Then,

$$\left(\frac{\int_a^b |h(t)| l(t) dg(t)}{\int_a^b |h(t)| dg(t)} \right)^\alpha \leq \frac{\int_a^b |h(t)| (l(t))^\alpha dg(t)}{\int_a^b |h(t)| dg(t)}$$

if $\alpha > 1$, *and*

$$\left(\frac{\int_a^b |h(t)| l(t) dg(t)}{\int_a^b |h(t)| dg(t)} \right)^\alpha \geq \frac{\int_a^b |h(t)| (l(t))^\alpha dg(s)}{\int_a^b |h(t)| dg(t)}$$

if $\alpha \in (0, 1)$.

Problem 3.2. Let $a, b \in \mathbb{R}$, $a \leq b$, $l, h \in \mathscr{C}((a, b))$ be such that

$$\int_a^b |h(t)| dg(t) > 0$$

and $l > 0$ on $[a, b]$, and

$$l(x) \geq \frac{\int_a^b |h(t)| l(t) dg(t)}{\int_a^b |h(t)| dg(t)}, \quad x \in [a, b].$$

Then, prove that

$$\log \left(\frac{\int_a^b |h(t)| l(t) dg(t)}{\int_a^b |h(t)| dg(t)} \right) \leq \frac{\int_a^b |h(t)| \log(l(t)) dg(t)}{\int_a^b |h(t)| dg(t)}.$$

Definition 3.1. A function $\phi : [0, \infty) \to \mathbb{R}$ is said to be superquadratic provided there exists a function $C \colon [0, \infty) \to \mathbb{R}$ such that

$$\phi(t) - \phi(s) - \phi(|t - s|) \geq C(s)(t - s), \quad s, t \geq 0.$$

Theorem 3.9. *Let* $a, b \in \mathbb{R}$, $a \leq b$, *and* $f \colon [a, b] \to [0, \infty)$ *be such that* $f \in \mathscr{C}([a, b])$, *and* $\phi \colon [0, \infty) \to \mathbb{R}$ *be continuous and superquadratic. Then,*

$$\phi \left(\frac{\int_a^b f(t) dg(t)}{g(b) - g(a)} \right) \leq \frac{1}{g(b) - g(a)} \int_a^b \left(\phi(f(s)) - \phi \left(\left| f(s) - \frac{\int_a^b f(t) dg(t)}{g(b) - g(a)} \right| \right) \right) dg(s).$$

Proof. Since $\phi : [0,\infty) \to \mathbb{R}$ is a superquadratic function, there exists a function $C : [0,\infty) \to \mathbb{R}$ such that

$$\phi(y) \geq \phi(x_0) + \phi(|y - x_0|) + C(x_0)(y - x_0), \quad x_0, y \geq 0.$$

Set

$$x_0 = \frac{1}{g(b) - g(a)} \int_a^b f(t)dg(t).$$

Then, for $s, x_0 \geq 0$,

$$\begin{aligned}\phi(f(s)) \geq \phi\left(\frac{\int_a^b f(t)dg(t)}{g(b) - g(a)}\right) &+ \phi\left(\left|f(s) - \frac{\int_a^b f(t)dg(t)}{g(b) - g(a)}\right|\right) \\ &+ C(x_0)(f(s) - x_0),\end{aligned}$$

i.e.,

$$\phi(f(s)) - \phi\left(\frac{\int_a^b f(t)dg(t)}{g(b) - g(a)}\right) - \phi\left(\left|f(s) - \frac{\int_a^b f(t)dg(t)}{g(b) - g(a)}\right|\right) \geq C(x_0)(f(s) - x_0).$$

Now, integrating the last inequality, in the sense of Stieltjes, from a to b, we find

$$\begin{aligned}\int_a^b &\left(\phi(f(s)) - \phi\left(\frac{\int_a^b f(t)dg(t)}{g(b) - g(a)}\right) - \phi\left(\left|f(s) - \frac{\int_a^b f(t)dg(t)}{g(b) - g(a)}\right|\right)\right)dg(s) \\ &\geq C(x_0)\int_a^b (f(s) - x_0)dg(s) \\ &= C(x_0)\left(\int_a^b f(s)dg(s) - x_0(g(b) - g(a))\right) \\ &= C(x_0)\left(\int_a^b f(s)dg(s) - \frac{\int_a^b f(s)dg(s)}{g(b) - g(a)}(g(b) - g(a))\right) \\ &= C(x_0)\left(\int_a^b f(s)dg(s) - \int_a^b f(s)dg(s)\right) \\ &= 0.\end{aligned}$$

Therefore,

$$\int_a^b \phi(f(s))dg(s) - (g(b)-g(a)) \int_a^b \left(\phi\left(\frac{\int_a^b f(t)dg(t)}{g(b)-g(a)} \right) - \phi\left(\left| f(s) - \frac{\int_a^b f(t)dg(t)}{g(b)-g(a)} \right| \right) dg(s) \right) \geq 0,$$

$$\int_a^b \phi(f(s))dg(s) \geq (g(b)-g(a)) \int_a^b \left(\phi\left(\frac{\int_a^b f(t)dg(t)}{g(b)-g(a)} \right) + \phi\left(\left| f(s) - \frac{\int_a^b f(t)dg(t)}{g(b)-g(a)} \right| \right) \right) dg(s),$$

or

$$\frac{1}{g(b)-g(a)} \int_a^b \phi(f(s))dg(s) \geq \int_a^b \left(\phi\left(\frac{\int_a^b f(t)dg(t)}{g(b)-g(a)} \right) + \phi\left(\left| f(s) - \frac{\int_a^b f(t)dg(t)}{g(b)-g(a)} \right| \right) \right) dg(s).$$

This completes the proof. □

3.4 Stieltjes–Hölder Inequalities

In this section, we establish some Hölder inequalities for Stieltjes integrals.

Theorem 3.10. *Let $f, l \colon [a,b] \to (0,\infty)$ be such that $f, l \in \mathscr{C}([a,b])$, and $p, q \in (1,\infty)$ be such that $\frac{1}{p} + \frac{1}{q} = 1$. Then,*

$$\int_a^b |f(t)l(t)|dg(t) \leq \left(\int_a^b |f(t)|^p dg(t) \right)^{\frac{1}{p}} \left(\int_a^b |l(t)|^q dg(t) \right)^{\frac{1}{q}}.$$

Proof. Without any restrictions, assume that

$$\left(\int_a^b |f(t)|^p dg(t) \right)^{\frac{1}{p}} \left(\int_a^b |l(t)|^q dg(t) \right)^{\frac{1}{q}} \neq 0.$$

Let

$$a = \frac{|f(t)|}{\left(\int_a^b |f(t)|^p dg(t) \right)^{\frac{1}{p}}} \quad \text{and} \quad b = \frac{|l(t)|}{\left(\int_a^b |l(t)|^q dg(t) \right)^{\frac{1}{q}}}.$$

Now, applying the Young inequality, Theorem 3.2 for a and b, we get

$$\frac{|f(t)l(t)|}{\left(\int_a^b |f(t)|^p dg(t)\right)^{\frac{1}{p}} \left(\int_a^b |l(t)|^q dg(t)\right)^{\frac{1}{q}}} \leq \frac{|f(t)|^p}{p\int_a^b |f(t)|^p dg(t)} + \frac{|l(t)|^q}{q\int_a^b |l(t)|^q dg(t)},$$

$t \in [a,b]$. Integrating the last inequality, in the sense of Stieltjes, from a to b, we arrive at the inequality

$$\int_a^b \frac{|f(t)l(t)|}{\left(\int_a^b |f(t)|^p dg(t)\right)^{\frac{1}{p}} \left(\int_a^b |l(t)|^q dg(t)\right)^{\frac{1}{q}}} dg(t) \leq \int_a^b \frac{|f(t)|^p}{p\int_a^b |f(t)|^p dg(t)} dg(t) + \int_a^b \frac{|l(t)|^q}{q\int_a^b |l(t)|^q dg(t)} dg(t), \quad t \in [a,b],$$

or

$$\begin{aligned} \frac{\int_a^b |f(t)l(t)| dg(t)}{\left(\int_a^b |f(t)|^p dg(t)\right)^{\frac{1}{p}} \left(\int_a^b |l(t)|^q dg(t)\right)^{\frac{1}{q}}} &\leq \frac{\int_a^b |f(t)|^p dg(t)}{p\int_a^b |f(t)|^p dg(t)} + \frac{\int_a^b |l(t)|^q dg(t)}{q\int_a^b |l(t)|^q dg(t)} \\ &= \frac{1}{p} + \frac{1}{q} \\ &= 1, \end{aligned}$$

whereupon

$$\int_a^b |f(t)l(t)| dg(t) \leq \left(\int_a^b |f(t)|^p dg(t)\right)^{\frac{1}{p}} \left(\int_a^b |l(t)|^q dg(t)\right)^{\frac{1}{q}}.$$

This completes the proof. □

Theorem 3.11. *Let $f, l, h\colon [a,b] \to [0,\infty)$ be such that $f, l, h \in \mathscr{C}([a,b])$, and $p, q \in (1,\infty)$ be such that $\frac{1}{p} + \frac{1}{q} = 1$. Then,*

$$\int_a^b h(t)f(t)l(t) dg(t) \leq \left(\int_a^b h(t)(f(t))^p dg(t)\right)^{\frac{1}{p}} \left(\int_a^b h(t)(l(t))^q dg(t)\right)^{\frac{1}{q}}.$$

Proof. Applying Theorem 3.10 for $(h(t))^{\frac{1}{p}} f(t)$ and $(h(t))^{\frac{1}{q}} l(t)$, $t \in [a, b]$, we get

$$\begin{aligned}\int_a^b h(t)f(t)l(t)dg(t) &= \int_a^b (h(t))^{\frac{1}{p}} f(t)(h(t))^{\frac{1}{q}} l(t)dg(t)\\ &\le \left(\int_a^b h(t)(f(t))^p dg(t)\right)^{\frac{1}{p}} \left(\int_a^b h(t)(l(t))^q dg(t)\right)^{\frac{1}{q}}.\end{aligned}$$

This completes the proof. □

Theorem 3.12. *Let $f, l \colon [a, b] \to (0, \infty)$ be such that $f, l \in \mathscr{C}([a, b])$ and*

$$0 < m < \frac{f}{l} \le M < \infty \quad on \quad [a, b]. \tag{3.3}$$

Then,

$$\begin{aligned}\int_a^b (f(t))^{\frac{1}{p}} (l(t))^{\frac{1}{q}} dg(t) &\le \frac{M^{\frac{1}{p^2}}}{m^{\frac{1}{q^2}}} \int_a^b (f(t))^{\frac{1}{q}} (l(t))^{\frac{1}{p}} dg(t)\\ &\le \frac{M^{\frac{1}{p^2}}}{m^{\frac{1}{q^2}}} \left(\int_a^b f(t)dg(t)\right)^{\frac{1}{q}} \left(\int_a^b l(t)dg(t)\right)^{\frac{1}{p}}.\end{aligned}$$

Proof. By Theorem 3.10, we find

$$\begin{aligned}\int_a^b (f(t))^{\frac{1}{p}} (l(t))^{\frac{1}{q}} dg(t) &\le \left(\int_a^b f(t)dg(t)\right)^{\frac{1}{p}} \left(\int_a^b l(t)dg(t)\right)^{\frac{1}{q}}\\ &= \left(\int_a^b (f(t))^{\frac{1}{p}} (f(t))^{\frac{1}{q}} dg(t)\right)^{\frac{1}{p}} \\ &\quad \times \left(\int_a^b (l(t))^{\frac{1}{p}} (l(t))^{\frac{1}{q}} dg(t)\right)^{\frac{1}{q}}.\end{aligned} \tag{3.4}$$

From the inequalities (3.3), we get

$$(f(t))^{\frac{1}{p}} \le M^{\frac{1}{p}} (l(t))^{\frac{1}{p}} \quad \text{and} \quad (f(t))^{\frac{1}{q}} m^{-\frac{1}{q}} > (l(t))^{\frac{1}{q}}, \quad t \in [a, b].$$

From this and (3.4), we arrive at

$$\begin{aligned}
&\int_a^b (f(t))^{\frac{1}{p}}(l(t))^{\frac{1}{q}}dg(t)\\
&\quad\le \frac{M^{\frac{1}{p^2}}}{m^{\frac{1}{q^2}}}\left(\int_a^b (l(t))^{\frac{1}{p}}(f(t))^{\frac{1}{q}}dg(t)\right)^{\frac{1}{p}}\times\left(\int_a^b (f(t))^{\frac{1}{q}}(l(t))^{\frac{1}{p}}dg(t)\right)^{\frac{1}{q}}\\
&\quad= \frac{M^{\frac{1}{p^2}}}{m^{\frac{1}{q^2}}}\left(\int_a^b (f(t))^{\frac{1}{q}}(l(t))^{\frac{1}{p}}dg(t)\right)\\
&\quad\le \frac{M^{\frac{1}{p^2}}}{m^{\frac{1}{q^2}}}\left(\int_a^b f(t)dg(t)\right)^{\frac{1}{q}}\left(\int_a^b l(t)dg(t)\right)^{\frac{1}{p}}.
\end{aligned}$$

This completes the proof. □

Theorem 3.13. *Let $f, l\colon [a,b]\to(0,\infty)$ be such that $f, l\in\mathscr{C}([a,b])$ and*

$$0 < m < \frac{f^p}{l^q} \le M < \infty \quad on \quad [a,b], \tag{3.5}$$

and $p, q\in(1,\infty)$ be such that $\frac{1}{p}+\frac{1}{q}=1$. Then,

$$\left(\int_a^b (f(t))^p dg(t)\right)^{\frac{1}{p}}\left(\int_a^b (l(t))^q dg(t)\right)^{\frac{1}{q}} \le \left(\frac{M}{m}\right)^{\frac{1}{pq}}\int_a^b f(t)l(t)dg(t).$$

Proof. Since $\frac{1}{p}+\frac{1}{q}=1$, we have $p+q=pq$. From (3.5), we get

$$l(t) \ge \frac{(f(t))^{\frac{p}{q}}}{M^{\frac{1}{q}}}, \quad t\in[a,b],$$

and then

$$\begin{aligned}
f(t)l(t) &\ge \frac{1}{M^{\frac{1}{q}}}(f(t))(f(t))^{\frac{p}{q}}\\
&= \frac{M^{\frac{1}{p^2}}}{m^{\frac{1}{q^2}}}(f(t))^{1+\frac{p}{q}}
\end{aligned}$$

$$= \frac{M^{\frac{1}{p^2}}}{m^{\frac{1}{q^2}}}(f(t))^{\frac{p+q}{q}}$$
$$= \frac{M^{\frac{1}{p^2}}}{m^{\frac{1}{q^2}}}(f(t))^{p}, \quad t \in [a,b],$$

whereupon $(f(t))^p \le M^{\frac{1}{q}} f(t)l(t)$, $t \in [a,b]$, and

$$\left(\int_a^b (f(t))^p dg(t)\right)^{\frac{1}{p}} \le M^{\frac{1}{pq}} \left(\int_a^b f(t)l(t)dg(t)\right)^{\frac{1}{p}}. \tag{3.6}$$

Now, applying (3.5), we have $f(t) \ge m^{\frac{1}{p}}(l(t))^{\frac{q}{p}}$, $t \in [a,b]$, and

$$f(t)l(t) \ge m^{\frac{1}{p}} l(t)(l(t))^{\frac{q}{p}}$$
$$= m^{\frac{1}{p}} l(t)^{1+\frac{1q}{p}}$$
$$= m^{\frac{1}{p}}(l(t))^{\frac{p+q}{p}}$$
$$= m^{\frac{1}{p}}(l(t))^{q}, \quad t \in [a,b],$$

whereupon

$$\int_a^b f(t)l(t)dg(t) \ge m^{\frac{1}{p}} \int_a^b (l(t))^q dg(t)$$

and

$$\left(\int_a^b f(t)l(t)dg(t)\right)^{\frac{1}{q}} \ge m^{\frac{1}{pq}} \left(\int_a^b (l(t))^q dg(t)\right)^{\frac{1}{q}},$$

i.e.,

$$\left(\int_a^b (l(t))^q dg(t)\right)^{\frac{1}{q}} \le \frac{1}{m^{\frac{1}{pq}}} \left(\int_a^b f(t)l(t)dg(t)\right)^{\frac{1}{q}}.$$

From this and (3.6), we obtain

$$\begin{aligned}
&\left(\int_a^b (f(t))^p dg(t)\right)^{\frac{1}{p}} \left(\int_a^b (l(t))^q dg(t)\right)^{\frac{1}{q}} \\
&\leq M^{\frac{1}{pq}} \left(\int_a^b f(t)l(t)dg(t)\right)^{\frac{1}{p}} \times \frac{1}{m^{\frac{1}{pq}}} \left(\int_a^b f(t)l(t)dg(t)\right)^{\frac{1}{q}} \\
&= \left(\frac{M}{m}\right)^{\frac{1}{pq}} \left(\int_a^b f(t)l(t)dg(t)\right).
\end{aligned}$$

This completes the proof. □

If we replace f^p with f and l^q with l in Theorem 3.13, we find the following result.

Corollary 3.2. *Let $f, l\colon [a,b] \to (0,\infty)$ be such that $f, l \in \mathscr{C}([a,b])$ and*

$$0 < m < \frac{f}{l} \leq M < \infty \quad on \quad [a,b],$$

and $p, q \in (1,\infty)$ be such that $\frac{1}{p} + \frac{1}{q} = 1$. Then,

$$\left(\int_a^b f(t)dg(t)\right)^{\frac{1}{p}} \left(\int_a^b l(t)dg(t)\right)^{\frac{1}{q}} \leq \left(\frac{M}{m}\right)^{\frac{1}{pq}} \int_a^b (f(t))^{\frac{1}{p}} (l(t))^{\frac{1}{q}} dg(t).$$

Theorem 3.14. *Let $f, l\colon [a,b] \to (0,\infty)$ be such that $f, l \in \mathscr{C}([a,b])$ and*

$$0 < m \leq f^p \leq M < \infty \quad on \quad [a,b],$$

for $p, q \in (1,\infty)$, so that $\frac{1}{p} + \frac{1}{q} = 1$. Then,

$$\left(\int_a^b (f(t))^{\frac{1}{p}} dg(t)\right)^p \geq (g(b)-g(a))^{\frac{p+1}{q}} \left(\frac{m}{M}\right)^{\frac{p+1}{pq}} \left(\int_a^b (f(t))^p dg(t)\right)^{\frac{1}{p}}.$$

Proof. First, we apply Theorem 3.13 for $l \equiv 1$ on $[a, b]$ and find

$$\left(\int_a^b (f(t))^p dg(t)\right)^{\frac{1}{p}} (g(b) - g(a))^{\frac{1}{q}} \le \left(\frac{M}{m}\right)^{\frac{1}{pq}} \int_a^b f(t)dg(t),$$

whereupon

$$\left(\int_a^b (f(t))^p dg(t)\right)^{\frac{1}{p}} \le \left(\frac{m}{M}\right)^{-\frac{1}{pq}} (g(b) - g(a))^{-\frac{1}{q}} \int_a^b f(t)dg(t). \tag{3.7}$$

Now, we apply Corollary 3.2 for $l \equiv 1$ on $[a, b]$ and obtain

$$\left(\int_a^b f(t)dg(t)\right)^{\frac{1}{p}} \le \left(\frac{m}{M}\right)^{-\frac{1}{pq}} (g(b) - g(a))^{-\frac{1}{q}} \int_a^b (f(t))^{\frac{1}{p}} dg(t),$$

whereupon

$$\int_a^b f(t)dg(t) \le \left(\frac{m}{M}\right)^{-\frac{1}{q}} (g(b) - g(a))^{-\frac{p}{q}} \left(\int_a^b (f(t))^{\frac{1}{p}} dg(t)\right)^p.$$

From this and (3.7), we arrive at

$$\begin{aligned}\left(\int_a^b (f(t))^p dg(t)\right)^{\frac{1}{p}} &\le \left(\frac{m}{M}\right)^{-\frac{1}{pq}} \left(\frac{m}{M}\right)^{-\frac{1}{q}} (g(b) - g(a))^{-\frac{p}{q}} \left(\int_a^b (f(t))^{\frac{1}{p}} dg(t)\right)^p \\ &= \left(\frac{m}{M}\right)^{-\frac{p+1}{pq}} (g(b) - g(a))^{-\frac{p+1}{q}} \left(\int_a^b (f(t))^{\frac{1}{p}} dg(t)\right)^p,\end{aligned}$$

whereupon we find

$$\left(\int_a^b (f(t))^{\frac{1}{p}} dg(t)\right)^p \ge (g(b) - g(a))^{\frac{p+1}{q}} \left(\frac{m}{M}\right)^{\frac{p+1}{pq}} \left(\int_a^b (f(t))^p dg(t)\right)^{\frac{1}{p}}.$$

This completes the proof. □

3.5 Stieltjes–Minkowski Inequalities

Theorem 3.15. *Let $f, l \in \mathscr{C}([a, b])$ and $p > 1$. Then,*

$$\left(\int_a^b |f(t) + l(t)|^p dg(t)\right)^{\frac{1}{p}} \le \left(\int_a^b |f(t)|^p dg(t)\right)^{\frac{1}{p}} + \left(\int_a^b |l(t)|^p dg(t)\right)^{\frac{1}{p}}.$$

Proof. Let $q > 1$ be such that $\frac{1}{p} + \frac{1}{q} = 1$. Then, $(p-1)q = p$. Without any restrictions, assume that

$$\int_a^b |f(t) + l(t)|^p dg(t) \neq 0.$$

Then, applying the Stieltjes–Hölder inequality, Theorem 3.10, we find

$$\begin{aligned}
\int_a^b |f(t) + l(t)|^p dg(t) &= \int_a^b |f(t) + l(t)|^{p-1} |f(t) + l(t)| dg(t) \\
&\leq \int_a^b |f(t) + l(t)|^{p-1} (|f(t)| + |l(t)|) dg(t) \\
&= \int_a^b |f(t) + l(t)|^{p-1} |f(t)| dg(t) \\
&\quad + \int_a^b |f(t) + l(t)|^{p-1} |l(t)| dg(t) \\
&\leq \left(\int_a^b |f(t) + l(t)|^{(p-1)q} dg(t) \right)^{\frac{1}{q}} \left(\int_a^b |f(t)|^p dg(t) \right)^{\frac{1}{p}} \\
&\quad + \left(\int_a^b |f(t) + l(t)|^{(p-1)q} dg(t) \right)^{\frac{1}{q}} \left(\int_a^b |l(t)|^p dg(t) \right)^{\frac{1}{p}} \\
&= \left(\int_a^b |f(t) + l(t)|^{(p-1)q} dg(t) \right)^{\frac{1}{q}} \\
&\quad \times \left(\left(\int_a^b |f(t)|^p dg(t) \right)^{\frac{1}{p}} + \left(\int_a^b |l(t)|^p dg(t) \right)^{\frac{1}{p}} \right) \\
&= \left(\int_a^b |f(t) + l(t)|^p dg(t) \right)^{\frac{1}{q}} \\
&\quad \times \left(\left(\int_a^b |f(t)|^p dg(t) \right)^{\frac{1}{p}} + \left(\int_a^b |l(t)|^p dg(t) \right)^{\frac{1}{p}} \right),
\end{aligned}$$

whereupon

$$\left(\int_a^b |f(t)+l(t)|^p dg(t)\right)^{1-\frac{1}{q}} \le \left(\int_a^b |f(t)|^p dg(t)\right)^{\frac{1}{p}} + \left(\int_a^b |l(t)|^p dg(t)\right)^{\frac{1}{p}}$$

or

$$\left(\int_a^b |f(t)+l(t)|^p dg(t)\right)^{\frac{1}{p}} \le \left(\int_a^b |f(t)|^p dg(t)\right)^{\frac{1}{p}} + \left(\int_a^b |l(t)|^p dg(t)\right)^{\frac{1}{p}}.$$

This completes the proof. □

Theorem 3.16. *Let* $f, l, h \in \mathscr{C}([a,b])$ *and* $p > 1$. *Then,*

$$\left(\int_a^b |h(t)||f(t)+l(t)|^p dg(t)\right)^{\frac{1}{p}}$$
$$\le \left(\int_a^b |h(t)||f(t)|^p dg(t)\right)^{\frac{1}{p}} + \left(\int_a^b |h(t)||l(t)|^p dg(t)\right)^{\frac{1}{p}}.$$

Proof. Let $q > 1$ be such that $\frac{1}{p}+\frac{1}{q} = 1$. Without any restrictions, assume that

$$\int_a^b |h(t)||f(t)+l(t)|^p dg(t) \neq 0.$$

Then, applying Theorem 3.11, we find

$$\int_a^b |h(t)||f(t)+l(t)|^p dg(t)$$
$$= \int_a^b |h(t)||f(t)+l(t)|^{p-1}|f(t)+l(t)|dg(t)$$
$$\le \int_a^b |h(t)||f(t)+l(t)|^{p-1}(|f(t)|+|l(t)|)dg(t)$$
$$= \int_a^b |h(t)||f(t)+l(t)|^{p-1}|f(t)|dg(t)$$
$$+ \int_a^b |h(t)||f(t)+l(t)|^{p-1}|l(t)|dg(t)$$

$$\leq \left(\int_a^b |h(t)||f(t)+l(t)|^{(p-1)q}dg(t)\right)^{\frac{1}{q}} \left(\int_a^b |h(t)||f(t)|^p dg(t)\right)^{\frac{1}{p}}$$

$$+ \left(\int_a^b |h(t)||f(t)+l(t)|^{(p-1)q}dg(t)\right)^{\frac{1}{q}} \left(\int_a^b |h(t)||l(t)|^p dg(t)\right)^{\frac{1}{p}}$$

$$= \left(\int_a^b |h(t)||f(t)+l(t)|^{(p-1)q}dg(t)\right)^{\frac{1}{q}}$$

$$\times \left(\left(\int_a^b |h(t)||f(t)|^p dg(t)\right)^{\frac{1}{p}} + \left(\int_a^b |h(t)||l(t)|^p dg(t)\right)^{\frac{1}{p}}\right)$$

$$= \left(\int_a^b |h(t)||f(t)+l(t)|^p dg(t)\right)^{\frac{1}{q}}$$

$$\times \left(\left(\int_a^b |h(t)||f(t)|^p dg(t)\right)^{\frac{1}{p}} + \left(\int_a^b |h(t)||l(t)|^p dg(t)\right)^{\frac{1}{p}}\right),$$

whereupon

$$\left(\int_a^b |h(t)||f(t)+l(t)|^p dg(t)\right)^{1-\frac{1}{q}}$$

$$\leq \left(\int_a^b |h(t)||f(t)|^p dg(t)\right)^{\frac{1}{p}} + \left(\int_a^b |h(t)||l(t)|^p dg(t)\right)^{\frac{1}{p}},$$

i.e.,

$$\left(\int_a^b |h(t)||f(t)+l(t)|^p dg(t)\right)^{\frac{1}{p}}$$

$$\leq \left(\int_a^b |h(t)||f(t)|^p dg(t)\right)^{\frac{1}{p}} + \left(\int_a^b |h(t)||l(t)|^p dg(t)\right)^{\frac{1}{p}}.$$

This completes the proof. □

Theorem 3.17. *Let $f, l\colon [a,b] \to (0,\infty)$ be such that $f, l \in \mathscr{C}([a,b])$, and*

$$0 < m \le \frac{f^p}{(f+l)^q} \le M < \infty,$$
$$0 < m \le \frac{l^p}{(f+l)^q} \le M < \infty \quad \text{on} \quad [a,b],$$

for $p, q \in (1,\infty)$, such that $\frac{1}{p} + \frac{1}{q} = 1$. Then,

$$\left(\int_a^b (f(t))^p dg(t)\right)^{\frac{1}{p}} + \left(\int_a^b (l(t))^p dg(t)\right)^{\frac{1}{p}}$$
$$\le \left(\frac{M}{m}\right)^{\frac{1}{pq}} \left(\int_a^b (f(t)+l(t))^p dg(t)\right)^{\frac{1}{p}}.$$

Proof. To prove this result, we apply Theorem 3.12. We have

$$\begin{aligned}
&\int_a^b (f(t)+l(t))^p dg(t)\\
&= \int_a^b (f(t)+l(t))^{p-1}(f(t)+l(t))dg(t)\\
&= \int_a^b (f(t)+l(t))^{p-1} f(t)dg(t) + \int_a^b (f(t)+l(t))^{p-1} l(t)dg(t)\\
&\ge \left(\frac{m}{M}\right)^{\frac{1}{pq}} \left(\int_a^b (f(t)+l(t))^{(p-1)q} dg(t)\right)^{\frac{1}{q}} \left(\int_a^b (f(t))^p dg(t)\right)^{\frac{1}{p}}\\
&\quad + \left(\frac{m}{M}\right)^{\frac{1}{pq}} \left(\int_a^b (f(t)+l(t))^{(p-1)q} dg(t)\right)^{\frac{1}{q}} \left(\int_a^b (l(t))^p dg(t)\right)^{\frac{1}{p}}\\
&= \left(\frac{m}{M}\right)^{\frac{1}{pq}} \left(\int_a^b (f(t)+l(t))^p dg(t)\right)^{\frac{1}{q}} \left(\int_a^b (f(t))^p dg(t)\right)^{\frac{1}{p}}\\
&\quad + \left(\frac{m}{M}\right)^{\frac{1}{pq}} \left(\int_a^b (f(t)+l(t))^p dg(t)\right)^{\frac{1}{q}} \left(\int_a^b (l(t))^p dg(t)\right)^{\frac{1}{p}}
\end{aligned}$$

$$= \left(\frac{m}{M}\right)^{\frac{1}{pq}} \left(\int_a^b (f(t)+l(t))^p dg(t)\right)^{\frac{1}{q}} \left(\left(\int_a^b (f(t))^p dg(t)\right)^{\frac{1}{p}} + \left(\int_a^b (l(t))^p dg(t)\right)^{\frac{1}{p}}\right),$$

whereupon

$$\left(\int_a^b (f(t)+l(t))^p dg(t)\right)^{1-\frac{1}{q}} \geq \left(\frac{m}{M}\right)^{\frac{1}{pq}} \left(\left(\int_a^b (f(t))^p dg(t)\right)^{\frac{1}{p}} + \left(\int_a^b (l(t))^p dg(t)\right)^{\frac{1}{p}}\right),$$

i.e.,

$$\left(\int_a^b (f(t)+l(t))^p dg(t)\right)^{\frac{1}{p}} \geq \left(\frac{m}{M}\right)^{\frac{1}{pq}} \left(\left(\int_a^b (f(t))^p dg(t)\right)^{\frac{1}{p}} + \left(\int_a^b (l(t))^p dg(t)\right)^{\frac{1}{p}}\right).$$

This completes the proof. □

3.6 Stieltjes–Steffensen Inequalities

Theorem 3.18. *Let $f, h \in \mathscr{C}([a,b])$ be such that f is of one sign and decreasing, and $0 \leq h(t) \leq 1$, $t \in [a,b]$. Assume that $\gamma, l \in [a,b]$ are such that*

$$g(b) - g(l) \leq \int_a^b h(t)dg(t) \leq g(\gamma) - g(a) \quad \textit{if} \quad f(t) > 0, \quad t \in [a,b],$$

and

$$g(\gamma) - g(a) \leq \int_a^b h(t)dg(t) \leq g(b) - g(l) \quad \textit{if} \quad f(t) < 0, \quad t \in [a,b].$$

Then,

$$\int_l^b f(t)dg(t) \le \int_a^b f(t)h(t)dg(t) \le \int_a^\gamma f(t)dg(t).$$

Proof. Without any restrictions, suppose that $f(t) > 0$, $t \in [a, b]$. We have

$$\begin{aligned}
\int_a^b f(t)h(t)dg(t) - \int_l^b f(t)dg(t)& \\
&= \int_a^l f(t)h(t)dg(t) + \int_l^b f(t)h(t)dg(t) - \int_l^b f(t)dg(t) \\
&= \int_a^l f(t)h(t)dg(t) - \int_l^b f(t)(1 - h(t))dg(t) \\
&\ge \int_a^l f(t)h(t)dg(t) - f(l)\int_l^b (1 - h(t))dg(t) \\
&= \int_a^l f(t)h(t)dg(t) - f(l)(g(b) - g(l)) + f(l)\int_l^b h(t)dg(t) \\
&\ge \int_a^l f(t)h(t)dg(t) - f(l)\int_a^b h(t)dg(t) + f(l)\int_l^b h(t)dg(t) \\
&= \int_a^l f(t)h(t)dg(t) - f(l)\left(\int_a^b h(t)dg(t) - \int_l^b h(t)dg(t)\right) \\
&= \int_a^l f(t)h(t)dg(t) - f(l)\int_a^l h(t)dg(t) \\
&= \int_a^l (f(t) - f(l))h(t)dg(t) \\
&\ge 0
\end{aligned}$$

and

$$\begin{aligned}
&\int_a^\gamma f(t)dg(t) - \int_a^b f(t)h(t)dg(t) \\
&\quad = \int_a^\gamma f(t)dg(t) - \int_a^\gamma f(t)h(t)dg(t) - \int_\gamma^b f(t)h(t)dg(t)
\end{aligned}$$

$$
\begin{aligned}
&= \int_a^{\gamma} f(t)(1 - h(t))dg(t) - \int_{\gamma}^{b} f(t)h(t)dg(t) \\
&\geq f(\gamma) \int_a^{\gamma} (1 - h(t))dg(t) - f(\gamma) \int_{\gamma}^{b} h(t)dg(t) \\
&= f(\gamma)(g(\gamma) - g(a)) - f(\gamma) \int_a^{\gamma} h(t)dg(t) - f(\gamma) \int_{\gamma}^{b} h(t)dg(t) \\
&\geq f(\gamma) \int_a^{b} h(t)dg(t) - f(\gamma) \left(\int_a^{\gamma} h(t)dg(t) + \int_{\gamma}^{b} h(t)dg(t) \right) \\
&= f(\gamma) \int_a^{b} h(t)dg(t) - f(\gamma) \int_a^{b} h(t)dg(t) \\
&= 0.
\end{aligned}
$$

This completes the proof. □

The following result is used to prove the next version of the Stieltjes–Steffensen inequality.

Theorem 3.19. *Let $f, r, h \in \mathscr{C}([a,b])$ and $l, \gamma \in [a,b]$ be such that*

$$
\int_a^{\gamma} h(t)dg(t) = \int_a^{b} r(t)dg(t) = \int_l^{b} h(t)dg(t).
$$

Then,

$$
\begin{aligned}
\int_a^b f(t)r(t)dg(t) = &\int_{\gamma}^{b} (f(t) - f(\gamma))r(t)dg(t) \\
&+ \int_a^{\gamma} (f(t)h(t) - (f(t) - f(\gamma))(h(t) - r(t)))dg(t)
\end{aligned}
$$

and

$$
\begin{aligned}
\int_a^b f(t)r(t)dg(t) = &\int_a^{l} (f(t) - f(l))dg(t) \\
&+ \int_l^{b} (f(t)h(t) - (f(t) - f(l))(h(t) - r(t)))dg(t).
\end{aligned}
$$

Proof. We have

$$\int_a^\gamma (f(t)h(t) - (f(t) - f(\gamma))(h(t) - r(t)))dg(t) - \int_a^b f(t)r(t)dg(t)$$

$$= \int_a^\gamma (f(t)h(t) - f(t)r(t) - (f(t) - f(\gamma))(h(t) - r(t)))dg(t)$$

$$+ \int_a^\gamma f(t)r(t)dg(t) - \int_a^b f(t)r(t)dg(t)$$

$$= \int_a^\gamma f(\gamma)(h(t) - r(t))dg(t) + \int_a^\gamma f(t)r(t)dg(t) - \int_a^b f(t)r(t)dg(t)$$

$$= \int_a^\gamma f(\gamma)(h(t) - r(t))dg(t) - \int_\gamma^b f(t)r(t)dg(t)$$

$$= f(\gamma)\int_a^\gamma h(t)dg(t) - f(\gamma)\int_a^\gamma r(t)dg(t) - \int_\gamma^b f(t)r(t)dg(t)$$

$$= f(\gamma)\int_a^b r(t)dg(t) - f(\gamma)\int_a^\gamma g(t)dg(t) - \int_\gamma^b f(t)r(t)dg(t)$$

$$= f(\gamma)\left(\int_a^b r(t)dg(t) - \int_a^\gamma r(t)dg(t)\right) - \int_\gamma^b f(t)r(t)dg(t)$$

$$= f(\gamma)\int_\gamma^b r(t)dg(t) - \int_\gamma^b f(t)r(t)dg(t)$$

$$= \int_\gamma^b (f(\gamma) - f(t))r(t)dg(t)$$

and

$$\int_l^b (f(t)h(t) - (f(t) - f(l))(h(t) - r(t)))dg(t) - \int_a^b f(t)r(t)dg(t)$$

$$= \int_l^b (f(t)h(t) - f(t)r(t) - (f(t) - f(l))(h(t) - r(t)))dg(t)$$

$$
\begin{aligned}
&+\int_l^b f(t)r(t)dg(t)-\int_a^b f(t)r(t)dg(t)\\
&=\int_l^b f(l)(h(t)-r(t))dg(t)+\int_l^b f(t)r(t)dg(t)-\int_a^b f(t)r(t)dg(t)\\
&=\int_l^b f(l)(h(t)-r(t))dg(t)-\int_a^l f(t)r(t)dg(t)\\
&=f(l)\int_l^b h(t)dg(t)-f(l)\int_l^b r(t)dg(t)-\int_a^l f(t)r(t)dg(t)\\
&=f(l)\int_a^b r(t)dg(t)-f(l)\int_l^b r(t)dg(t)-\int_a^l f(t)r(t)dg(t)\\
&=f(l)\left(\int_a^b r(t)dg(t)-\int_l^b r(t)dg(t)\right)-\int_a^l f(t)r(t)dg(t)\\
&=f(l)\int_a^l r(t)dg(t)-\int_a^l f(t)r(t)dg(t)\\
&=\int_a^r (f(l)-f(t))r(t)dg(t).
\end{aligned}
$$

This completes the proof. □

Theorem 3.20. *Let $f, r, h \in \mathscr{C}([a,b])$ be such that f is of one sign and decreasing, and $0 \le r(t) \le h(t)$, $t \in [a,b]$. Assume that $\gamma, l \in [a,b]$ are such that*

$$
\int_l^b h(t)dg(t) \le \int_a^b r(t)dg(t) \le \int_a^\gamma h(t)dg(t) \quad \textit{if} \quad f(t) > 0, \quad t \in [a,b],
$$

and

$$
\int_a^\gamma h(t)dg(t) \le \int_a^b r(t)dg(t) \le \int_l^b h(t)dg(t) \quad \textit{if} \quad f(t) < 0, \quad t \in [a,b].
$$

Then,

$$
\int_l^b f(t)h(t)dg(t) \le \int_a^b f(t)r(t)dg(t) \le \int_a^\gamma f(t)h(t)dg(t).
$$

Proof. Without any restrictions, suppose that $f(t) > 0$, $t \in [a, b]$. We have

$$\begin{aligned}
&\int_a^b f(t)r(t)dg(t)\\
&\quad - \int_l^b f(t)h(t)dg(t)\\
&= \int_a^l f(t)r(t)dg(t) + \int_l^b f(t)r(t)dg(t) - \int_l^b f(t)h(t)dg(t)\\
&= \int_a^l f(t)r(t)dg(t) - \int_l^b f(t)(h(t) - r(t))dg(t)\\
&\geq \int_a^l f(t)r(t)dg(t) - f(l)\int_l^b (h(t) - r(t))dg(t)\\
&= \int_a^l f(t)r(t)dg(t) - f(l)\int_l^b h(t)dg(t) + f(l)\int_l^b r(t)dg(t)\\
&\geq \int_a^l f(t)r(t)dg(t) - f(l)\int_a^b r(t)h(t)dg(t) + f(l)\int_l^b r(t)h(t)dg(t)\\
&= \int_a^l f(t)r(t)dg(t) - f(l)\left(\int_a^b r(t)dg(t) - \int_l^b r(t)dg(t)\right)\\
&= \int_a^l f(t)r(t)dg(t) - f(l)\int_a^l r(t)dg(t)\\
&= \int_a^l (f(t) - f(l))r(t)dg(t)\\
&\geq 0
\end{aligned}$$

and

$$\begin{aligned}
&\int_a^\gamma f(t)h(t)dg(t) - \int_a^b f(t)r(t)dg(t)\\
&= \int_a^\gamma f(t)h(t)dg(t) - \int_a^\gamma f(t)r(t)dg(t) - \int_\gamma^b f(t)r(t)dg(t)
\end{aligned}$$

$$
\begin{aligned}
&= \int_a^\gamma f(t)(h(t) \quad r(t))dg(t) \quad \int_\gamma^b f(t)r(t)dg(t) \\
&\geq f(\gamma)\int_a^\gamma (h(t)-r(t))dg(t) - f(\gamma)\int_\gamma^b r(t)dg(t) \\
&= f(\gamma)\int_a^\gamma h(t)dg(t) - f(\gamma)\int_a^\gamma r(t)dg(t) - f(\gamma)\int_\gamma^b r(t)dg(t) \\
&\geq f(\gamma)\int_a^b r(t)dg(t) - f(\gamma)\left(\int_a^\gamma r(t)dg(t) + \int_\gamma^b r(t)dg(t)\right) \\
&= f(\gamma)\int_a^b r(t)dg(t) - f(\gamma)\int_a^b r(t)dg(t) \\
&= 0.
\end{aligned}
$$

This completes the proof. □

Theorem 3.21. *Let $f, r, h \in \mathscr{C}([a,b])$ be such that f is of one sign and decreasing on $[a,b]$, and $0 \leq r(t) \leq h(t)$, $t \in [a,b]$. Assume that $l, \gamma \in [a,b]$ are such that*

$$\int_a^\gamma h(t)dg(t) = \int_a^b r(t)dg(t) = \int_l^b h(t)dg(t).$$

Then,

$$
\begin{aligned}
\int_a^b f(t)h(t)dg(t) &\leq \int_l^b (f(t)h(t) - (f(t)-f(l))(h(t)-r(t)))dg(t) \\
&\leq \int_a^b f(t)r(t)dg(t) \\
&\leq \int_a^\gamma (f(t)h(t) - (f(t)-f(\gamma))(h(t)-r(t)))dg(t) \\
&\leq \int_a^\gamma f(t)h(t)dg(t).
\end{aligned}
$$

Proof. Without any restrictions, assume that $f > 0$ on $[a,b]$. Since f is decreasing on $[a,b]$ and $0 \leq r(t) \leq h(t)$, $t \in [a,b]$, we have the following set of inequalities:

$$\int_a^l (f(t) - f(l))r(t)dg(t) \geq 0,$$

$$\int_l^b (f(l) - f(t))(h(t) - r(t))dg(t) \geq 0,$$

$$\int_\gamma^b (f(\gamma) - f(t))r(t)dg(t) \geq 0,$$

$$\int_a^\gamma (f(t) - f(\gamma))(h(t) - r(t))dg(t) \geq 0.$$

By the first equality of Theorem 3.19, we find

$$\begin{aligned}\int_a^b f(t)r(t)dg(t) &= \int_\gamma^b (f(t) - f(\gamma))r(t)dg(t)\\ &\quad + \int_a^\gamma (f(t)h(t) - (f(t) - f(\gamma))(h(t) - r(t)))dg(t)\\ &\leq \int_a^\gamma (f(t)h(t) - (f(t) - f(\gamma))(h(t) - r(t)))dg(t)\\ &\leq \int_a^\gamma f(t)h(t)dg(t).\end{aligned}$$

By the second equality of Theorem 3.19, we obtain

$$\begin{aligned}\int_a^b f(t)r(t)dg(t) &= \int_a^l (f(t) - f(l))r(t)dg(t)\\ &\quad + \int_l^b (f(t)h(t) - (f(t) - f(l))(h(t) - r(t)))dg(t)\\ &\geq \int_l^b (f(t)h(t) - (f(t) - f(l))(h(t) - r(t)))dg(t)\\ &\geq \int_l^b f(t)h(t)dg(t).\end{aligned}$$

This completes the proof. □

Theorem 3.22. *Let $f, r, h, \phi \in \mathscr{C}([a,b])$ be such that f is of one sign and decreasing, and*

$$0 \leq \phi(t) \leq r(t) \leq h(t) - \phi(t), \quad t \in [a,b]. \tag{3.8}$$

Assume that $l, \gamma \in [a,b]$ are such that

$$\int_a^\gamma h(t)dg(t) = \int_a^b r(t)dg(t) = \int_l^b h(t)dg(t).$$

Then,

$$\begin{aligned}\int_l^b f(t)h(t)dg(t) &+ \int_a^b |f(t) - f(l)|\phi(t)dg(t) \\ &\leq \int_a^b f(t)r(t)dg(t) \\ &\leq \int_a^\gamma f(t)h(t)dg(t) - \int_a^\gamma |f(t) - f(\gamma)|\phi(t)dg(t).\end{aligned}$$

Proof. Keeping in mind (3.8) and the fact that f is decreasing, we get the following inequalities:

$$\begin{aligned}&\int_a^\gamma (f(t) - f(\gamma))(h(t) - r(t))dg(t) + \int_\gamma^b (f(\gamma) - f(t))r(t)dg(t) \\ &\quad = \int_a^\gamma |f(t) - f(\gamma)|(h(t) - r(t))dg(t) + \int_\gamma^b |f(t) - f(\gamma)|r(t)dg(t) \\ &\quad \geq \int_a^\gamma |f(t) - f(\gamma)|\phi(t)dg(t) + \int_\gamma^b |f(t) - f(\gamma)|\phi(t)dg(t) \\ &\quad = \int_a^b |f(t) - f(\gamma)|\phi(t)dg(t)\end{aligned}$$

and

$$\begin{aligned}&\int_a^l (f(t) - f(l))r(t)dg(t) + \int_l^b (f(l) - f(t))(h(t) - r(t))dg(t) \\ &\quad = \int_a^l |f(t) - f(l)|r(t)dg(t) + \int_l^b |f(t) - f(l)|(h(t) - r(t))dg(t)\end{aligned}$$

$$\geq \int_a^l |f(t) - f(l)|\phi(t)dg(t) + \int_l^b |f(t) - f(l)|\phi(t)dg(t)$$
$$= \int_a^b |f(t) - f(l)|\phi(t)dg(t).$$

Now, applying the first equality of Theorem 3.19, we get

$$\begin{aligned}
\int_a^b f(t)r(t)dg(t) &= \int_\gamma^b (f(t) - f(\gamma))r(t)dg(t) \\
&\quad + \int_a^\gamma (f(t)h(t) - (f(t) - f(\gamma))(h(t) - r(t)))dg(t) \\
&= -\int_\gamma^b (f(\gamma) - f(t))r(t)dg(t) \\
&\quad - \int_a^\gamma (f(t) - f(\gamma))(h(t) - r(t))dg(t) \\
&\quad + \int_a^\gamma f(t)h(t)dg(t) \\
&\leq -\int_a^b |f(t) - f(\gamma)|\phi(t)dg(t) + \int_a^\gamma f(t)h(t)dg(t).
\end{aligned}$$

By the second equality of Theorem 3.19, we find

$$\begin{aligned}
\int_a^b f(t)r(t)dg(t) &= \int_a^l (f(t) - f(l))r(t)dg(t) \\
&\quad + \int_l^b (f(t)h(t) - (f(t) - f(l))(h(t) - r(t)))dg(t) \\
&= \int_a^l |f(t) - f(l)|r(t)dg(t) \\
&\quad + \int_l^b (f(l) - f(t))(h(t) - r(t))dg(t) \\
&\quad + \int_l^b f(t)h(t)dg(t) \\
&\geq \int_a^b |f(t) - f(l)|\phi(t)dg(t) + \int_l^b f(t)h(t)dg(t).
\end{aligned}$$

This completes the proof. □

3.7 Stieltjes–Hermite–Hadamard Inequalities

In this section, we derive some Stieltjes analogs of Hermite–Hadamard inequalities. We start with the first one.

Theorem 3.23. *Let $f \in \mathscr{C}_g^1([a,b])$ and $m \le f_g' \le M$ on $[a,b]$. If there exist $l, \gamma \in [a,b]$ such that*

$$g(\gamma) - g(a) \le \frac{f(b) - f(a) - m(g(b) - g(a))}{M - m} \le g(b) - g(l),$$

then

$$\begin{aligned} mh_2(a,b) + (M-m)h_2(a,\gamma) &\le (g(b) - g(a))f(b) - \int_a^b f(t)dg(t) \\ &\le Mh_2(a,b) + (m-M)h_2(a,l), \end{aligned}$$

where $h_1(\cdot,\cdot)$ and $h_2(\cdot,\cdot)$ are Stieltjes–Taylor monomials.

Proof. Let

$$r(t) = \frac{f(t) - m(g(t) - g(a))}{M - m}, \quad F(t) = h_1(a,t),$$
$$G(t) = r_g'(t) = \frac{f_g'(t) - m}{M - m} \in [0,1].$$

We find that F is decreasing and nonpositive, and

$$\begin{aligned} \int_a^b G(t)dg(t) &= \int_a^b \frac{f_g'(t) - m}{M - m} dg(t) \\ &= \frac{1}{M-m} f(t)\Big|_{t=a}^{t=b} - \frac{m(g(b) - g(a))}{M - m} \\ &= \frac{f(b) - f(a)}{M - m} - \frac{m(g(b) - g(a))}{M - m} \\ &= \frac{f(b) - f(a) - m(g(b) - g(a))}{M - m} \in [g(\gamma) - g(a), g(b) - g(l)], \end{aligned}$$

$$\begin{aligned}
\int_l^b F(t)dg(t) &= \int_l^b h_1(a,t)dg(t) \\
&= -h_2(a,t)\Big|_{t=l}^{t=b} \\
&= -h_2(a,b) + h_2(a,l), \\
\int_a^\gamma F(t)dg(t) &= \int_a^\gamma h_1(a,t)dg(t) \\
&= -h_2(a,t)\Big|_{t=a}^{t=\gamma} \\
&= -h_2(a,\gamma),
\end{aligned}$$

and

$$\begin{aligned}
&\int_a^b F(t)G(t)dg(t) \\
&\quad = \int_a^b F(t)r_g'(t)dg(t) \\
&\quad = \int_a^b h_1(a,t)r_g'(t)dg(t) \\
&\quad = h_1(a,t)r(t)\Big|_{t=a}^{t=b} - \int_a^b h_{1g}'(a,t)r(t)dg(t) \\
&\quad = h_1(a,b)r(b) + \int_a^b r(t)dg(t) \\
&\quad = (g(a)-g(b))\frac{f(b)}{M-m} + \int_a^b \frac{f(t)-m(g(t)-g(b))}{M-m}dg(t) \\
&\quad = \frac{1}{M-m}\left(-(g(b)-g(a))f(b) + \int_a^b f(t)dg(t) + m\int_a^b (g(b)-g(t))dg(t)\right) \\
&\quad = \frac{1}{M-m}\left(-(g(b)-g(a))f(b) + \int_a^b f(t)dg(t) + mh_2(a,b)\right).
\end{aligned}$$

By the Stieltjes–Steffensen inequality, Theorem 3.18, we have

$$\int_l^b F(t)dg(t) \leq \int_a^b F(t)G(t)dg(t) \leq \int_a^\gamma F(t)dg(t),$$

whereupon

$$\begin{aligned}
&-h_2(a,b)+h_2(a,l)\\
&\quad\leq \frac{1}{M-m}\left(-(g(b)-g(a))f(b)+\int_a^b f(t)dg(t)+mh_2(a,b)\right)\\
&\quad\leq -h_2(a,\gamma),\\
&-(M-m)h_2(a,b)+(M-m)h_2(a,l)\\
&\quad\leq -(g(b)-g(a))f(b)+\int_a^b f(t)dg(t)+mh_2(a,b)\\
&\quad\leq -(M-m)h_2(a,\gamma),\\
&(M-m)h_2(a,b)-(M-m)h_2(a,l)\\
&\quad\geq (g(b)-g(a))f(b)-\int_a^b f(t)dg(t)-mh_2(a,b)\\
&\quad\geq (M-m)h_2(a,\gamma),
\end{aligned}$$

or

$$\begin{aligned}
Mh_2(a,b)-(M-m)h_2(a,l) &\geq (g(b)-g(a))f(b)-\int_a^b f(t)dg(t)\\
&\geq (M-m)h_2(a,\gamma)+mh_2(a,b).
\end{aligned}$$

This completes the proof. □

Theorem 3.24. *Let $f \in \mathscr{C}([a,b])$ be convex and monotonic. Assume that $l, \gamma \in [a,b]$ are such that*

$$l \geq a + \frac{h_2(b,a)}{g(b)-g(a)}, \quad \gamma \geq b - \frac{h_2(b,a)}{g(b)-g(a)}$$

if f is decreasing, and

$$l \leq a + \frac{h_2(b,a)}{g(b)-g(a)}, \quad \gamma \leq b - \frac{h_2(b,a)}{g(b)-g(a)}$$

if f is increasing. Then,

$$f(\gamma) \leq \frac{1}{g(b)-g(a)} \int_a^b f(t)dg(t) \leq f(a) + f(b) - f(l).$$

Proof. Assume that f is decreasing and convex. Then, $f_g'' \geq 0$ and $f_g' \leq 0$. Set

$$F(t) = -f_g'(t), \quad G(t) = \frac{g(b)-g(t)}{g(b)-g(a)}, \quad t \in [a,b].$$

Then,

$$F_g'(t) = -f_g''(t) \leq 0, \quad t \in [a,b].$$

Therefore, F is decreasing. Moreover, $0 \leq G(t) \leq 1$, $t \in [a,b]$. Now, the inequality

$$g(b) - g(l) \leq \int_a^b G(t)dg(t) \leq g(\gamma) - g(a)$$

can be rewritten in the form

$$g(b) - g(l) \leq \frac{1}{g(b)-g(a)} \int_a^b (g(b)-g(t))dg(t) \leq g(\gamma) - g(a). \tag{3.9}$$

Observe that

$$\begin{aligned}\frac{1}{g(b)-g(a)} \int_a^b (g(b)-g(t))dg(t) &= \frac{1}{g(b)-g(a)} \int_a^b (g(b)-g(a)+g(a)-g(t))dg\\ &= g(b)-g(a) - \frac{1}{g(b)-g(a)} \int_a^b (g(t)-g(a))dg\\ &= g(b)-g(a) - \frac{h_2(b,a)}{g(b)-g(a)}.\end{aligned}$$

Thus, (3.9) takes the form

$$g(b) - g(l) \leq g(b) - g(a) - \frac{h_2(b,a)}{g(b)-g(a)} \leq g(\gamma) - g(a).$$

Then,

$$g(b) - g(l) \le g(b) - g(a) - \frac{h_2(b,a)}{g(b) - g(a)},$$

$$g(\gamma) - g(a) \ge g(b) - g(a) - \frac{h_2(b,a)}{g(b) - g(a)},$$

i.e.,

$$g(l) \ge g(a) + \frac{h_2(b,a)}{g(b) - g(a)},$$

$$g(\gamma) \ge g(b) - \frac{h_2(b,a)}{g(b) - g(a)},$$

which is true. Therefore, the functions F and G satisfy all conditions of the Stieltjes–Steffensen inequality, Theorem 3.18. We get

$$\int_l^b F(t)dg(t) \le \int_a^b F(t)G(t)dg(t) \le \int_a^\gamma F(t)dg(t). \tag{3.10}$$

We have

$$-\int_l^b f_g'(t)dg(t) = f(l) - f(b),$$

$$\int_a^\gamma f_g'(t)dg(t) = f(a) - f(\gamma),$$

and

$$\begin{aligned}\int_a^b F(t)G(t)dg(t) &= -\int_a^b f_g'(t)\frac{g(b) - g(t)}{g(b) - g(a)}dg(t)\\ &= -f(t)\frac{g(b) - g(t)}{g(b) - g(a)}\bigg|_{t=a}^{t=b} - \frac{1}{g(b) - g(a)}\int_a^b f(t)dg(t)\\ &= f(a) - \frac{1}{g(b) - g(a)}\int_a^b f(t)dg(t).\end{aligned}$$

Hence, with (3.10), we find

$$f(l) - f(b) \le f(a) - \frac{1}{g(b) - g(a)}\int_a^b f(t)dg(t) \le f(a) - f(\gamma),$$

or

$$f(l) - f(b) \leq f(a) - \frac{1}{g(b) - g(a)} \int_a^b f(t)dg(t),$$

$$f(a) - f(\gamma) \geq f(a) - \frac{1}{g(b) - g(a)} \int_a^b f(t)dg(t),$$

or

$$\frac{1}{g(b) - g(a)} \int_a^b f(t)dg(t) \leq \frac{f(a)}{g(b) - g(a)} + f(b) - f(l),$$

$$\frac{1}{g(b) - g(a)} \int_a^b f(t)dg(t) \geq f(\gamma).$$

This completes the proof. □

3.8 Advanced Practical Problems

Problem 3.3. Let $a, b \in \mathbb{R}$, $a \leq b$, be such that $h_1(x, a)(g(b) - g(a)) \geq h_2(b, a)$, $x \in [a, b]$. Then, for $n \in \mathbb{N}$, prove that

$$\left(\frac{h_2(b, a)}{g(b) - g(a)} \right)^{2n} \leq \frac{1}{g(b) - g(a)} \int_a^b (h_1(s, a))^{2n} dg(s).$$

Problem 3.4. Let $p, q \in (1, \infty)$ be such that $\frac{1}{p} + \frac{1}{q} = 1$. Then, prove that

$$\int_a^b h_2(t, a)h_3(t, a)dg(t) \leq \left(\int_a^b (h_2(t, a))^p dg(t) \right)^{\frac{1}{p}} \left(\int_a^b (h_3(t, a))^q dg(t) \right)^{\frac{1}{q}}.$$

Problem 3.5. Let $p, q \in (1, \infty)$ be such that $\frac{1}{p} + \frac{1}{q} = 1$. Then, prove that

$$\int_a^b h_4(t, a)h_6(t, a)h_8(t, a)dg(t)$$
$$\leq \left(\int_a^b (h_4(t, a))^p h_8(t, a)dg(t) \right)^{\frac{1}{p}} \left(\int_a^b (h_6(t, a))^q h_8(t, a)dg(t) \right)^{\frac{1}{q}}.$$

Problem 3.6. Let $a, b \in \mathbb{R}$, $a \leq b$, and $p, q \in (1, \infty)$ be such that $\frac{1}{p} + \frac{1}{q} = 1$. Then, prove that there exists a constant $C > 0$ such that

$$\left(\int_a^b (h_3(t,a))^p dg(t)\right)^{\frac{1}{p}} \left(\int_a^b (h_5(t,a))^q dg(t)\right)^{\frac{1}{q}} \leq C \int_a^b h_3(t,a) h_5(t,a) dg(t).$$

Problem 3.7. Let $a, b \in \mathbb{R}$, $a \leq b$, and $p, q \in (1, \infty)$ be such that $\frac{1}{p} + \frac{1}{q} = 1$. Then, prove that there exists a constant $C > 0$ such that

$$\left(\int_a^b (h_7(t,a))^p dg(t)\right)^{\frac{1}{p}} \left(\int_a^b (h_{11}(t,a))^q dg(t)\right)^{\frac{1}{q}} \leq C \int_a^b h_7(t,a) h_{11}(t,a) dg(t).$$

Problem 3.8. Let $a, b \in \mathbb{R}$, $a \leq b$, and $p, q \in (1, \infty)$ be such that $\frac{1}{p} + \frac{1}{q} = 1$. Then, prove that there exists a constant $C > 0$ such that

$$\left(\int_a^b (h_4(t,a))^{\frac{1}{p}} dg(t)\right)^p \geq C \left(\int_a^b (h_4(t,a))^p dg(t)\right)^{\frac{1}{p}}.$$

Problem 3.9. Let $a, b \in \mathbb{R}$, $a \leq b$, and $p \in (1, \infty)$. Then, prove that

$$\left(\int_a^b (h_2(t,a) + h_4(t,a))^p dg(t)\right)^{\frac{1}{p}} \leq \left(\int_a^b (h_2(t,a))^p dg(t)\right)^{\frac{1}{p}} + \left(\int_a^b (h_4(t,a))^p dg(t)\right)^{\frac{1}{p}}.$$

Chapter 4

The Stieltjes–Wirtinger-Type Inequalities

In this chapter, we derive some Wirtinger-type inequalities for Stieltjes calculus. Some applications of the Stieltjes–Wirtinger-type inequalities are given.

Suppose that $I \subset \mathbb{R}$ and $g\colon I \to \mathbb{R}$ is a monotone nondecreasing function that is continuous from the left everywhere on I. Let $a, t_0 \in I$, $a \geq t_0$ and $J = [t_0, a]$.

4.1 Auxiliary Inequalities

In this section, we provide some auxiliary inequalities that will be used in the following sections. The proofs of these results can be found in our earlier book (Georgiev and Tikare, 2025). We start with the following important inequality.

Lemma 4.1. *For any $r, x, y > 0$, we have the following:*

$$(x+y)^r \leq \begin{cases} 2^r(x^r + y^r) & \text{for } r > 0, \\ x^r + y^r & \text{for } 0 < r \leq 1, \\ 2^{r-1}(x^r + y^r) & \text{for } r > 1. \end{cases}$$

The above lemma can be generalized as follows.

Theorem 4.1. *Let $k \geq 2$ be an integer and $x_j \geq 0$, $j \in \{1, \ldots, k\}$. Then,*

$$\left(\sum_{j=1}^{k} x_j\right)^{\alpha} \leq \begin{cases} \sum_{j=1}^{k} x_j^{\alpha} & \text{if } 0 \leq \alpha \leq 1, \\ 2^{(k-1)(\alpha-1)} \sum_{j=1}^{k} x_j^{\alpha} & \text{if } \alpha \geq 1. \end{cases} \tag{4.1}$$

Theorem 4.2. *Let $p > 2$ be an integer and $u, v \geq 0$. Then, there exists a positive constant $A(p)$ that does not depend on u and v such that*

$$(u+v)^p \leq u^p + pu^{p-1}v + A(p)(v^p + u^{p-2}v^2). \tag{4.2}$$

Theorem 4.3. *For any $p > 2$ and $u \geq 0$, we have the inequality*

$$(1+u)^p \geq pu + \frac{p(p-1)}{2}u^2. \tag{4.3}$$

Theorem 4.4. *For any integer p and $u, v \geq 0$, we have the inequality*

$$u^p + (p-1)v^p - puv^{p-1} \geq 0. \tag{4.4}$$

4.2 The Stieltjes–Wirtinger-Type Inequality I

The main result in this section reads as follows.

Theorem 4.5 (The Stieltjes–Wirtinger Inequality I). *Suppose that $f \in \mathscr{C}_g^1(J)$, $f > 0$ on J and satisfies either $f_g' > 0$ for $f_g' < 0$ on J. Set*

$$h = \left(\sup_{t\in J}\left(\frac{f(t)}{f(t+)}\right)\right)^{\frac{1}{2}} + \left(\sup_{t\in J}\left(\frac{(\Delta g(t))f_g'(t)}{f(t)}\right) + \sup_{t\in J}\left(\frac{f(t)}{f(t+)}\right)\right)^{\frac{1}{2}}.$$

Then, for any $y \in \mathscr{C}_g^1(J)$ for which $y(t_0) = y(a) = 0$, we have

$$\int_{t_0}^{a} \frac{f(t)f(t+)}{|f_g'(t)|}\left(y_g'(t)\right)^2 dg(t) \geq \frac{1}{h^2}\int_{t_0}^{a} |f_g'(t)|(y(t+))^2 dg(t). \tag{4.5}$$

Proof. Without any restrictions, assume that $f_g'(t) > 0$, $t \in J$. We leave the case $f_g' < 0$ on J to the reader as an exercise. Set

$$\alpha = \left(\sup_{t\in J}\left(\frac{f(t)}{f(t+)}\right)\right)^{\frac{1}{2}}, \quad \beta = \sup_{t\in J}\left(\frac{(\Delta g(t))f_g'(t)}{f(t+)}\right),$$

$$I_1 = \int_{t_0}^{a} |f_g'(t)|(y(t+))^2 dg(t), \quad I_2 = \int_{t_0}^{a} \frac{f(t)f(t+)}{|f_g'(t)|}\left(y_g'(t)\right)^2 dg(t).$$

Note that $\alpha+\sqrt{\alpha^2+\beta}=h$. We have the following estimations for I_1, using integration by parts in the sense of Stieltjes:

$$\begin{aligned}
I_1 &= \int_{t_0}^{a} |f_g'(t)|(y(t+))^2 dg(t) \\
&= f(t)y(t)\Big|_{t=t_0}^{t=a} - \int_{t_0}^{a} f(t)((y(t))^2)_g' dg(t) \\
&= -\int_{t_0}^{a} f(t)\left(y(t)y_g'(t)+y(t+)y_g'(t)\right) dg(t) \\
&= -\int_{t_0}^{a} f(t)y_g'(t)(y(t)+y(t+))dg(t) \\
&= -\int_{t_0}^{a} f(t)y_g'(t)(y(t+)-(\Delta g(t))y_g'(t)+y(t+))dg(t) \\
&= -\int_{t_0}^{a} f(t)y_g'(t)(2y(t+)-(\Delta g(t))y_g'(t))dg(t) \\
&\leq 2\int_{t_0}^{a} f(t)|y(t+)||y_g'(t)|dg(t) + \int_{t_0}^{a} (\Delta g(t))f(t)(y_g'(t))^2 dg(t) \\
&= 2\int_{t_0}^{a} \sqrt{\frac{f(t)f(t+)}{f_g'(t)}}|y_g'(t)|\sqrt{\frac{f(t)}{f(t+)}f_g'(t)}|y(t+)|dg(t) \\
&\quad + \int_{t_0}^{a} \frac{f(t)f(t+)}{f_g'(t)}\frac{(\Delta g(t))f_g'(t)}{f(t+)}(y_g'(t))^2 dg(t) \\
&\leq 2\int_{t_0}^{a} \sqrt{\frac{f(t)f(t+)}{f_g'(t)}}|y_g'(t)|\sqrt{\frac{f(t)}{f(t+)}f_g'(t)}|y(t+)|dg(t) \\
&\quad + \sup_{t\in J}\left(\frac{(\Delta g(t))f_g'(t)}{f(t+)}\right)\int_{t_0}^{a} \frac{f(t)f(t+)}{f_g'(t)}(y_g'(t))^2 dg(t) \\
&= 2\int_{t_0}^{a} \sqrt{\frac{f(t)f(t+)}{f_g'(t)}}|y_g'(t)|\sqrt{\frac{f(t)}{f(t+)}f_g'(t)}|y(t+)|dg(t) + \beta I_2,
\end{aligned}$$

i.e.,

$$I_1 \leq 2\int_{t_0}^{a} \sqrt{\frac{f(t)f(t+)}{f_g'(t)}}|y_g'(t)|\sqrt{\frac{f(t)}{f(t+)}f_g'(t)}|y(t+)|dg(t) + \beta I_2.$$

Now, applying the Stieltjes–Hölder inequality, we get

$$I_1 \leq 2\Big(\int_{t_0}^{a} \frac{f(t)f(t+)}{f_g'(t)}(y_g'(t))^2 dg(t)\Big)^{2}\Big(\int_{t_0}^{a} \frac{f(t)}{f(t+)} f_g'(t)(y(t+))^2 dg(t)\Big)^{\frac{1}{2}} + \beta I_2$$

$$\leq 2\left(\sup_{t\in J}\left(\frac{f(t)}{f(t+)}\right)\right)^{\frac{1}{2}} I_2^{\frac{1}{2}}\left(\int_{t_0}^{a} f_g'(t)(y(t+))^2 dg(t)\right)^{\frac{1}{2}} + \beta I_2$$

$$= 2\alpha I_1^{\frac{1}{2}} I_2^{\frac{1}{2}} + \beta I_2,$$

i.e.,

$$I_1 \leq 2\alpha I_1^{\frac{1}{2}} I_2^{\frac{1}{2}} + \beta I_2.$$

Dividing the last inequality by $I_1^{\frac{1}{2}} I_2^{\frac{1}{2}}$, we find

$$\frac{I_1^{\frac{1}{2}}}{I_2^{\frac{1}{2}}} \leq 2\alpha + \beta \frac{I_2^{\frac{1}{2}}}{I_1^{\frac{1}{2}}}.$$

Set

$$I_3 = \frac{I_1^{\frac{1}{2}}}{I_2^{\frac{1}{2}}}.$$

Then, we arrive at the inequality

$$I_3 \leq 2\alpha + \beta \frac{1}{I_3}$$

or

$$I_3^2 \leq 2\alpha I_3 + \beta,$$

and

$$(I_3 - \alpha)^2 \leq \alpha^2 + \beta.$$

From here, we get the inequalities

$$-\sqrt{\alpha^2+\beta} \leq I_3 - \alpha \leq \sqrt{\alpha^2+\beta}.$$

Since $I_3 \geq 0$, we obtain

$$I_3 \leq \alpha + \sqrt{\alpha^2+\beta}$$

and

$$I_3^2 \le \left(\alpha + \sqrt{\alpha^2 + \beta}\right)^2 = h^2.$$

Then,

$$I_1 \le h^2 I_2$$

or

$$I_2 \ge \frac{1}{h^2} I_1,$$

whereupon we find (4.5). This completes the proof. □

4.3 Stieltjes–Wirtinger-Type Inequality II

In this section, we present the other Stieltjes analogues of the Wirtinger-type inequality. For $\gamma \ge 1$ and a function $f \in \mathscr{C}_g^1(J)$ for which $f > 0$ on J, we set

$$\alpha = \sup_{t \in J} \left(\frac{f(t+)}{f(t)}\right)^{\frac{\gamma}{\gamma+1}} \quad \text{and} \quad \beta = \sup_{t \in J} \left(\frac{(\Delta g(t))|f_g'(t)|}{f(t)}\right)^{\gamma}.$$

The main result in this section is as follows.

Theorem 4.6 (The Stieltjes–Wirtinger-Type Inequality II). *Let $\gamma \in \mathbb{N}$ and $f \in \mathscr{C}_g^1(J)$, $f > 0$ on J, satisfies either $f_g' > 0$ or $f_g' < 0$ J. Also, let $h(\alpha, \beta, \gamma)$ be the largest root of the equation*

$$x^{\gamma+1} - 2^{\gamma-1}(\gamma+1)\alpha x^{\gamma} - 2^{\gamma-1}\beta = 0. \tag{4.6}$$

Then, for any $y \in \mathscr{C}_g^1(J)$ for which $y > 0$ on J and $y(t_0) = y(a) = 0$, we have

$$\int_{t_0}^{a} \frac{(f(t))^{\gamma} f(t+)}{|f_g'(t)|^{\gamma}} (y_g'(t))^{\gamma+1} dg(t) \ge \frac{1}{(h(\alpha, \beta, \gamma))^{\gamma+1}} \int_{t_0}^{a} |f_g'(t)| (y(t))^{\gamma+1} dg(t).$$

Proof. Let $\phi(x) = x^{\gamma+1} - 2^{\gamma-1}(\gamma+1)\alpha x^{\gamma} - 2^{\gamma-1}\beta$, $x \ge 0$. Then, $\phi \in \mathscr{C}([0, \infty))$. Moreover,

$$\phi(0) = -2^{\gamma-1}\beta \le 0$$

and

$$\lim_{x \to \infty} \phi(x) = \infty.$$

Therefore, Equation (4.6) has positive roots which are at most $\gamma+1$. Consequently, $h(\alpha,\beta,\gamma)$ exists. Without any restrictions, assume that $f_g'(t)>0$, $t\in J$. We leave the case $f_g'<0$ on J to the reader as an exercise. Let

$$I_1=\int_{t_0}^{a} f_g'(t)(y(t))^{\gamma+1}dg(t) \quad \text{and} \quad I_2=\int_{t_0}^{a} \frac{(f(t))^{\gamma} f(t+)}{(f_g'(t))^{\gamma}}(y_g'(t))^{\gamma+1}dg(t).$$

For I_1, we have the following estimate:

$$\begin{aligned} I_1 &= \int_{t_0}^{a} f_g'(t)(y(t))^{\gamma+1}dg(t) \\ &= f(t)(y(t))^{\gamma+1}\Big|_{t=t_0}^{t=a} - \int_{t_0}^{a} f(t+)((y_g')^{\gamma+1})(t)dg(t) \\ &= -\int_{t_0}^{a} f(t+)((y_g')^{\gamma+1})(t)dg(t) \\ &\le \int_{t_0}^{a} f(t+)|((y_g')^{\gamma+1})(t)|dg(t), \end{aligned}$$

i.e.,

$$I_1 \le \int_{t_0}^{a} f(t+)|((y_g')^{\gamma+1})(t)|dg(t). \tag{4.7}$$

Applying the chain rule, Theorem 1.42 and Lemma 4.1, we get

$$\begin{aligned} |((y_g')^{\gamma+1})(t)| &= (\gamma+1)\left|y_g'(t)\int_0^1 (y(t)+(\Delta g(t))hy_g'(t))^{\gamma}dh\right| \\ &= (\gamma+1)\,|y_g'(t)|\left|\int_0^1 (y(t)+(\Delta g(t))hy_g'(t))^{\gamma}dh\right| \\ &\le (\gamma+1)\,|y_g'(t)|\int_0^1 |y(t)+(\Delta g(t))hy_g'(t)|^{\gamma}dh \\ &\le (\gamma+1)\,|y_g'(t)|\int_0^1 (|y(t)|+(\Delta g(t))h|y_g'(t)|)^{\gamma}dh \\ &\le 2^{\gamma-1}(\gamma+1)|y_g'(t)|\int_0^1 \left(|y(t)|^{\gamma}+(\Delta g(t))^{\gamma}h^{\gamma}|y_g'(t)|^{\gamma}\right)dh \\ &= 2^{\gamma-1}(\gamma+1)|y_g'(t)|\left(\int_0^1 |y(t)|^{\gamma}dg(t)+(\Delta g(t))|y_g'(t)|^{\gamma}\int_0^1 h^{\gamma}dh\right) \\ &= 2^{\gamma-1}(\gamma+1)|y_g'(t)||y(t)|^{\gamma}+2^{\gamma-1}|y_g'(t)|(\Delta g(t))^{\gamma}|y_g'(t)|^{\gamma}, \quad t\in J. \end{aligned}$$

Hence, with (4.7), we arrive at the following estimate for I_1:

$$\begin{aligned}
I_1 &\leq 2^{\gamma-1}(\gamma+1)\int_{t_0}^{a} f(t+)|y_g'(t)||y(t)|^{\gamma}dg(t)\\
&\quad + 2^{\gamma-1}\int_{t_0}^{a} f(t+)|y_g'(t)|^{\gamma+1}(\Delta g(t))^{\gamma}dg(t)\\
&= 2^{\gamma-1}(\gamma+1)\int_{t_0}^{a}\left(\frac{f(t+)(f(t))^{\gamma}}{(f_g'(t))^{\gamma}}\right)^{\frac{1}{\gamma+1}}|y_g'(t)|\left(\frac{f(t+)f_g'(t)}{f(t)}\right)^{\frac{\gamma}{\gamma+1}}\\
&\quad\times|y(t)|^{\gamma}dg(t) + 2^{\gamma-1}\int_{t_0}^{a}\left(\frac{(\Delta g(t))f_g'(t)}{f(t)}\right)^{\gamma}\frac{f(t+)(f(t))^{\gamma}}{(f_g'(t))^{\gamma}}|y_g'(t)|^{\gamma+1}dg(t)\\
&= 2^{\gamma-1}(\gamma+1)\int_{t_0}^{a}\left(\frac{f(t+)(f(t))^{\gamma}}{(f_g'(t))^{\gamma}}|y_g'(t)|^{\gamma+1}\right)^{\frac{1}{\gamma+1}}\\
&\quad\times\left(\frac{f(t+)f_g'(t)}{f(t)}|y(t)|^{\gamma+1}\right)^{\frac{\gamma}{\gamma+1}}dg(t)\\
&\quad + 2^{\gamma-1}\int_{t_0}^{a}\left(\frac{(\Delta g(t))f_g'(t)}{f(t)}\right)^{\gamma}\frac{f(t+)(f(t))^{\gamma}}{(f_g'(t))^{\gamma}}|y_g'(t)|^{\gamma+1}dg(t)\\
&\leq 2^{\gamma-1}(\gamma+1)\left(\int_{t_0}^{a}\frac{f(t+)(f(t))^{\gamma}}{(f_g'(t))^{\gamma}}|y_g'(t)|^{\gamma+1}dg(t)\right)^{\frac{1}{\gamma+1}}\\
&\quad\times\left(\int_{t_0}^{a}\frac{f(t+)f_g'(t)}{f(t)}|y(t)|^{\gamma+1}dg(t)\right)^{\frac{\gamma}{\gamma+1}}\\
&\quad + 2^{\gamma-1}\int_{t_0}^{a}\left(\frac{(\Delta g(t))f_g'(t)}{f(t)}\right)^{\gamma}\frac{f(t+)(f(t))^{\gamma}}{(f_g'(t))^{\gamma}}|y_g'(t)|^{\gamma+1}dg(t)\\
&\leq 2^{\gamma-1}(\gamma+1)\alpha\left(\int_{t_0}^{a}\frac{f(t+)(f(t))^{\gamma}}{(f_g'(t))^{\gamma}}|y_g'(t)|^{\gamma+1}dg(t)\right)^{\frac{1}{\gamma+1}}\\
&\quad\times\left(\int_{t_0}^{a}f_g'(t)|y(t)|^{\gamma+1}dg(t)\right)^{\frac{\gamma}{\gamma+1}} + 2^{\gamma-1}\beta\int_{t_0}^{a}\frac{f(t+)(f(t))^{\gamma}}{(f_g'(t))^{\gamma}}|y_g'(t)|^{\gamma+1}dg(t)\\
&= 2^{\gamma-1}(\gamma+1)\alpha I_1^{\frac{\gamma}{\gamma+1}}I_2^{\frac{1}{\gamma+1}} + 2^{\gamma-1}\beta I_2\\
&= 2^{\gamma-1}\left((\gamma+1)\alpha I_1^{\frac{\gamma}{\gamma+1}}I_2^{\frac{1}{\gamma+1}} + \beta I_2\right),
\end{aligned}$$

i.e.,

$$I_1 \leq 2^{\gamma-1}\left((\gamma+1)\alpha I_1^{\frac{\gamma}{\gamma+1}} I_2^{\frac{1}{\gamma+1}} + \beta I_2\right).$$

Dividing the last inequality by $I_1^{\frac{\gamma}{\gamma+1}} I_2^{\frac{1}{\gamma+1}}$, we find

$$\frac{I_1^{\frac{1}{\gamma+1}}}{I_2^{\frac{1}{\gamma+1}}} \leq 2^{\gamma-1}\left((\gamma+1)\alpha + \beta\frac{I_2^{\frac{\gamma}{\gamma+1}}}{I_1^{\frac{\gamma}{\gamma+1}}}\right).$$

Set

$$I_3 = \frac{I_1^{\frac{1}{\gamma+1}}}{I_2^{\frac{1}{\gamma+1}}}.$$

Then,

$$I_3 \leq 2^{\gamma-1}\left((\gamma+1)\alpha + \frac{\beta}{I_3^{\gamma}}\right)$$

or

$$I_3^{\gamma+1} - 2^{\gamma-1}(\gamma+1)\alpha I_3^{\gamma} - 2^{\gamma-1}\beta \leq 0.$$

By definition of $h(\alpha,\beta,\gamma)$, we find

$$I_3 \leq h(\alpha,\beta,\gamma),$$

$$\frac{I_1^{\frac{1}{\gamma+1}}}{I_2^{\frac{1}{\gamma+1}}} \leq h(\alpha,\beta,\gamma),$$

$$I_2^{\frac{1}{\gamma+1}} \geq \frac{1}{h(\alpha,\beta,\gamma)} I_1^{\frac{1}{\gamma+1}},$$

or

$$I_2 \geq \frac{1}{(h(\alpha,\beta,\gamma))^{\gamma+1}} I_1.$$

This completes the proof. □

Corollary 4.1 (The Stieltjes–Wirtinger-Type Inequality III). *Let $f \in \mathscr{C}_g^1(J)$, $f > 0$ on J, satisfy either $f_g' > 0$ or $f_g' < 0$ on J. Then, for any $y \in \mathscr{C}_g^1(J)$ for which $y > 0$ on J and $y(t_0) = y(a) = 0$, we have*

$$\int_{t_0}^{a} \frac{f(t+)f(t)}{|f_g'(t)|}(y_g'(t))^2 dg(t) \geq \frac{1}{\left(\alpha + \sqrt{\alpha^2+\beta}\right)^2}\int_{t_0}^{a} |f_g'(t)|(y(t))^2 dg(t).$$

Proof. We apply Theorem 4.6 for $\gamma = 1$. □

4.4 Other Stieltjes–Wirtinger-Type Inequalities

In this section, we present some other Stieltjes–Wirtinger-type inequalities using the inequalities given in the previous section. We use the notations from the previous section.

Theorem 4.7. *Let $A(p)$ be the smallest positive constant for which (4.2) holds and $\gamma \in \mathbb{N}$, $\gamma \geq 2$ be such that $\gamma(\gamma+1) > 6A(p)$. Suppose that $f \in \mathscr{C}_g^1(J)$ is a positive function for which either $f_g' > 0$ or $f_g' < 0$ and*

$$\begin{aligned} \alpha &< \frac{A(p)}{2^\gamma(\gamma+1)^2}, \\ \beta &< \frac{A(p)}{2^\gamma}. \end{aligned} \tag{4.8}$$

Moreover, let $l(\alpha,\beta,\gamma)$ be the largest positive root of the equation

$$\begin{aligned} &-(\gamma+1)\left(1+\alpha 2^{\gamma-1}(\gamma+1)\right)x^\gamma - A(p)x^{\gamma-1} + (\gamma+1)x \\ &\quad + \frac{\gamma(\gamma+1)}{2}x^2 - A(p) - 2^{\gamma-1}\beta = 0. \end{aligned} \tag{4.9}$$

Then, for any $y \in \mathscr{C}_g^1(J)$ for which $y(t_0) = y(a) = 0$, we have the following inequality:

$$\int_{t_0}^a \frac{(f(t))^\gamma f(t+)}{(f_g'(t))^\gamma}(y_g'(t))^{\gamma+1}dg(t) \geq \frac{1}{(l(\alpha,\beta,\gamma))^{\gamma+1}}\int_{t_0}^a |f_g'(t)|(y(t))^{\gamma+1}dg(t).$$

Proof. Without any restrictions, assume that $f_g'(t) > 0$, $t \in J$. We leave the case $f_g'(t) < 0$ to the reader as an exercise. Let

$$\begin{aligned} \phi(x) &= -(\gamma+1)\left(1+\alpha 2^{\gamma-1}(\gamma+1)\right)x^\gamma - A(p)x^{\gamma-1} + (\gamma+1)x \\ &\quad + \frac{\gamma(\gamma+1)}{2}x^2 - A(p) - 2^{\gamma-1}\beta, \quad x \in [0,\infty). \end{aligned}$$

We have $\phi \in \mathscr{C}(J)$ and $\phi(0) = -A(p) - 2^{\gamma-1}\beta < 0$, and applying (4.8), we find

$$\begin{aligned} \phi(1) &= -(\gamma+1)\left(1+\alpha 2^{\gamma-1}(\gamma+1)\right) - A(p) + (\gamma+1) + \frac{\gamma(\gamma+1)}{2} \\ &\quad - A(p) - 2^{\gamma-1}\beta \end{aligned}$$

$$
\begin{aligned}
&= -\alpha(\gamma+1)^2 2^{\gamma-1} - 2A(p) - 2^{\gamma-1}\beta + \frac{\gamma(\gamma+1)}{2} \\
&= \frac{1}{2}\left(\gamma(\gamma+1) - \alpha(\gamma+1)^2 2^{\gamma} - 2^{\gamma}\beta - 4A(p)\right) \\
&> \frac{1}{2}\left(\gamma(\gamma+1) - \frac{A(p)}{(\gamma+1)^2 2^{\gamma}}(\gamma+1)^2 2^{\gamma} - 2^{\gamma}\frac{A(p)}{2^{\gamma}} - 4A(p)\right) \\
&= \frac{1}{2}\left(\gamma(\gamma+1) - 6A(p)\right) \\
&> 0.
\end{aligned}
$$

Thus, Equation (4.9) has positive roots and they are at most γ. In the following, we use the notations used in the proof of Theorem 4.6. By the proof of Theorem 4.6, we have

$$
I_3^{\gamma+1} \leq 2^{\gamma-1}(\gamma+1)\alpha I_3^{\gamma} + 2^{\gamma-1}\beta. \tag{4.10}
$$

Now, we apply (4.2) for $u = I_3$, $v = 1$, and $p = \gamma + 1$ and obtain

$$
(1+I_3)^{\gamma+1} \leq I_3^{\gamma+1} + (\gamma+1)I_3^{\gamma} + A(p)(1 + I_3^{\gamma-1}),
$$

whereupon

$$
I_3^{\gamma+1} \geq (1+I_3)^{\gamma+1} - (\gamma+1)I_3^{\gamma} - A(p)I_3^{\gamma-1} - A(p).
$$

Hence, applying (4.10), we arrive at the inequality

$$
(1+I_3)^{\gamma+1} - (\gamma+1)I_3^{\gamma} - A(p)I_3^{\gamma-1} - A(p) \leq 2^{\gamma-1}(\gamma+1)\alpha I_3^{\gamma} + 2^{\gamma-1}\beta
$$

or

$$
(1+I_3)^{\gamma+1} - (\gamma+1)\left(1 + \alpha 2^{\gamma-1}(\gamma+1)\right) I_3^{\gamma} - A(p)I_3^{\gamma-1} - A(p) - 2^{\gamma-1}\beta \leq 0. \tag{4.11}
$$

Now, we apply the inequality (4.3) and find

$$
(1+I_3)^{\gamma+1} \geq (\gamma+1)I_3 + \frac{\gamma(\gamma+1)}{2} I_3^2.
$$

From this and (4.11), we get

$$
\begin{aligned}
-(\gamma+1)\left(1 + \alpha 2^{\gamma-1}(\gamma+1)\right) I_3^{\gamma} - A(p)I_3^{\gamma-1} + (\gamma+1)I_3 + \frac{\gamma(\gamma+1)}{2} I_3^2 \\
- A(p) - 2^{\gamma-1}\beta \leq 0.
\end{aligned}
$$

Therefore,

$$I_3 \leq l(\alpha, \beta, \gamma),$$
$$I_2^{\frac{1}{\gamma+1}} \geq \frac{1}{l(\alpha, \beta, \gamma)} I_1^{\frac{1}{\gamma+1}},$$

or

$$I_2 \geq \frac{1}{(l(\alpha, \beta, \gamma))^{\gamma+1}} I_1.$$

This completes the proof. □

Theorem 4.8. *Let $A(p)$ be the smallest positive constant for which* (4.2) *holds and $\gamma \in \mathbb{N}$, $\gamma \geq 3$ be such that $(\gamma - 2)(\gamma + 1) > 6A(p)$. Suppose that $f \in \mathscr{C}_g^1(J)$ is a positive function for which either $f_g' > 0$ or $f_g' < 0$, and* (4.8) *holds. Moreover, let $p(\alpha, \beta, \gamma)$ be the largest positive root of the equation*

$$\begin{aligned} &-(\gamma+1)\left(1+\alpha 2^{\gamma-1}(\gamma+1)\right)x^\gamma - A(p)x^{\gamma-1} \\ &\quad + \frac{\gamma(\gamma+1)}{2}x^2 - A(p) - 2^{\gamma-1}\beta = 0. \end{aligned} \tag{4.12}$$

Then, for any $y \in \mathscr{C}_g^1(J)$ for which $y(t_0) = y(a) = 0$, we have the following inequality:

$$\begin{aligned} &\int_{t_0}^{a} \frac{(f(t))^\gamma f(t+)}{(f_g'(t))^\gamma}(y_g'(t))^{\gamma+1} dg(t) \\ &\quad \geq \frac{1}{(p(\alpha, \beta, \gamma))^{\gamma+1}} \int_{t_0}^{a} |f_g'(t)|(y(t))^{\gamma+1} dg(t). \end{aligned}$$

Proof. Without any restrictions, assume that $f_g'(t) > 0$, $t \in J$. We leave the case $f_g' < 0$ to the reader as an exercise. Let

$$\begin{aligned} \phi(x) &= -(\gamma+1)\left(1+\alpha 2^{\gamma-1}(\gamma+1)\right)x^\gamma - A(p)x^{\gamma-1} \\ &\quad + \frac{\gamma(\gamma+1)}{2}x^2 - A(p) - 2^{\gamma-1}\beta, \quad x \in [0, \infty). \end{aligned}$$

We have $\phi \in \mathscr{C}(J)$ and $\phi(0) = -A(p) - 2^{\gamma-1}\beta < 0$, and applying (4.8), we find

$$\begin{aligned}\phi(1) &= -(\gamma+1)\left(1+\alpha 2^{\gamma-1}(\gamma+1)\right) - A(p) + \frac{\gamma(\gamma+1)}{2} - A(p) - 2^{\gamma-1}\beta \\ &= -\alpha(\gamma+1)^2 2^{\gamma-1} - 2A(p) - 2^{\gamma-1}\beta + \frac{(\gamma-2)(\gamma+1)}{2} \\ &= \frac{1}{2}\left((\gamma-2)(\gamma+1) - \alpha(\gamma+1)^2 2^{\gamma} - 2^{\gamma}\beta - 4A(p)\right) \\ &> \frac{1}{2}\left((\gamma-2)(\gamma+1) - \frac{A(p)}{(\gamma+1)^2 2^{\gamma}}(\gamma+1)^2 2^{\gamma} - 2^{\gamma}\frac{A(p)}{2^{\gamma}} - 4A(p)\right) \\ &= \frac{1}{2}\left((\gamma-2)(\gamma+1) - 6A(p)\right) \\ &> 0.\end{aligned}$$

Thus, Equation (4.12) has positive roots and they are at most γ. In the following, we use the notations used in the proof of Theorem 4.6. Now, we apply the inequality (4.3) and find

$$(1+I_3)^{\gamma+1} \geq \frac{\gamma(\gamma+1)}{2} I_3^2.$$

From this and (4.11), we get

$$-(\gamma+1)\left(1+\alpha 2^{\gamma-1}(\gamma+1)\right) I_3^{\gamma} - A(p) I_3^{\gamma-1} + \frac{\gamma(\gamma+1)}{2} I_3^2 - A(p) - 2^{\gamma-1}\beta \leq 0.$$

Therefore,

$$I_3 \leq p(\alpha,\beta,\gamma),$$
$$I_2^{\frac{1}{\gamma+1}} \geq \frac{1}{p(\alpha,\beta,\gamma)} I_1^{\frac{1}{\gamma+1}},$$

or

$$I_2 \geq \frac{1}{(p(\alpha,\beta,\gamma))^{\gamma+1}} I_1.$$

This completes the proof. □

Theorem 4.9. *Let $A(p)$ be the smallest positive constant for which* (4.2) *holds and $\gamma \in \mathbb{N}$, $\gamma \geq 2$ be such that*

$$\gamma + 1 > 20A(p). \tag{4.13}$$

Suppose that $f \in \mathscr{C}_g^1(J)$ is a positive function for which either $f_g' > 0$ or $f_g' < 0$, and

$$\begin{aligned} \alpha &< \frac{A(p)}{2^{\gamma-1}(\gamma+1)^2}, \\ \beta &< \frac{A(p)}{2^{\gamma-1}}. \end{aligned} \tag{4.14}$$

Moreover, let $q(\alpha,\beta,\gamma)$ be the largest positive root of the equation

$$\begin{aligned} -(\gamma+1)\left(1+\alpha 2^{\gamma-1}(\gamma+1)\right)x^\gamma - A(p)x^{\gamma-1} + (\gamma+1)x \\ - A(p) - 2^{\gamma-1}\beta = 0. \end{aligned} \tag{4.15}$$

Then, for any $y \in \mathscr{C}_g^1(J)$ for which $y(t_0) = y(a) = 0$, we have the following inequality:

$$\int_{t_0}^a \frac{(f(t))^\gamma f(t+)}{(f_g'(t))^\gamma}(y_g'(t))^{\gamma+1}dg(t) \geq \frac{1}{(q(\alpha,\beta,\gamma))^{\gamma+1}}\int_{t_0}^a |f_g'(t)|(y(t))^{\gamma+1}dg(t).$$

Proof. Without any restrictions, assume that $f_g'(t) > 0$, $t \in J$. We leave the case $f_g' < 0$ to the reader as an exercise. Let

$$\begin{aligned} \phi(x) = -(\gamma+1)\left(1+\alpha 2^{\gamma-1}(\gamma+1)\right)x^\gamma - A(p)x^{\gamma-1} + (\gamma+1)x \\ - A(p) - 2^{\gamma-1}\beta, \quad x \in [0,\infty). \end{aligned}$$

We have $\phi \in \mathscr{C}(J)$ and $\phi(0) = -A(p) - 2^{\gamma-1}\beta < 0$. From (4.13), we get

$$\frac{1}{5}(\gamma+1) > 4A(p). \tag{4.16}$$

Note that $3^\gamma > \frac{15}{2}$ yields $\frac{1}{3} - \frac{1}{3^\gamma} > \frac{1}{5}$. Also, $4 > \frac{1}{3^\gamma} + \frac{1}{3^{\gamma-1}} + 2$. Hence, applying (4.16), we arrive at

$$\begin{aligned} \left(\frac{1}{3} - \frac{1}{3^\gamma}\right)(\gamma+1) &> \frac{1}{5}(\gamma+1) \\ &> 4A(p) \\ &> \left(\frac{1}{3^\gamma} + \frac{1}{3^{\gamma-1}} + 2\right)A(p), \end{aligned}$$

i.e.,

$$\left(\frac{1}{3} - \frac{1}{3^\gamma}\right)(\gamma+1) > \left(\frac{1}{3^\gamma} + \frac{1}{3^{\gamma-1}} + 2\right)A(p).$$

From this and (4.16), we find

$$\begin{aligned}\phi\left(\frac{1}{3}\right) &= \frac{1}{3}(\gamma+1) - \frac{1}{3^\gamma}(\gamma+1) - \frac{1}{3^\gamma}(\gamma+1)^2\alpha 2^{\gamma-1} \\ &\quad - \frac{1}{3^{\gamma-1}}A(p) - A(p) - 2^{\gamma-1}\beta \\ &= \left(\frac{1}{3} - \frac{1}{3^\gamma}\right)(\gamma+1) - \frac{1}{3^\gamma}(\gamma+1)^2\alpha 2^{\gamma-1} \\ &\quad - \frac{1}{3^{\gamma-1}}A(p) - A(p) - 2^{\gamma-1}\beta \\ &> \left(\frac{1}{3} - \frac{1}{3^\gamma}\right)(\gamma+1) - \frac{1}{3^\gamma}(\gamma+1)^2 2^{\gamma-1}\frac{A(p)}{(\gamma+1)^2 2^{\gamma-1}} \\ &\quad - \frac{1}{3^{\gamma-1}}A(p) - A(p) - 2^{\gamma-1}\frac{A(p)}{2^{\gamma-1}} \\ &= \left(\frac{1}{3} - \frac{1}{3^\gamma}\right)(\gamma+1) - \left(\frac{1}{3^\gamma} + \frac{1}{3^{\gamma-1}} + 2\right)A(p) \\ &> 0.\end{aligned}$$

Thus, Equation (4.15) has positive roots and they are at most γ. In the following, we use the notations used in the proof of Theorem 4.7. We apply the inequality (4.3) and find

$$(1+I_3)^{\gamma+1} \geq (\gamma+1)I_3.$$

From this and (4.11), we get

$$-(\gamma+1)\left(1+\alpha 2^{\gamma-1}(\gamma+1)\right)I_3^\gamma - A(p)I_3^{\gamma-1} + (\gamma+1)I_3 - A(p) - 2^{\gamma-1}\beta \leq 0.$$

Therefore,

$$I_3 \leq q(\alpha,\beta,\gamma),$$

$$I_2^{\frac{1}{\gamma+1}} \geq \frac{1}{q(\alpha,\beta,\gamma)}I_1^{\frac{1}{\gamma+1}},$$

or

$$I_2 \geq \frac{1}{(q(\alpha,\beta,\gamma))^{\gamma+1}}I_1.$$

This completes the proof. □

Theorem 4.10. *Let $\gamma \in \mathbb{N}$, $\gamma \geq 2$. Suppose that $f \in \mathscr{C}_g^1(J)$ is a positive function for which either $f_g' > 0$ or $f_g' < 0$, and*

$$\alpha < \frac{1}{2^\gamma(\gamma+1)^2},$$
$$\beta < \frac{1}{2^\gamma}. \tag{4.17}$$

Moreover, let $r(\alpha, \beta, \gamma)$ be the largest positive root of the equation

$$-2^{\gamma-1}(\gamma+1)\alpha x^\gamma + (\gamma+1)x - 2^{\gamma-1}\beta - \gamma = 0. \tag{4.18}$$

Then, for any $y \in \mathscr{C}_g^1(J)$ for which $y(t_0) = y(a) = 0$, we have the following inequality:

$$\int_{t_0}^a \frac{(f(t))^\gamma f(t+)}{(f_g'(t))^\gamma}(y_g'(t))^{\gamma+1}dg(t) \geq \frac{1}{(r(\alpha,\beta,\gamma))^{\gamma+1}}\int_{t_0}^a |f_g'(t)|(y(t))^{\gamma+1}dg(t).$$

Proof. Without any restrictions, assume that $f_g'(t) > 0$, $t \in J$. We leave the case $f_g' < 0$ to the reader as an exercise. Let

$$\phi(x) = -2^{\gamma-1}(\gamma+1)\alpha x^\gamma + (\gamma+1)x - 2^{\gamma-1}\beta - \gamma, \quad x \in [0,\infty).$$

We have $\phi \in \mathscr{C}(J)$ and $\phi(0) = -2^{\gamma-1}\beta - \gamma < 0$. Applying (4.17), we find

$$\begin{aligned}
\phi(1) &= -2^{\gamma-1}(\gamma+1)\alpha + \gamma + 1 - 2^{\gamma-1}\beta - \gamma \\
&= 1 - 2^{\gamma-1}(\gamma+1)\alpha - 2^{\gamma-1}\beta \\
&> 1 - 2^{\gamma-1}(\gamma+1)\frac{1}{2^\gamma(\gamma+1)} - 2^{\gamma-1}\frac{1}{2^\gamma} \\
&= 1 - \frac{1}{2} - \frac{1}{2} \\
&= 0.
\end{aligned}$$

Thus, Equation (4.18) has positive roots and they are at most γ. In the following, we use the notations used in the proof of Theorem 4.6. We apply the inequality (4.4) for $u = I_3$ and $v = 1$, and find

$$I_3^{\gamma+1} + \gamma - (\gamma+1)I_3 \geq 0,$$

i.e.,

$$I_3^{\gamma+1} \geq (\gamma+1)I_3 - \gamma.$$

Hence, applying (4.10), we obtain

$$(\gamma+1)I_3-\gamma \le 2^{\gamma-1}(\gamma+1)\alpha I_3^{\gamma}+2^{\gamma-1}\beta$$

or

$$-2^{\gamma-1}(\gamma+1)\alpha I_3^{\gamma}+(\gamma+1)I_3-\gamma-2^{\gamma-1}\beta \le 0.$$

Therefore,

$$I_3 \le r(\alpha,\beta,\gamma),$$
$$I_2^{\frac{1}{\gamma+1}} \ge \frac{1}{r(\alpha,\beta,\gamma)} I_1^{\frac{1}{\gamma+1}},$$

or

$$I_2 \ge \frac{1}{(r(\alpha,\beta,\gamma))^{\gamma+1}} I_1.$$

This completes the proof. □

4.5 Advanced Practical Problems

Problem 4.1. Let $\gamma > 2$, $\gamma \in \mathbb{N}$ and f, g be positive Stieltjes integrable functions on J. Then, prove that there is a positive constant $A(\gamma)$ that does not depend on f and g such that

$$\begin{aligned}\int_{t_0}^{a}(f(t)+g(t))^{\gamma}dg(t) &\le \int_{t_0}^{a}(f(t))^{\gamma}dg(t)+\gamma\int_{t_0}^{a}(f(t))^{\gamma-1}g(t)dg(t)\\ &\quad + A(p)\left(\int_{t_0}^{a}(g(t))^{\gamma}dg(t)\right.\\ &\quad \left.+\int_{t_0}^{a}(f(t))^{\gamma-2}(g(t))^{2}dg(t)\right).\end{aligned}$$

Problem 4.2. Let $\gamma > 2$, $\gamma \in \mathbb{N}$ and g be a positive Stieltjes integrable function on J. Then, prove that there is a positive constant $A(\gamma)$ that does not depend on f and g such that

$$\begin{aligned}\int_{t_0}^{a}(1+g(t))^{\gamma}dg(t) &\le (a-t_0)+\gamma\int_{t_0}^{a}g(t)dg(t)\\ &\quad + A(p)\left(\int_{t_0}^{a}(g(t))^{\gamma}dg(t)+\int_{t_0}^{a}(g(t))^{2}dg(t)\right).\end{aligned}$$

Problem 4.3. Let $\gamma > 2$, $\gamma \in \mathbb{N}$ and g be a positive Stieltjes integrable function on J. Then, prove that there is a positive constant $A(\gamma)$ that does not depend on f and g such that

$$\int_{t_0}^{a} (g(t))^{\gamma} dg(t) \leq \int_{t_0}^{a} (g(t))^{\gamma} dg(t) + \gamma \int_{t_0}^{a} (g(t))^{\gamma-1} dg(t) + A(p) \left((a - t_0) + \int_{t_0}^{a} (g(t))^{\gamma-2} dg(t) \right).$$

Problem 4.4. Let $\gamma \in \mathbb{N}$ and f, g be positive Stieltjes integrable functions on J. Then, prove that

$$\int_{t_0}^{a} (f(t))^{\gamma} dg(t) + (\gamma - 1) \int_{t_0}^{a} (g(t))^{\gamma} dg(t) - \gamma \int_{t_0}^{a} f(t)(g(t))^{\gamma-1} dg(t) \geq 0.$$

Problem 4.5. Let $\gamma \in \mathbb{N}$ and g be a positive Stieltjes integrable function on J. Then, prove that

$$a - t_0 + (\gamma - 1) \int_{t_0}^{a} (g(t))^{\gamma} dg(t) - \gamma \int_{t_0}^{a} (g(t))^{\gamma-1} dg(t) \geq 0.$$

Problem 4.6. Let $\gamma \in \mathbb{N}$ and g be a positive Stieltjes integrable function on J. Then, prove that

$$\int_{t_0}^{a} (g(t))^{\gamma} dg(t) + (\gamma - 1)(a - t_0) - \gamma \int_{t_0}^{a} g(t) dg(t) \geq 0.$$

Problem 4.7. Let $y \in \mathscr{C}_g^1(J)$ and $y(t_0) = y(a) = 0$. Then, prove that

$$\int_{t_0}^{a} (y_g'(t))^{\gamma+1} dg(t) \geq \frac{1}{(h(\alpha, \beta, \gamma))^{\gamma+1}} \int_{t_0}^{a} (y(t))^{\gamma+1} dg(t),$$

where $\gamma \in \mathbb{N}$ and $h(\alpha, \beta, \gamma)$ is the largest positive root of Equation (4.6).

Chapter 5

Stieltjes–Opial-Type Inequalities

Suppose that $I \subset \mathbb{R}$ and $g \colon I \to \mathbb{R}$ is a monotone nondecreasing function that is continuous from the left everywhere on I. Assume that $t_0, a \in I$ be such that $t_0 \leq a$ and $J = [t_0, a]$.

5.1 Auxiliary Inequalities

In this section, we present some auxiliary inequalities needed to prove some of the main results in this chapter. We start with the following useful inequality.

Theorem 5.1. *For any $p > 0$ and $y \geq 0$, we have the inequality*

$$y^{p+1} - (p+1)y + p \geq 0. \tag{5.1}$$

Proof. Let $h(y) = y^{p+1} - (p+1)y + p, \quad y \geq 0$. Then, we have

$$\begin{aligned} h'(y) &= (p+1)y^p - (p+1) \\ &= (p+1)(y^p - 1), \quad y \geq 0. \end{aligned}$$

We have the following cases:

(1) Suppose $y > 1$. Then, $h'(y) > 0$ and g is nondecreasing on $(1, \infty)$. Hence,

$$\begin{aligned} h(y) &> h(1) \\ &= 1 - (p+1) + p \\ &= 0. \end{aligned}$$

(2) Suppose $y \in [0,1]$. Then, $h'(y) \leq 0$ and g is nonincreasing on $[0,1]$. Thus,

$$\begin{aligned} h(y) &\geq h(1) \\ &= 0. \end{aligned}$$

This completes the proof. $\square$

Theorem 5.2. *For any $a \neq b > 0$ and $p > 0$, we have the inequality*

$$a^{p+1} + pb^{p+1} - (p+1)ab^p \geq 0. \tag{5.2}$$

Proof. In the inequality (5.1), we take $y = \frac{a}{b}$. Then, we get

$$\left(\frac{a}{b}\right)^{p+1} - (p+1)\frac{a}{b} + p \geq 0,$$

whereupon

$$a^{p+1} - (p+1)ab^p + pb^{p+1} \geq 0.$$

This completes the proof. □

5.2 Stieltjes–Opial-Type Inequalities I

Theorem 5.3. *Let $y \in \mathscr{C}_g^1(J)$ and $y(t_0) = 0$. Then,*

$$\int_{t_0}^{a} |y(t) + y(t+)||y_g'(t)|dg(t) \leq (g(a) - g(t_0)) \int_{t_0}^{a} |y_g'(t)|^2 dg(t).$$

Proof. Set

$$z(t) = \int_{t_0}^{t} |y_g'(s)|dg(s), \quad t \in J.$$

Then,

$$\begin{aligned} z(t) &\geq \left| \int_{t_0}^{t} y_g'(s)dg(s) \right| \\ &= \left| y(s) \Big|_{s=t_0}^{s=t} \right| \\ &= |y(t)|, \quad t \in J, \end{aligned}$$

and

$$\begin{aligned} z_g'(t) &= |y(t)|, \\ z(t_0) &= 0. \end{aligned}$$

Hence, applying the Cauchy–Schwartz inequality, we arrive at

$$\begin{aligned}\int_{t_0}^{a} |y(t)+y(t+)||y_g'(t)|dg(t) &\leq \int_{t_0}^{a} (|y(t)|+|y(t+)|)z_g'(t)dg(t)\\&\leq \int_{t_0}^{a} (z(t)+z(t+))z_g'(t)dg(t)\\&= \int_{t_0}^{a} (z_g')^2(t)dg(t)\\&= (z(t))^2\Big|_{t=t_0}^{t=a}\\&= (z(a))^2\\&= \left(\int_{t_0}^{a} |y_g'(t)|dg(t)\right)^2\\&\leq (g(a)-g(t_0))\int_{t_0}^{a} (y_g'(t))^2 dg(t).\end{aligned}$$

This completes the proof. □

Theorem 5.4. *For $y \in \mathscr{C}_g^1(J)$, we have the inequality*

$$\begin{aligned}&\int_{t_0}^{a} |y(t)+y(t+)||y_g'(t)|dg(t)\\&\quad\leq (g(a)-g(t_0))\int_{t_0}^{a} (y_g'(t))^2 dg(t) + \max\{|y(t_0)|, |y(a)|\}\int_{t_0}^{a} |y_g'(t)|dg(t).\end{aligned}$$

Proof. Let

$$y_1(t) = \int_{t_0}^{t} |y_g'(s)|dg(s) \quad \text{and} \quad y_2(t) = \int_{t}^{a} |y_g'(s)|dg(s), \quad t \in J.$$

Then,

$$y_{1g}'(t) = |y_g'(t)| \quad \text{and} \quad y_{2g}'(t) = -|y_g'(t)|, \quad t \in J.$$

Hence,

$$\begin{aligned}|y(t)| &= |y(t)-y(t_0)+y(t_0)|\\&\leq |y(t)-y(t_0)| + |y(t_0)|\end{aligned}$$

$$
\begin{aligned}
&= \left| \int_{t_0}^{t} y_g'(s)dg(s) \right| + |y(t_0)| \\
&\leq \int_{t_0}^{t} |y_g'(s)|dg(s) + |y(t_0)| \\
&= y_1(t) + |y(t_0)|, \quad t \in J,
\end{aligned}
$$

and

$$
\begin{aligned}
|y(t)| &= |y(t) - y(a) + y(a)| \\
&\leq |y(t) - y(a)| + |y(a)| \\
&= \left| \int_{t}^{a} y_g'(s)dg(s) \right| + |y(a)| \\
&\leq \int_{t}^{a} |y_g'(s)|dg(s) + |y(a)| \\
&= y_2(t) + |y(a)|, \quad t \in J.
\end{aligned}
$$

Next, applying the Cauchy–Schwartz inequality, we find

$$
\begin{aligned}
&\int_{t_0}^{t} |y(s) + y(s+)||y_g'(s)|dg(s) \\
&\quad \leq \int_{t_0}^{t} (|y(s)| + |y(s+)|)y_{1g}'(s)dg(s) \\
&\quad \leq \int_{t_0}^{t} (y_1(s) + y_1(s+) + 2|y(t_0)|)y_{1g}'(s)dg(s) \\
&\quad = \int_{t_0}^{t} (y_1(s) + y_1(s+))y_{1g}'(s)dg(s) + 2|y(t_0)| \int_{t_0}^{t} y_{1g}'(s)dg(s) \\
&\quad = \int_{t_0}^{t} (y_{1g}')^2(s)dg(s) + 2|y(t_0)|y_1(s)\Big|_{s=t_0}^{s=t} \\
&\quad = (y_1(s))^2\Big|_{s=t_0}^{s=t} + 2|y(t_0)|y_1(t) \\
&\quad = (y_1(t))^2 + 2|y(t_0)|y_1(t)
\end{aligned}
$$

$$= \left(\int_{t_0}^{t} |y_g'(s)|dg(s)\right)^2 + 2|y(t_0)|\int_{t_0}^{t} |y_g'(s)|dg(s)$$
$$\leq (g(a) - g(t_0))\int_{t_0}^{t} (y_g'(s))^2 dg(s) + 2|y(t_0)|\int_{t_0}^{t} |y_g'(s)|dg(s), \quad t \in J,$$

i.e.,

$$\int_{t_0}^{t} |y(s) + y(s+)||y_g'(s)|dg(s)$$
$$\leq (g(a) - g(t_0))\int_{t_0}^{t} (y_g'(s))^2 dg(s) + 2|y(t_0)|\int_{t_0}^{t} |y_g'(s)|dg(s). \tag{5.3}$$

Moreover,

$$\int_{t}^{a} |y(s) + y(s+)||y_g'(s)|dg(s)$$
$$\leq -\int_{t}^{a} (|y(s)| + |y(s+)|)y_{2g}'(s)dg(s)$$
$$\leq -\int_{t}^{a} (y_2(s) + y_2(s+) + 2|y(a)|)y_{2g}'(s)dg(s)$$
$$= -\int_{t}^{a} (y_2(s) + y_2(s+))y_{2g}'(s)dg(s) - 2|y(a)|\int_{t}^{a} y_{2g}'(s)dg(s)$$
$$= -\int_{t}^{a} (y_{2g}')^2(s)dg(s) + 2|y(a)|y_2(s)\Big|_{s=t}^{s=a}$$
$$= -(y_2(s))^2\Big|_{s=t}^{s=a} + 2|y(t_0)|y_2(t)$$
$$= (y_2(t))^2 + 2|y(a)|y_2(t)$$
$$= \left(\int_{t}^{a} |y_g'(s)|dg(s)\right)^2 + 2|y(a)|\int_{t}^{a} |y_g'(s)|dg(s)$$
$$\leq (g(a) - g(t_0))\int_{t}^{a} (y_g'(s))^2 dg(s) + 2|y(a)|\int_{t}^{a} |y_g'(s)|dg(s), \quad t \in J,$$

i.e.,

$$\int_{t}^{a} |y(s) + y(s+)||y_g'(s)|dg(s)$$
$$\leq (g(a) - g(t_0))\int_{t}^{a} (y_g'(s))^2 dg(s) + 2|y(a)|\int_{t}^{a} |y_g'(s)|dg(s), \quad t \in J.$$

Adding the last inequality and the inequality (5.3), we arrive at

$$\begin{aligned}
&\int_{t_0}^{t} |y(s)+y(s+)||y'_g(s)|dg(s) + \int_{t}^{a} |y(s)+y(s+)||y'_g(s)|dg(s)\\
&\quad \le (g(a)-g(t_0))\int_{t_0}^{t} (y'_g(s))^2 dg(s) + 2|y(t_0)|\int_{t_0}^{t} |y'_g(s)|dg(s)\\
&\qquad + (g(a)-g(t_0))\int_{t}^{a} (y'_g(s))^2 dg(s) + 2|y(a)|\int_{t}^{a} |y'_g(s)|dg(s)
\end{aligned}$$

or

$$\begin{aligned}
&\int_{t_0}^{a} |y(s)+y(s+)||y'_g(s)|dg(s)\\
&\quad \le (g(a)-g(t_0))\int_{t_0}^{a} (y'_g(s))^2 dg(s) + 2\max\{|y(t_0)|, y(a)\}\int_{t_0}^{t} |y'_g(s)|dg(s)\\
&\qquad + 2\max\{|y(t_0)|, |y(a)|\}\int_{t}^{a} |y'_g(s)|dg(s)\\
&\quad \le (g(a)-g(t_0))\int_{t_0}^{a} (y'_g(s))^2 dg(s) + 2\max\{|y(t_0)|, y(a)\}\int_{t_0}^{a} |y'_g(s)|dg(s).
\end{aligned}$$

This completes the proof. □

Corollary 5.1. *For $y \in \mathscr{C}_g^1(J)$ such that $y(t_0) = y(a) = 0$, we have the inequality*

$$\int_{t_0}^{a} |y(t)+y(t+)||y'_g(t)|dg(t) \le (g(a)-g(t_0))\int_{t_0}^{a} (y'_g(t))^2 dg(t).$$

Theorem 5.5. *Let $w \in \mathscr{C}(J)$ be a positive function such that*

$$\int_{t_0}^{a} (w(t))^{1-q} dg(t) < \infty$$

for some $q > 1$, and let $y \in \mathscr{C}_g^1(J)$ be such that $y(t_0) = 0$. Then,

$$\begin{aligned}
&\int_{t_0}^{a} |y(t)+y(t+)||y'_g(t)|dg(t)\\
&\quad \le \left(\int_{t_0}^{a} (w(t))^{1-q} dg(t)\right)^{\frac{2}{q}} \left(\int_{t_0}^{a} w(a)|y'_g(t)|^p dg(t)\right)^{\frac{2}{p}}.
\end{aligned}$$

Proof. Set

$$y_1(t) = \int_{t_0}^{t} |y'_g(s)|dg(s), \quad t \in J.$$

As in the proof of Theorem 5.3, we have

$$y'_{1g}(t) = |y'_g(t)|, \quad y_1(t) \geq |y(t)|, \quad t \in J, \quad \text{and} \quad y_1(t_0) = 0.$$

Hence, applying the Stieltjes–Hölder inequality, we obtain

$$\begin{aligned}
&\int_{t_0}^{a} |y(t) + y(t+)||y'_g(t)|dg(t) \\
&\quad \leq \int_{t_0}^{a} (|y(t)| + |y(t+)|)y'_{1g}(t)dg(t) \\
&\quad \leq \int_{t_0}^{a} (y_1(t) + y_1(t+))y'_{1g}(t)dg(t) \\
&\quad = \int_{t_0}^{t} (y'_{1g})^2(t)dg(t) \\
&\quad = (y_1(t))^2\Big|_{t=t_0}^{t=a} \\
&\quad = (y_1(t))^2 \\
&\quad = \left(\int_{t_0}^{a} |y'_g(t)|dg(t)\right)^2 \\
&\quad = \left(\int_{t_0}^{a} (w(t))^{-\frac{1}{p}}(w(t))^{\frac{1}{p}}|y'_g(t)|dg(t)\right)^2 \\
&\quad \leq \left(\int_{t_0}^{a} (w(t))^{-\frac{q}{p}}dg(t)\right)^{\frac{2}{q}} \left(\int_{t_0}^{t} w(t)|y'_g(t)|^p dg(t)\right)^{\frac{2}{p}} \\
&\quad = \left(\int_{t_0}^{a} (w(t))^{1-q}dg(t)\right)^{\frac{2}{q}} \left(\int_{t_0}^{a} w(t)|y'_g(t)|^p dg(t)\right)^{\frac{2}{p}}.
\end{aligned}$$

This completes the proof. □

Theorem 5.6. *Let $w \in \mathscr{C}(J)$ be a positive function such that*

$$\int_{t_0}^{a} (w(t))^{1-q}dg(t) < \infty$$

for some $q > 1$. *Then, for any* $y \in \mathscr{C}_g^1(J)$, *we have the inequality*

$$\int_{t_0}^{a} |y(t) + y(t+)||y'_g(t)|dg(t) \leq 2v(t)\left(\int_{t_0}^{a} w(t)|y'_g(t)|^p dg(t)\right)^{\frac{2}{p}} + 2\max\{y(t_0), y(a)\}\int_{t_0}^{a} |y'_g(t)|dg(t),$$

where $\frac{1}{p} + \frac{1}{q} = 1$ *and*

$$v(t) = 2\max\left\{\left(\int_{t_0}^{t} (w(s))^{1-q} dg(s)\right)^{\frac{2}{q}}, \left(\int_{t}^{a} (w(s))^{1-q} dg(s)\right)^{\frac{2}{q}}\right\}, \quad t \in J.$$

Proof. Let

$$y_1(t) = \int_{t_0}^{t} |y'_g(s)|dg(s) \quad \text{and} \quad y_2(t) = \int_{t}^{a} |y'_g(s)|dg(s), \quad t \in J.$$

Then,

$$y'_{1g}(t) = |y'_g(t)| \quad \text{and} \quad y'_{2g}(t) = -|y'_g(t)|, \quad t \in J.$$

Hence,

$$\begin{aligned} |y(t)| &= |y(t) - y(t_0) + y(t_0)| \\ &\leq |y(t) - y(t_0)| + |y(t_0)| \\ &= \left|\int_{t_0}^{t} y'_g(s)dg(s)\right| + |y(t_0)| \\ &\leq \int_{t_0}^{t} |y'_g(s)|dg(s) + |y(t_0)| \\ &= y_1(t) + |y(t_0)|, \quad t \in J, \end{aligned}$$

and

$$\begin{aligned} |y(t)| &= |y(t) - y(a) + y(a)| \\ &\leq |y(t) - y(a)| + |y(a)| \\ &= \left|\int_{t}^{a} y'_g(s)dg(s)\right| + |y(a)| \\ &\leq \int_{t}^{a} |y'_g(s)|dg(s) + |y(a)| \\ &= y_2(t) + |y(a)|, \quad t \in J. \end{aligned}$$

Next, applying the Stieltjes–Hölder inequality, we find

$$
\begin{aligned}
&\int_{t_0}^{t} |y(s) + y(s+)||y_g'(s)|dg(s)\\
&\quad\le \int_{t_0}^{t} (|y(s)| + |y(s+)|)y_{1g}'(s)dg(s)\\
&\quad\le \int_{t_0}^{t} (y_1(s) + y_1(s+) + 2|y(t_0)|)y_{1g}'(s)dg(s)\\
&\quad= \int_{t_0}^{t} (y_1(s) + y_1(s+))y_{1g}'(s)dg(s) + 2|y(t_0)| \int_{t_0}^{t} y_{1g}'(s)dg(s)\\
&\quad= \int_{t_0}^{t} (y_{1g}')^2(s)dg(s) + 2|y(t_0)|y_1(s)\Big|_{s=t_0}^{s=t}\\
&\quad= (y_1(s))^2\Big|_{s=t_0}^{s=t} + 2|y(t_0)|y_1(t)\\
&\quad= (y_1(t))^2 + 2|y(t_0)|y_1(t)\\
&\quad= \left(\int_{t_0}^{t} |y_g'(s)|dg(s)\right)^2 + 2|y(t_0)| \int_{t_0}^{t} |y_g'(s)|dg(s)\\
&\quad= \left(\int_{t_0}^{t} (w(s))^{-\frac{1}{p}}(w(s))^{\frac{1}{p}}|y_g'(s)|dg(s)\right)^2 + 2|y(t_0)| \int_{t_0}^{t} |y_g'(s)|dg(s)\\
&\quad\le \left(\int_{t_0}^{t} (w(s))^{-\frac{q}{p}}dg(s)\right)^{\frac{2}{q}} \left(\int_{t_0}^{t} w(s)|y_g'(s)|^p dg(s)\right)^{\frac{2}{p}}\\
&\qquad + 2|y(t_0)| \int_{t_0}^{t} |y_g'(s)|dg(s)\\
&\quad\le v(t)\left(\int_{t_0}^{a} w(s)|y_g'(s)|^p dg(s)\right)^{\frac{2}{p}} + 2|y(t_0)| \int_{t_0}^{t} |y_g'(s)|dg(s), \quad t \in J,
\end{aligned}
$$

i.e.,

$$
\begin{aligned}
&\int_{t_0}^{t} |y(s) + y(s+)||y_g'(s)|dg(s)\\
&\quad\le v(t)\left(\int_{t_0}^{a} w(s)|y_g'(s)|^p dg(s)\right)^{\frac{2}{p}} + 2|y(t_0)| \int_{t_0}^{t} |y_g'(s)|dg(s), \quad t \in J.
\end{aligned}
\tag{5.4}
$$

Moreover,

$$\begin{aligned}
&\int_t^a |y(s)+y(s+)||y_g'(s)|dg(s)\\
&\quad\le -\int_t^a (|y(s)|+|y(s+)|)y_{2g}'(s)dg(s)\\
&\quad\le -\int_t^a (y_2(s)+y_2(s+)+2|y(a)|)y_{2g}'(s)dg(s)\\
&\quad= -\int_t^a (y_2(s)+y_2(s+))y_{2g}'(s)dg(s)-2|y(a)|\int_t^a y_{2g}'(s)dg(s)\\
&\quad= -\int_t^a (y_{2g}')^2(s)dg(s)-2|y(a)|y_2(s)\Big|_{s=t}^{s=a}\\
&\quad= -(y_2(s))^2\Big|_{s=t}^{s=a}+2|y(a)|y_2(t)\\
&\quad= (y_2(t))^2+2|y(a)|y_2(t)\\
&\quad= \left(\int_t^a |y_g'(s)|dg(s)\right)^2+2|y(a)|\int_t^a |y_g'(s)|dg(s)\\
&\quad= \left(\int_t^a (w(s))^{-\frac{1}{p}}(w(s))^{\frac{1}{p}}|y_g'(s)|dg(s)\right)^2+2|y(a)|\int_t^a |y_g'(s)|dg(s)\\
&\quad\le \left(\int_t^a (w(s))^{-\frac{q}{p}}dg(s)\right)^{\frac{2}{q}}\left(\int_t^a w(s)|y_g'(s)|^p dg(s)\right)^{\frac{2}{p}}\\
&\qquad+2|y(a)|\int_t^a |y_g'(s)|dg(s)\\
&\quad\le v(t)\left(\int_{t_0}^a w(s)|y_g'(s)|^p dg(s)\right)^{\frac{2}{p}}+2|y(a)|\int_t^a |y_g'(s)|dg(s),\quad t\in J,
\end{aligned}$$

i.e.,

$$\begin{aligned}
&\int_t^a |y(s)+y(s+)||y_g'(s)|dg(s)\\
&\quad\le v(t)\left(\int_t^a w(s)|y_g'(s)|^p dg(s)\right)^{\frac{2}{p}}+2|y(a)|\int_t^a |y_g'(s)|dg(s),\quad t\in J.
\end{aligned}$$

Adding the last inequality and the inequality (5.4), we arrive at

$$\int_{t_0}^{t} |y(s)+y(s+)||y_g'(s)|dg(s) + \int_{t}^{a} |y(s)+y(s+)||y_g'(s)|dg(s)$$
$$\le v(t)\left(\int_{t_0}^{a} w(s)|y_g'(s)|^p dg(s)\right)^{\frac{2}{p}} + 2|y(t_0)|\int_{t_0}^{t} |y_g'(s)|dg(s)$$
$$+ v(t)\left(\int_{t_0}^{a} w(s)|y_g'(s)|^p dg(s)\right)^{\frac{2}{p}} + 2|y(a)|\int_{t}^{a} |y_g'(s)|dg(s), \quad t\in J,$$

or

$$\int_{t_0}^{a} |y(s)+y(s+)||y_g'(s)|dg(s)$$
$$\le 2v(t)\left(\int_{t_0}^{a} w(s)|y_g'(s)|^p dg(s)\right)^{\frac{2}{p}}$$
$$+ 2\max\{|y(t_0)|, |y(a)|\}\int_{t}^{a} |y_g'(s)|dg(s), \quad t\in J.$$

This completes the proof. □

Corollary 5.2. *Let $w\in\mathscr{C}(J)$ be a positive function such that*

$$\int_{t_0}^{a} (w(t))^{1-q} dg(t) < \infty$$

for some $q>1$. Then, for any $y\in\mathscr{C}_g^1(J)$, we have the inequality

$$\int_{t_0}^{a} |y(t)+y(t+)||y_g'(t)|dg(t)$$
$$\le 2\left(\int_{t_0}^{a} (w(s))^{1-q} dg(s)\right)^{\frac{2}{q}}\left(\int_{t_0}^{a} w(t)|y_g'(t)|^p dg(t)\right)^{\frac{2}{p}}$$
$$+ 2\max\{y(t_0), y(a)\}\int_{t_0}^{a} |y_g'(t)|dg(t).$$

Proof. We have

$$v(t) \le \left(\int_{t_0}^{a} (w(s))^{1-q} dg(s)\right)^{\frac{2}{q}}, \quad t\in J.$$

Hence, the result follows immediately from Theorem 5.6. □

Corollary 5.3. *Let $w \in \mathscr{C}(J)$ be a positive function such that*

$$\int_{t_0}^{a} (w(t))^{1-q} dg(t) < \infty$$

for some $q > 1$. Then, for any $y \in \mathscr{C}_g^1(J)$ for which $y(t_0) = y(a) = 0$, we have the inequality

$$\int_{t_0}^{a} |y(t) + y(t+)||y_g'(t)|dg(t) \leq 2v(t) \left(\int_{t_0}^{a} w(t)|y_g'(t)|^p dg(t) \right)^{\frac{2}{p}},$$

where $\frac{1}{p} + \frac{1}{q} = 1$ and

$$v(t) = 2 \max \left\{ \left(\int_{t_0}^{t} (w(s))^{1-q} dg(s) \right)^{\frac{2}{q}}, \left(\int_{t}^{a} (w(s))^{1-q} dg(s) \right)^{\frac{2}{q}} \right\}, \quad t \in J.$$

Corollary 5.4. *Let $w \in \mathscr{C}(J)$ be a positive function such that*

$$\int_{t_0}^{a} (w(t))^{1-q} dg(t) < \infty$$

for some $q > 1$. Then, for any $y \in \mathscr{C}_g^1(J)$ for which $y(t_0) = y(a) = 0$, we have the inequality

$$\int_{t_0}^{a} |y(t) + y(t+)||y_g'(t)|dg(t)$$
$$\leq 2 \left(\int_{t_0}^{a} (w(t))^{1-q} dg(t) \right)^{\frac{2}{q}} \left(\int_{t_0}^{a} w(t)|y_g'(t)|^p dg(t) \right)^{\frac{2}{p}},$$

where $\frac{1}{p} + \frac{1}{q} = 1$.

Theorem 5.7. *Let $y, f \in \mathscr{C}_g^1(J)$ and $y(t_0) = 0$. Set*

$$K_f(a, t_0) = 2 \left(\int_{t_0}^{a} (f(t))^2 (g(t) - g(t_0)) dg(t) \right)^{\frac{1}{2}}.$$

Then,

$$\int_{t_0}^{a} f(t)|y(t) + y(t+)||y_g'(t)|dg(t) \leq K_f(a, t_0) \int_{t_0}^{a} (y_g'(t))^2 dg(t).$$

Proof. Note that

$$y(t) = \int_{t_0}^{t} y'_g(s)dg(s)$$
$$\leq \left((g(t) - g(t_0)) \int_{t_0}^{t} (y'_g(s))^2 dg(s)\right)^{\frac{1}{2}}, \quad t \in J.$$

Define

$$h(t) = \int_{t_0}^{t} (y'_g(s))^2 dg(s), \quad t \in J.$$

Then, $y(t) \leq ((g(t) - g(t_0))h(s))^{\frac{1}{2}}$ and $h'_g(t) = (y'_g(t))^2$, $t \in J$, whereupon

$$(h'_g(t))^{\frac{1}{2}} = y'_g(t), \quad t \in J.$$

Now, using the inequality $\alpha^{\frac{1}{2}} + \beta^{\frac{1}{2}} \leq (\alpha + \beta)^{\frac{1}{2}}$, $\alpha, \beta \geq 0$ and applying the Stieltjes–Hölder inequality, we arrive at

$$\int_{t_0}^{a} f(t)|y(t) + y(t+)||y'_g(t)|dg(t)$$
$$\leq \int_{t_0}^{a} |f(t)|(|y(t)| + |y(t+)|)|y'_g(t)|dg(t)$$
$$\leq \int_{t_0}^{a} |f(t)| \left(((g(t) - g(t_0))h(t))^{\frac{1}{2}} + ((g(t) - g(t_0))(h(t+)))^{\frac{1}{2}}\right) |y'_g(t)|dg(t)$$
$$\leq \int_{t_0}^{a} |f(t)|(g(t) - g(t_0))^{\frac{1}{2}} \left((h(t))^{\frac{1}{2}} + (h(t+))^{\frac{1}{2}}\right) |y'_g(t)|dg(t)$$
$$= \int_{t_0}^{a} |f(t)|(g(t) - g(t_0))^{\frac{1}{2}} \left((h(t))^{\frac{1}{2}} + (h(t+))^{\frac{1}{2}}\right) (h'_g(t))^{\frac{1}{2}} dg(t)$$
$$\leq 2 \int_{t_0}^{a} |f(t)|(g(t) - g(t_0))^{\frac{1}{2}} \left((h(t) + h(t+))h'_g(t)\right)^{\frac{1}{2}} dg(t)$$
$$= 2 \int_{t_0}^{a} |f(t)|(g(t) - g(t_0))^{\frac{1}{2}} \left((h^2(t))'_t\right)^{\frac{1}{2}} dg(t)$$
$$\leq 2 \left(\int_{t_0}^{a} (f(t))^2 (g(t) - g(t_0))dg(t)\right)^{\frac{1}{2}} \left(\int_{t_0}^{a} (h^2(t))'_g dg(t)\right)^{\frac{1}{2}}$$
$$= K_f(a, t_0) \left((h(t))^2 \Big|_{t=t_0}^{t=a} \right)^{\frac{1}{2}}$$

$$= K_f(a,t_0)h(a)$$
$$= K_f(a,t_0)\int_{t_0}^{a}(y_g'(t))^{\frac{1}{2}}dg(t).$$

This completes the proof. □

Theorem 5.8. *Let $y, f \in \mathscr{C}_g^1(J)$ and $y(a) = 0$. Set*

$$K_f(a,t_0) = 2\left(\int_{t_0}^{a}(f(t))^2(g(a)-g(t))dg(t)\right)^{\frac{1}{2}}.$$

Then,

$$\int_{t_0}^{a} f(t)|y(t)+y(t+)||y_g'(t)|dg(t) \le L_f(a,t_0)\int_{t_0}^{a}(y_g'(t))^2dg(t).$$

Proof. Note that

$$y(t) = \int_{t}^{a} y_g'(s)dg(s)$$
$$\le \left((g(a)-g(t))\int_{t}^{a}(y_g'(s))^2dg(s)\right)^{\frac{1}{2}}, \quad t \in J.$$

Define

$$h(t) = \int_{t}^{a}(y_g'(s))^2dg(s), \quad t \in J.$$

Then,

$$y(t) \le ((g(a)-g(t))h(t))^{\frac{1}{2}}, \quad t \in J,$$

and

$$h_g'(t) = -(y_g'(t))^2, \quad t \in J.$$

Hence, we find

$$\int_{t_0}^{a} f(t)|y(t)+y(t+)||y_g'(t)|dg(t)$$
$$\le \int_{t_0}^{a}|f(t)|(|y(t)|+|y(t+)|)|y_g'(t)|dg(t)$$
$$\le \int_{t_0}^{a}|f(t)|\left(((g(a)-g(t))h(t))^{\frac{1}{2}} + ((g(a)-g(t))h(t+))^{\frac{1}{2}}\right)|y_g'(t)|dg(t)$$

$$\leq \int_{t_0}^{a} |f(t)|(g(a)-g(t))^{\frac{1}{2}}\left((h(t))^{\frac{1}{2}}+(h(t+))^{\frac{1}{2}}\right)|y'_y(t)|dg(t)$$

$$= \int_{t_0}^{a} |f(t)|(g(a)-g(t))^{\frac{1}{2}}\left((h(t))^{\frac{1}{2}}+(h(t+))^{\frac{1}{2}}\right)(h'_g(t))^{\frac{1}{2}}dg(t)$$

$$\leq 2\int_{t_0}^{a} |f(t)|(g(a)-g(t))^{\frac{1}{2}}\left((h(t)+h(t+))h'_g(t)\right)^{\frac{1}{2}}dg(t)$$

$$= 2\int_{t_0}^{a} |f(t)|(g(a)-g(t))^{\frac{1}{2}}\left((h^2(t))'_g\right)^{\frac{1}{2}}dg(t)$$

$$\leq 2\left(\int_{t_0}^{a}(f(t))^2(g(a)-g(t))dg(t)\right)^{\frac{1}{2}}\left(\int_{t_0}^{a}(h^2(t))'_g dg(t)\right)^{\frac{1}{2}}$$

$$= K_f(a,t_0)\left((h(t))^2\Big|_{t=t_0}^{t=a}\right)^{\frac{1}{2}}$$

$$= K_f(a,t_0)h(a)$$

$$= K_f(a,t_0)\int_{t_0}^{a}(y'_g(t))^{\frac{1}{2}}dg(t).$$

This completes the proof. □

Theorem 5.9. *Let $f, y \in \mathscr{C}_g^1(J)$ and $y(t_0) = y(a) = 0$. Then,*

$$\int_{t_0}^{a} f(t)|y(t)+y(t+)||y'_g(t)|dg(t) \leq \max\{K_f(c,t_0)K_f(a,c)\}\int_{t_0}^{a}(y'_g(t))^2dg(t) \tag{5.5}$$

for any $c \in J$.

Proof. Let $c \in J$ be arbitrarily chosen. By Theorem 5.7, we find

$$\int_{t_0}^{c} f(t)|y(t)+y(t+)||y'_g(t)|dg(t) \leq K_f(c,t_0)\int_{t_0}^{c}(y'_g(t))^2dg(t) \leq K_f(c,t_0)\int_{t_0}^{a}(y'_g(t))^2dg(t). \tag{5.6}$$

By Theorem 5.8, we have

$$\int_c^a f(t)|y(t)+y(t+)||y'_g(t)|dg(t) \le K_f(a,c)\int_c^a (y'_g(t))^2 dg(t)$$
$$\le K_f(a,c)\int_{t_0}^a (y'_g(t))^2 dg(t).$$

Adding the last inequality and the inequality (5.6), we find

$$\begin{aligned}\int_{t_0}^a & f(t)|y(t)+y(t+)||y'_g(t)|dg(t)\\ &= \int_{t_0}^c f(t)|y(t)+y(t+)||y'_g(t)|dg(t)\\ &\quad + \int_c^a f(t)|y(t)+y(t+)||y'_g(t)|dg(t)\\ &\le K_f(c,t_0)\int_{t_0}^a (y'_g(t))^2 dg(t) + L_f(c,t_0)\int_{t_0}^a (y'_g(t))^2 dg(t)\\ &\le \max\{K_f(c,t_0), K_f(a,c)\}\int_{t_0}^a (y'_g(t))^2 dg(t).\end{aligned}$$

This completes the proof. □

Theorem 5.10. *Let* $f, y \in \mathscr{C}_g^1(J)$ *and* $y(t_0) = y(a) = 0$. *Then,*

$$\int_{t_0}^a f(t)|y(t)+y(t+)||y'_g(t)|dg(t)$$
$$\le \min_{c\in J}\left(\max\{K_f(c,t_0)K_f(a,c)\}\right)\int_{t_0}^a (y'_g(t))^2 dg(t).$$

Proof. Since (5.5) holds for any $c \in J$, we get the desired inequality. □

Theorem 5.11. *Let* $r, q \in \mathscr{C}(J)$ *be positive functions such that*

$$\int_{t_0}^a \frac{1}{r(t)} dg(t) < \infty$$

and q *be a nonincreasing function. Also, let* $y \in \mathscr{C}_g^1(J)$ *be such that* $y(t_0) = 0$. *Then,*

$$\int_{t_0}^a q(t)|(y(t)+y(t+))y'_g(t)|dg(t)$$
$$\le \left(\int_{t_0}^a \frac{1}{r(t)} dg(t)\right)\left(\int_{t_0}^a r(t)q(t)(y'_g(t))^2 dg(t)\right).$$

Proof. Set

$$z(t) = \int_{t_0}^{t} \sqrt{q(s)}|y'_g(s)|dg(s), \quad t \in J.$$

Then,

$$z'_g(t) = \sqrt{q(t)}|y'_g(t)|, \quad t \in J.$$

Note that, since q is a nonincreasing function, for any $s \in [t_0, t)$, we have $q(s) \geq q(t)$, $t \in J$. Then,

$$\begin{aligned}
|y(t)| &= \left| \int_{t_0}^{t} y'_g(s)dg(s) \right| \\
&\leq \int_{t_0}^{t} |y'_g(s)|dg(s) \\
&\leq \int_{t_0}^{t} \sqrt{\frac{q(s)}{q(t)}}|y'_g(s)|dg(s) \\
&= \frac{1}{\sqrt{q(t)}} \int_{t_0}^{t} \sqrt{q(s)}|y'_g(s)|dg(s) \\
&= \frac{1}{\sqrt{q(t)}} z(t), \quad t \in J.
\end{aligned}$$

Hence,

$$\begin{aligned}
&\int_{t_0}^{a} q(t)|(y(t) + y(t+))y'_g(t)|dg(t) \\
&\quad \leq \int_{t_0}^{a} q(t)(|y(t)| + |y(t+)|)|y'_g(t)|dg(t) \\
&\quad \leq \int_{t_0}^{a} q(t) \left(\frac{1}{q(t)} z(t) + \frac{1}{q(t)} z(t+) \right) \frac{z'_g(t)}{\sqrt{q(t)}} dg(t) \\
&\quad \leq \int_{t_0}^{a} (z(t) + z(t+)) z'_g(t)dg(t) \\
&\quad = \int_{t_0}^{a} (z^2)'_g(t)dg(t) \\
&\quad = (z(t))^2 \Big|_{t=t_0}^{t=a}
\end{aligned}$$

$$\begin{aligned}
&= (z(a))^2 \\
&= \left(\int_{t_0}^{a} \sqrt{q(t)}|y_g'(t)|dg(t)\right)^2 \\
&= \left(\int_{t_0}^{a} \frac{1}{\sqrt{r(t)}}\sqrt{r(t)q(t)}|y_g'(t)|dg(t)\right)^2 \\
&\leq \left(\int_{t_0}^{a} \frac{1}{r(t)}dg(t)\right)\left(\int_{t_0}^{a} r(t)q(t)(y_g'(t))^2 dg(t)\right).
\end{aligned}$$

This completes the proof. □

Theorem 5.12. *Let $s, r \in \mathscr{C}(J)$ be positive functions such that*

$$\int_{t_0}^{a} \frac{1}{r(t)}dg(t) < \infty.$$

Also, let $y \in \mathscr{C}_g^1(J)$ with $y(t_0) = 0$ and y_g' does not change its sign on J. Then,

$$\int_{t_0}^{a} s(t)|y(t) + y(t+)||y_g'(t)|dg(t) \leq K_1(a, t_0)\int_{t_0}^{a} r(t)(y_g'(t))^2 dg(t),$$

where

$$K_1(a, t_0) = \sqrt{2}\left(\int_{t_0}^{a} \frac{(s(t))^2}{r(t)}\left(\int_{t_0}^{t} \frac{1}{r(\tau)}dg(\tau)\right)dg(t)\right)^{\frac{1}{2}} + \sup_{t\in J}\left(\Delta g(t)\frac{s(t)}{r(t)}\right).$$

Proof. Note that

$$\begin{aligned}
|y(t)| &= \int_{t_0}^{t} |y_g'(\tau)|dg(\tau) \\
&= \int_{t_0}^{t} \frac{1}{\sqrt{r(\tau)}}\sqrt{r(\tau)}|y_g'(\tau)|dg(\tau) \\
&\leq \left(\int_{t_0}^{t} \frac{1}{r(\tau)}dg(\tau)\right)^{\frac{1}{2}}\left(\int_{t_0}^{t} r(\tau)(y_g'(\tau))^2 dg(\tau)\right)^{\frac{1}{2}}, \quad t \in J,
\end{aligned}$$

where we have used the fact that y_g' does not change its sign on J. Next,

$$\begin{aligned}
y(t) + y(t+) &= y(t) + y(t) + (\Delta g(t))y_g'(t) \\
&= 2y(t) + (\Delta g(t))y_g'(t), \quad t \in J.
\end{aligned}$$

We define

$$z(t) = \int_{t_0}^{t} r(\tau)(y'_g(\tau))^2 dg(\tau), \quad t \in J.$$

Then,

$$z(t_0) = 0 \quad \text{and} \quad z'_g(t) = r(t)(y'_g(t))^2 > 0, \quad t \in J.$$

Hence,

$$(y'_g(t))^2 = \frac{z'_g(t)}{r(t)}, \quad t \in J,$$

and

$$|y'_g(t)| = \left(\frac{z'_g(t)}{r(t)}\right)^{\frac{1}{2}}, \quad t \in J.$$

Now, observe that

$$\begin{aligned}
&s(t)|y(t) + y(t+)||y'_g(t)| \\
&\quad = s(t)|2y(t) + (\Delta g(t))y'_g(t)||y'_g(t)| \\
&\quad \leq s(t)(2|y(t)| + (\Delta g(t))|y'_g(t)|)|y'_g(t)| \\
&\quad = 2s(t)|y(t)||y'_g(t)| + s(t)(\Delta g(t))(y'_g(t))^2 \\
&\quad \leq 2s(t)\left(\int_{t_0}^{t} \frac{1}{r(\tau)} dg(\tau)\right)^{\frac{1}{2}} \left(\int_{t_0}^{t} r(\tau)(y'_g(\tau))^2 dg(\tau)\right)^{\frac{1}{2}} \left(\frac{z'_g(t)}{r(t)}\right)^{\frac{1}{2}} \\
&\qquad + (\Delta g(t))s(t)\frac{z'_g(t)}{r(t)} \\
&\quad = 2s(t)\frac{1}{(r(t))^{\frac{1}{2}}}\left(\int_{t_0}^{t} \frac{1}{r(\tau)} dg(\tau)\right)^{\frac{1}{2}} (z(t))^{\frac{1}{2}}(z'_g(t))^{\frac{1}{2}} \\
&\qquad + (\Delta g(t))s(t)\frac{z'_g(t)}{r(t)}, \quad t \in J,
\end{aligned}$$

i.e.,

$$\begin{aligned}
s(t)|y(t) + y((t+))||y'_g(t)| \leq 2s(t)\frac{1}{(r(t))^{\frac{1}{2}}}\left(\int_{t_0}^{x} \frac{1}{r(\tau)} dg(\tau)\right)^{\frac{1}{2}} (z(t))^{\frac{1}{2}}(z'_g(t))^{\frac{1}{2}} \\
+ (\Delta g(t))s(t)\frac{z'_g(t)}{r(t)}, \quad t \in J.
\end{aligned}$$

Integrating the last inequality, in the sense of Stieltjes, from t_0 to a, we get

$$\int_{t_0}^{a} s(t)|y(t)+y(t+)||y_g'(t)|dg(t)$$
$$\leq 2\int_{t_0}^{a} s(t)\frac{1}{(r(t))^{\frac{1}{2}}}\left(\int_{t_0}^{t}\frac{1}{r(\tau)}dg(\tau)\right)^{\frac{1}{2}}(z(t))^{\frac{1}{2}}(z_g'(t))^{\frac{1}{2}}dg(t)$$
$$+\int_{t_0}^{a}(\Delta g(t))s(t)\frac{z_g'(t)}{r(t)}dg(t)$$
$$\leq 2\int_{t_0}^{a} s(t)\frac{1}{(r(t))^{\frac{1}{2}}}\left(\int_{t_0}^{t}\frac{1}{r(\tau)}dg(\tau)\right)^{\frac{1}{2}}(z(t))^{\frac{1}{2}}(z_g'(t))^{\frac{1}{2}}dg(t)$$
$$+\sup_{t\in J}\left((\Delta g(t))\frac{s(t)}{r(t)}\right)\int_{t_0}^{a} z_g'(t)dg(t)$$
$$= 2\int_{t_0}^{a} s(t)\frac{1}{(r(t))^{\frac{1}{2}}}\left(\int_{t_0}^{t}\frac{1}{r(\tau)}dg(\tau)\right)^{\frac{1}{2}}(z(t))^{\frac{1}{2}}(z_g'(t))^{\frac{1}{2}}dg(t)$$
$$+\sup_{t\in J}\left((\Delta g(t))\frac{s(t)}{r(t)}\right)z(t)\Big|_{t=t_0}^{t=a}$$
$$= 2\int_{t_0}^{a} s(t)\frac{1}{(r(t))^{\frac{1}{2}}}\left(\int_{t_0}^{t}\frac{1}{r(\tau)}dg(\tau)\right)^{\frac{1}{2}}(z(t))^{\frac{1}{2}}(z_g'(t))^{\frac{1}{2}}dg(t)$$
$$+\sup_{t\in J}\left((\Delta g(t))\frac{s(t)}{r(t)}\right)z(a)$$
$$\leq 2\left(\int_{t_0}^{a}\frac{(s(t))^2}{r(t)}\left(\int_{t_0}^{t}\frac{1}{r(\tau)}dg(\tau)\right)dg(t)\right)^{\frac{1}{2}}\left(\int_{t_0}^{a} z(t)z_g'(t)dg(t)\right)^{\frac{1}{2}}$$
$$+\sup_{t\in J}\left((\Delta g(t))\frac{s(t)}{r(t)}\right)z(a)$$
$$\leq \sqrt{2}\left(\int_{t_0}^{a}\frac{(s(t))^2}{r(t)}\left(\int_{t_0}^{t}\frac{1}{r(\tau)}dg(\tau)\right)dg(t)\right)^{\frac{1}{2}}$$
$$\times\left(\int_{t_0}^{a}(z(t)+z(t+))z_g'(t)dg(t)\right)^{\frac{1}{2}}+\sup_{t\in J}\left((\Delta g(t))\frac{s(t)}{r(t)}\right)z(a)$$
$$= \sqrt{2}\left(\int_{t_0}^{a}\frac{(s(t))^2}{r(t)}\left(\int_{t_0}^{t}\frac{1}{r(\tau)}dg(\tau)\right)dg(t)\right)^{\frac{1}{2}}\left(\int_{t_0}^{a}(z_g')^2(t)dg(t)\right)^{\frac{1}{2}}$$

$$+\sup_{t\in J}\left((\Delta g(t))\frac{s(t)}{r(t)}\right)z(a)$$

$$=\sqrt{2}\left(\int_{t_0}^{a}\frac{(s(t))^2}{r(t)}\left(\int_{t_0}^{t}\frac{1}{r(\tau)}dg(\tau)\right)dg(t)\right)^{\frac{1}{2}}\left((z(t))^2\Big|_{t=t_0}^{t=a}\right)^{\frac{1}{2}}$$

$$+\sup_{t\in J}\left((\Delta g(t))\frac{s(t)}{r(t)}\right)z(a)$$

$$=\sqrt{2}\left(\int_{t_0}^{a}\frac{(s(t))^2}{r(t)}\left(\int_{t_0}^{t}\frac{1}{r(\tau)}dg(\tau)\right)dg(t)\right)^{\frac{1}{2}}z(a)$$

$$+\sup_{t\in J}\left((\Delta g(t))\frac{s(t)}{r(t)}\right)z(a)$$

$$=\left(\sqrt{2}\left(\int_{t_0}^{a}\frac{(s(t))^2}{r(t)}\left(\int_{t_0}^{t}\frac{1}{r(\tau)}dg(\tau)\right)dg(t)\right)^{\frac{1}{2}}\right.$$

$$\left.+\sup_{t\in J}\left((\Delta g(t))\frac{s(t)}{r(t)}\right)\right)z(a)$$

$$=K_1(a,t_0)z(a)$$

$$=K_1(a,t_0)\int_{t_0}^{a}r(t)(y_g'(t))^2dg(t).$$

In the above set of inequalities, we have used the fact that z is nondecreasing and $z(t)\le z(t+)$, $t\in J$. This completes the proof. □

Theorem 5.13. *Let $s,r\in\mathscr{C}(J)$ be positive functions such that*

$$\int_{t_0}^{a}\frac{1}{r(t)}dg(t)<\infty.$$

Also, let $y\in\mathscr{C}_g^1(J)$ with $y(a)=0$ and y_g' does not change its sign on J. Then,

$$\int_{t_0}^{a}s(t)|y(t)+y(t+)||y_g'(t)|dg(t)\le K_2(a,t_0)\int_{t_0}^{a}r(t)(y_g'(t))^2dg(t),$$

where

$$K_2(a,t_0)=\sqrt{2}\left(\int_{t_0}^{a}\frac{(s(t))^2}{r(t)}\left(\int_{t}^{a}\frac{1}{r(\tau)}dg(\tau)\right)dg(t)\right)^{\frac{1}{2}}+\sup_{t\in J}\left((\Delta g(t))\frac{s(t)}{r(t)}\right).$$

Proof. We have

$$
\begin{aligned}
|y(t)| &= \int_t^a |y_g'(\tau)|dg(\tau) \\
&= \int_t^a \frac{1}{\sqrt{r(\tau)}}\sqrt{r(\tau)}|y_g'(\tau)|dg(\tau) \\
&\le \left(\int_t^a \frac{1}{r(\tau)}dg(\tau)\right)^{\frac{1}{2}} \left(\int_t^a r(\tau)(y_g'(\tau))^2 dg(\tau)\right)^{\frac{1}{2}}, \quad t \in J,
\end{aligned}
$$

where we have used the fact that y_g' does not change its sign on J. Moreover,

$$
\begin{aligned}
y(t) + y(t+) &= y(t) + y(t) + (\Delta g(t))y_g'(t) \\
&= 2y(t) + (\Delta g(t))y_g'(t), \quad t \in J.
\end{aligned}
$$

Set

$$
z(t) = \int_t^a r(\tau)(y_g'(\tau))^2 dg(\tau), \quad t \in J.
$$

Then, $z(a) = 0$ and $z_g'(t) = -r(t)(y_g'(t))^2 < 0$, $t \in J$, i.e., z is a nonincreasing function on J and $z(t) \ge z(t+)$, $t \in J$. Then,

$$
(y_g'(t))^2 = -\frac{z_g'(t)}{r(t)} \quad \text{and} \quad |y_g'(t)| = \left(-\frac{z_g'(t)}{r(t)}\right)^{\frac{1}{2}}, \quad t \in J.
$$

Note that

$$
\begin{aligned}
& s(t)|y(t) + y(t+)||y_g'(t)| \\
&\quad = s(t)|2y(t) + (\Delta g(t))y_g'(t)||y_g'(t)| \\
&\quad \le s(t)(2|y(t)| + (\Delta g(t))|y_g'(t)|)|y_g'(t)| \\
&\quad = 2s(t)|y(t)||y_g'(t)| + s(t)(\Delta g(t))(y_g'(t))^2 \\
&\quad \le 2s(t)\left(\int_t^a \frac{1}{r(\tau)}dg(\tau)\right)^{\frac{1}{2}} \left(\int_t^a r(t)(y_g'(t))^2 dg(t)\right)^{\frac{1}{2}} \\
&\qquad \times \left(-\frac{z_g'(t)}{r(t)}\right)^{\frac{1}{2}} - (\Delta g(t))s(t)\frac{z_g'(t)}{r(t)} \\
&\quad = 2s(t)\frac{1}{(r(t))^{\frac{1}{2}}}\left(\int_t^a \frac{1}{r(\tau)}dg(\tau)\right)^{\frac{1}{2}} (z(t))^{\frac{1}{2}}(-z_g'(t))^{\frac{1}{2}} \\
&\qquad - (\Delta g(t))s(t)\frac{z_g'(t)}{r(t)}, \quad t \in J,
\end{aligned}
$$

i.e.,

$$s(t)|y(t)+y(t+)||y_g'(t)| \le 2s(t)\frac{1}{(r(t))^{\frac{1}{2}}}\left(\int_t^a \frac{1}{r(\tau)}dg(\tau)\right)^{\frac{1}{2}}(z(t))^{\frac{1}{2}}(-z_g'(t))^{\frac{1}{2}} - (\Delta g(t))s(t)\frac{z_g'(t)}{r(t)}, \quad t\in J.$$

Integrating the last inequality, in the sense of Stieltjes, from t_0 to a, we get

$$\begin{aligned}
&\int_{t_0}^a s(t)|y(t)+y(t+)||y_g'(t)|dg(t)\\
&\le 2\int_{t_0}^a s(t)\frac{1}{(r(t))^{\frac{1}{2}}}\left(\int_t^a \frac{1}{r(\tau)}dg(\tau)\right)^{\frac{1}{2}}(z(t))^{\frac{1}{2}}(-z_g'(t))^{\frac{1}{2}}dg(t)\\
&\quad -\int_{t_0}^a (\Delta g(t))s(t)\frac{z_g'(t)}{r(t)}dg(t)\\
&\le 2\int_{t_0}^a s(t)\frac{1}{(r(t))^{\frac{1}{2}}}\left(\int_t^a \frac{1}{r(\tau)}dg(\tau)\right)^{\frac{1}{2}}(z(t))^{\frac{1}{2}}(-z_g'(t))^{\frac{1}{2}}dg(t)\\
&\quad -\sup_{t\in J}\left((\Delta g(t))\frac{s(t)}{r(t)}\right)\int_{t_0}^a z_g'(t)dg(t)\\
&= 2\int_{t_0}^a s(t)\frac{1}{(r(t))^{\frac{1}{2}}}\left(\int_t^a \frac{1}{r(\tau)}dg(\tau)\right)^{\frac{1}{2}}(z(t))^{\frac{1}{2}}(-z_g'(t))^{\frac{1}{2}}dg(t)\\
&\quad -\sup_{t\in J}\left((\Delta g(t))\frac{s(t)}{r(t)}\right)z(t)\Big|_{t=t_0}^{t=a}\\
&= 2\int_{t_0}^a s(t)\frac{1}{(r(t))^{\frac{1}{2}}}\left(\int_t^a \frac{1}{r(\tau)}dg(\tau)\right)^{\frac{1}{2}}(z(t))^{\frac{1}{2}}(-z_g'(t))^{\frac{1}{2}}dg(t)\\
&\quad +\sup_{t\in J}\left((\Delta g(t))\frac{s(t)}{r(t)}\right)z(t_0)\\
&\le 2\left(\int_{t_0}^a \frac{(s(t))^2}{r(t)}\left(\int_t^a \frac{1}{r(\tau)}dg(\tau)\right)dg(t)\right)^{\frac{1}{2}}\left(-\int_{t_0}^a z(t)z_g'(t)dg(t)\right)^{\frac{1}{2}}\\
&\quad +\sup_{t\in J}\left((\Delta g(t))\frac{s(t)}{r(t)}\right)z(t_0)\\
&\le \sqrt{2}\left(\int_{t_0}^a \frac{(s(t))^2}{r(t)}\left(\int_t^a \frac{1}{r(\tau)}dg(\tau)\right)dg(t)\right)^{\frac{1}{2}}
\end{aligned}$$

$$\times\left(-\int_{t_0}^{a}(z(t)+z(t+))z_g'(t)dg(t)\right)^{\frac{1}{2}}+\sup_{t\in J}\left((\Delta g(t))\frac{s(t)}{r(t)}\right)z(t_0)$$

$$=\sqrt{2}\left(\int_{t_0}^{a}\frac{(s(t))^2}{r(t)}\left(\int_{t_0}^{x}\frac{1}{r(\tau)}dg(\tau)\right)dg(t)\right)^{\frac{1}{2}}\left(-\int_{t_0}^{a}(z^2)_g'(t)dg(t)\right)^{\frac{1}{2}}$$

$$+\sup_{t\in J}\left((\Delta g(t))\frac{s(t)}{r(t)}\right)z(t_0)$$

$$=\sqrt{2}\left(\int_{t_0}^{a}\frac{(s(t))^2}{r(t)}\left(\int_{t}^{a}\frac{1}{r(\tau)}dg(\tau)\right)dg(t)\right)^{\frac{1}{2}}\left(-(z(t))^2\Big|_{t=t_0}^{t=a}\right)^{\frac{1}{2}}$$

$$+\sup_{t\in J}\left((\Delta g(t))\frac{s(t)}{r(t)}\right)z(t_0)$$

$$=\sqrt{2}\left(\int_{t_0}^{a}\frac{(s(t))^2}{r(t)}\left(-\int_{t}^{a}\frac{1}{r(\tau)}dg(\tau)\right)dg(t)\right)^{\frac{1}{2}}z(t_0)$$

$$+\sup_{t\in J}\left((\Delta g(t))\frac{s(t)}{r(t)}\right)z(t_0)$$

$$=\left(\sqrt{2}\left(\int_{t_0}^{a}\frac{(s(t))^2}{r(t)}\left(\int_{t}^{a}\frac{1}{r(\tau)}dg(\tau)\right)dg(t)\right)^{\frac{1}{2}}\right.$$

$$\left.+\sup_{t\in J}\left((\Delta g(t))\frac{s(t)}{r(t)}\right)\right)z(t_0)$$

$$=K_2(a,t_0)z(t_0)$$

$$=K_2(a,t_0)\int_{t_0}^{a}r(t)(y_g'(t))^2dg(t).$$

This completes the proof. □

Theorem 5.14. *Let $s, r \in \mathscr{C}(J)$ be positive functions such that*

$$\int_{t_0}^{a}\frac{1}{r(t)}dg(t)<\infty.$$

Also, let $y \in \mathscr{C}_g^1(J)$ with $y(t_0) = y(a) = 0$ and y_g' does not change its sign on J. Then, for any $c \in J$, we have

$$\int_{t_0}^{a}s(t)|y(t)+y(t+)||y_g'(t)|dg(t)$$

$$\leq 2\max\{K_1(c,t_0),K_2(a,c)\}\int_{t_0}^{a}r(t)(y_g'(t))^2dg(t).$$

Proof. By Theorem 5.12, we have

$$\int_{t_0}^{c} s(t)|y(t)+y(t+)||y_g'(t)|dg(t) \le K_1(c,t_0)\int_{t_0}^{c} r(t)(y_g'(t))^2 dg(t)$$
$$\le K_1(c,t_0)\int_{t_0}^{a} r(t)(y_g'(t))^2 dg(t).$$

On the other hand, applying Theorem 5.13, we get

$$\int_{c}^{a} s(t)|y(t)+y(t+)||y_g'(t)|dg(t) \le K_2(a,c)\int_{c}^{a} r(t)(y_g'(t))^2 dg(t)$$
$$\le K_2(a,c)\int_{t_0}^{a} r(t)(y_g'(t))^2 dg(t).$$

Hence,

$$\int_{t_0}^{a} s(t)|y(t)+y(t+)||y_g'(t)|dg(t)$$
$$= \int_{t_0}^{c} s(t)|y(t)+y(t+)||y_g'(t)|dg(t) + \int_{c}^{a} s(t)|y(t)+y(t+)||y_g'(t)|dg(t)$$
$$\le K_1(c,t_0)\int_{t_0}^{a} r(t)(y_g'(t))^2 dg(t) + K_2(a,c)\int_{t_0}^{a} r(t)(y_g'(t))^2 dg(t)$$
$$\le 2\max\{K_1(c,t_0),K_2(a,c)\}\int_{t_0}^{a} r(t)(y_g'(t))^2 dg(t).$$

This completes the proof. □

Theorem 5.15. *Let $s, r \in \mathscr{C}(J)$ be positive functions such that*

$$\int_{t_0}^{a} \frac{1}{r(t)} dg(t) < \infty.$$

Also, let $y \in \mathscr{C}_g^1(J)$ with $y(t_0) = y(a) = 0$ and y_g' does not change its sign on J. Then,

$$\int_{t_0}^{a} s(t)|y(t)+y(t+)||y_g'(t)|dg(t)$$
$$\le 2\min_{c\in J}\left(\max\{K_1(c,t_0),K_2(a,c)\}\right)\int_{t_0}^{a} r(t)(y_g'(t))^2 dg(t).$$

Proof. Since Theorem 5.14 holds for any $c \in J$, we get the desired result. This completes the proof. □

Example 5.1. By the chain rule, Theorem 1.42, we have

$$\begin{aligned}((t-t_0)^2)'_g &= 2\int_0^1 (h((t-t_0)+)+(1-h)(t-t_0))dh\\ &\geq 2\int_0^1 (h(t-t_0)+(1-h)(t-t_0))dh\\ &= 2(t-t_0), \quad t\in J.\end{aligned}$$

Hence,

$$t-t_0 \leq \frac{1}{2}((t-t_0)^2)'_g, \quad t\in J,$$

and

$$\begin{aligned}\int_{t_0}^a (t-t_0)dg(t) &\leq \frac{1}{2}\int_{t_0}^a ((t-t_0)^2)'_g dg(t)\\ &= \frac{1}{2}(t-t_0)^2\Big|_{t=t_0}^{t=a}\\ &= \frac{1}{2}(a-t_0)^2.\end{aligned}$$

Let $r(t)=s(t)=1$, $t\in J$. Then,

$$\begin{aligned}K_1(a,t_0) &= \sqrt{2}\left(\int_{t_0}^a (t-t_0)dg(t)\right)^{\frac{1}{2}} + \sup_{t\in J}(\Delta g(t))\\ &\leq \sqrt{2}\left(\frac{1}{2}(a-t_0)^2\right)^{\frac{1}{2}} + \sup_{t\in J}(\Delta g(t))\\ &= \sqrt{2}\frac{1}{\sqrt{2}}(a-t_0) + \sup_{t\in J}(\Delta g(t))\\ &= a-t_0+\sup_{t\in J}(\Delta g(t)).\end{aligned}$$

Then, applying Theorem 5.12, we get the following estimate:

$$\int_{t_0}^a |y(t)+y(t+)||y'_g(t)|dg(t) \leq (a-t_0+\sup_{t\in J}(\Delta g(t)))\int_{t_0}^a (y'_g(t))^2 dg(t).$$

Problem 5.1. For $t\in J$, prove the following:

(1) $(t-t_0)^2 \leq \frac{1}{3}((t-t_0)^3)'_g$.
(2) $\int_{t_0}^a (t-t_0)^2 dg(t) \leq \frac{1}{3}(a-t_0)^3$.

(3) $K_1(a, t_0) \leq \sqrt{\frac{2}{3}}(a-t_0)^{\frac{3}{2}} + g(a)$ for $r(t) = 1$ and $s(t) = (t-t_0)^{\frac{1}{2}}$, $t \in J$.

(4)
$$\int_{t_0}^{a} (t-t_0)^{\frac{1}{2}} |y(t) + y(t+)||y_g'(t)|dg(t)$$
$$\leq \left(\sqrt{\frac{2}{3}}(a-t_0)^{\frac{3}{2}} + \sup_{t\in J}\left((\Delta g(t))(t-t_0)^{\frac{1}{2}}\right)\right) \times \int_{t_0}^{a} (y_g'(t))^2 dg(t).$$

Problem 5.2. Generalize Exercise 5.1 for $r(t) = 1$ and $s(t) = (t-t_0)^{\frac{n}{2}}$, $t \in J$, where $n \in \mathbb{N}$.

5.3 An Application

In this section, we find an estimate of the smallest eigenvalue of the Stieltjes–Sturm–Liouville eigenvalue problem

$$\begin{aligned} -y_g''(t) + a(t)y(t+) &= \lambda y(t+), \quad t \in J, \\ y(t_0) &= y(a) = 0, \end{aligned} \tag{5.7}$$

where $a \in \mathscr{C}(J)$ and $a(t) = A_g'(t) + \gamma$, $t \in J$, for some $A \in \mathscr{C}_g^1(J)$ and for some $\gamma \in \mathbb{R}$. Suppose that λ_0 is the smallest eigenvalue of the problem (5.7) and $y \in \mathscr{C}_g^2(J)$ is its eigenvalue. Multiplying the first equation of (5.7) with $y(t+)$, $t \in J$, we find

$$-y_g''(t)y(t+) + a(t)(y(t+))^2 = \lambda(y(t+))^2, \quad t \in J,$$

which we integrate in the sense of Stieltjes, from t_0 to a, and obtain

$$\begin{aligned} \lambda \int_{t_0}^{a} (y(t+))^2 dg(t) &= -\int_{t_0}^{a} y_g''(t)y(t+)dg(t) + \int_{t_0}^{a} a(t)(y(t+))^2 dg(t) \\ &= -y_g'(t)y(t)\Big|_{t=t_0}^{t=a} + \int_{t_0}^{a} (y_g'(t))^2 dg(t) \\ &\quad + \int_{t_0}^{a} \left(A_g'(t) + \gamma\right)(y(t+))^2 dg(t) \\ &= \int_{t_0}^{a} (y_g'(t))^2 dg(t) + \int_{t_0}^{a} A_g'(t)(y(t+))^2 dg(t) \\ &\quad + \gamma \int_{t_0}^{a} (y(t+))^2 dg(t), \end{aligned}$$

whereupon we get

$$
\begin{aligned}
(\lambda_0 - \gamma) & \int_{t_0}^{a} (y(t+))^2 dg(t) \\
&= \int_{t_0}^{a} (y_g'(t))^2 dg(t) + \int_{t_0}^{a} A_g'(t)(y(t+))^2 dg(t) \\
&= \int_{t_0}^{a} (y_g'(t))^2 dg(t) + A(t)(y(t))^2 \Big|_{t=t_0}^{t=a} \\
&\quad - \int_{t_0}^{a} A(t)(y^2)_g'(t) dg(t) \\
&= \int_{t_0}^{a} (y_g'(t))^2 dg(t) - \int_{t_0}^{a} A(t)(y(t) + y(t+))y_g'(t) dg(t) \\
&\geq \int_{t_0}^{a} (y_g'(t))^2 dg(t) - \int_{t_0}^{a} |A(t)||y(t) + y(t+)||y_g'(t)| dg(t),
\end{aligned}
$$

i.e.,

$$
\begin{aligned}
(\lambda_0 - \gamma) & \int_{t_0}^{a} (y(t+))^2 dg(t) \\
&\geq \int_{t_0}^{a} (y_g'(t))^2 dg(t) - \int_{t_0}^{a} |A(t)||y(t) + y(t+)||y_g'(t)| dg(t).
\end{aligned} \tag{5.8}
$$

Now, applying Theorem 5.12 for $s(t) = |A(t)|$ and $r(t) = 1$, $t \in J$, we find

$$
\begin{aligned}
\int_{t_0}^{a} & |A(t)||y(t) + y(t+)||y_g'(t)| dg(t) \\
&\leq \left(\sqrt{2} \left(\int_{t_0}^{a} (t - t_0)(A(t))^2 dg(t) \right)^{\frac{1}{2}} + \sup_{t \in J}((\Delta g(t))|A(t)|) \right) \\
&\quad \times \int_{t_0}^{a} (y_g'(t))^2 dg(t).
\end{aligned}
$$

Let

$$
q(A) = \sqrt{2} \left(\int_{t_0}^{a} (t - t_0)(A(t))^2 dg(t) \right)^{\frac{1}{2}} + \sup_{t \in J}((\Delta g(t))|A(t)|).
$$

Then,

$$
\int_{t_0}^{a} |A(t)||y(t) + y(t+)||y_g'(t)| dg(t) \geq q(A) \int_{t_0}^{a} (y_g'(t))^2 dg(t).
$$

Hence, with (5.8), we find

$$(\lambda_0-\gamma)\int_{t_0}^{a}(y(t+))^2dg(t)\geq\int_{t_0}^{a}(y_g'(t))^2dg(t)-q(A)\int_{t_0}^{a}(y_g'(t))^2dg(t)$$
$$=(1-q(A))\int_{t_0}^{a}(y_g'(t))^2dg(t)$$

and

$$|\lambda_0-\gamma|\int_{t_0}^{a}(y(t+))^2dg(t)\geq(1-q(A))\int_{t_0}^{a}(y_g'(t))^2dg(t). \tag{5.9}$$

Let

$$h=2+\sup_{t\in J}\left(\frac{\Delta g(t)}{t}\right).$$

Applying the Stieltjes–Wirtinger inequality for $f(t)=t$, $t\in J$, we find

$$\int_{t_0}^{a}t^2(y_g'(t))^2dg(t)\geq\frac{1}{h^2}\int_{t_0}^{a}(y(t+))^2dg(t),$$

whereupon

$$\int_{t_0}^{a}(y(t+))^2dg(t)\leq h^2\int_{t_0}^{a}t^2(y_g'(t))^2dg(t)$$
$$\leq h^2a^2\int_{t_0}^{a}(y_g'(t))^2dg(t).$$

From this and (5.9), we find

$$|\lambda_0-\gamma|h^2a^2\int_{t_0}^{a}(y_g'(t))^2dg(t)\geq(1-q(A))\int_{t_0}^{a}(y_g'(t))^2dg(t).$$

Therefore,

$$|\lambda_0-\gamma|\geq\frac{1-q(A)}{h^2a^2}.$$

5.4 Stieltjes–Opial-Type Inequalities II

Theorem 5.16. *Let $r\in\mathscr{C}(J)$ be a nonnegative nonincreasing function and $p\geq 0$, $q\geq 1$. Also, let $y\in\mathscr{C}_g^1(J)$ and $y(t_0)=0$. Then,*

$$\int_{t_0}^{a}r(t)|y(t)|^p|y_g'(t)|^qdg(t)\leq\frac{q(g(a)-g(t_0))^p}{p+q}\int_{t_0}^{a}r(t)|y_g'(t)|^{p+q}dg(t).$$

Proof. Introduce the function

$$h(t) = \int_{t_0}^{t} (r(s))^{\frac{q}{p+q}} |y'_g(s)|^q dg(s), \quad t \in J.$$

Then,

$$h(t_0) = 0 \quad \text{and} \quad h'_g(t) = (r(t))^{\frac{q}{p+q}} |y'_g(t)|^q \geq 0, \quad t \in J.$$

Thus, h is a nondecreasing function on J. Then, applying the Stieltjes–Hölder inequality for $\frac{q}{q-q}$ and q, we get

$$\begin{aligned}
|y(t)| &= \left| \int_{t_0}^{t} y'_g(s) dg(s) \right| \\
&\leq \int_{t_0}^{t} |y'_g(s)| dg(s) \\
&= \int_{t_0}^{t} (r(s))^{-\frac{1}{p+q}} (r(s))^{\frac{1}{p+q}} |y'_g(s)| dg(s) \\
&\leq \left(\int_{t_0}^{t} (r(s))^{-\frac{1}{p+q}\frac{q}{q-1}} dg(s) \right)^{\frac{q-1}{q}} \left(\int_{t_0}^{t} (r(s))^{\frac{q}{p+q}} |y'_g(s)|^q dg(s) \right)^{\frac{1}{q}} \\
&\leq (r(t))^{-\frac{1}{p+q}} (g(t) - g(t_0))^{\frac{q-1}{q}} \left(\int_{t_0}^{t} (r(s))^{\frac{q}{p+q}} |y'_g(s)|^q dg(s) \right)^{\frac{1}{q}}, \quad t \in J,
\end{aligned}$$

whereupon

$$(r(t))^{\frac{1}{p+q}} |y(t)| \leq (g(t) - g(t_0))^{\frac{q-1}{q}} \left(\int_{t_0}^{t} (r(s))^{\frac{q}{p+q}} |y'_g(s)|^q dg(s) \right)^{\frac{1}{q}}, \quad t \in J,$$

and

$$\begin{aligned}
(r(t))^{\frac{p}{p+q}} |y(t)|^p &\leq (g(t) - g(t_0))^{\frac{p(q-1)}{q}} \left(\int_{t_0}^{t} (r(s))^{\frac{q}{p+q}} |y'_g(s)|^q dg(s) \right)^{\frac{p}{q}} \\
&\leq (g(t) - g(t_0))^{\frac{p(q-1)}{q}} (h(t))^{\frac{p}{q}}, \quad t \in J.
\end{aligned}$$

Hence,

$$\begin{aligned}
\int_{t_0}^{a} r(t)|y(t)|^p |y'_g(t)|^q dg(t) &= \int_{t_0}^{a} (r(t))^{\frac{p}{p+q}} |y(t)|^p (r(t))^{\frac{q}{p+q}} |y'_g(t)|^q dg(t) \\
&= \int_{t_0}^{a} (g(t) - g(t_0))^{\frac{p(q-1)}{q}} (h(t))^{\frac{p}{q}} h'_g(t) dg(t) \\
&\leq (g(a) - g(t_0))^{\frac{p(q-1)}{q}} \int_{t_0}^{a} (h(t))^{\frac{p}{q}} h'_g(t) dg(t),
\end{aligned}$$

i.e.,

$$\int_{t_0}^{a} r(t)|y(t)|^p|y_g'(t)|^q dg(t) \le (g(a)-g(t_0))^{\frac{p(q-1)}{q}} \int_{t_0}^{a} (h(t))^{\frac{p}{q}} h_g'(t)dg(t). \tag{5.10}$$

By the chain rule, Theorem 1.42, and using that g is nondecreasing, we get

$$\begin{aligned}\left(h^{\frac{p+q}{q}}(t)\right)_g' &= \frac{p+q}{q} h_g'(t)\int_0^1 (h(mt+(1-m)t))^{\frac{p}{q}}\, dm\\ &\ge \frac{p+q}{q} h_g'(t)\int_0^1 (h(mt+(1-m)t))^{\frac{p}{q}} dm\\ &= \frac{p+q}{q} h_g'(t)(h(t))^{\frac{p}{q}}, \quad t\in J.\end{aligned}$$

From this and (5.10), we find

$$\begin{aligned}\int_{t_0}^{a} r(t)|h(t)|^p|y_g'(t)|dg(t) &\le \frac{p}{p+q}(g(a)-g(t_0))^{\frac{p(q-1)}{q}} \int_{t_0}^{a} (h^{\frac{p+q}{q}}(t))_g' dg(t)\\ &= \frac{q}{p+q}(g(a)-g(t_0))^{\frac{p(q-1)}{q}} (h(t))^{\frac{p+q}{q}}\Big|_{t=t_0}^{t=a}\\ &= \frac{q}{p+q}(g(a)-g(t_0))^{\frac{p(q-1)}{q}} (h(a))^{\frac{p+q}{q}},\end{aligned}$$

i.e.,

$$\int_{t_0}^{a} r(t)|h(t)|^p|y_g'(t)|dg(t) \le \frac{q}{p+q}(g(a)-g(t_0))^{\frac{p(q-1)}{q}} (h(a))^{\frac{p+q}{q}}. \tag{5.11}$$

Now, applying the definition for h and the Stieltjes–Hölder inequality, we find

$$\begin{aligned}h(t) &= \int_{t_0}^{t} (r(s))^{\frac{q}{p+q}}|y_g'(s)|^q dg(s)\\ &\le (g(t)-g(t_0))^{\frac{p}{p+q}}\left(\int_{t_0}^{t} r(s)|y_g'(s)|^{p+q}dg(s)\right)^{\frac{q}{p+q}}, \quad t\in J.\end{aligned}$$

From here, we obtain

$$h(a) \le (g(a)-g(t_0))^{\frac{p}{p+q}}\left(\int_{t_0}^{a} r(s)|y_g'(s)|^{p+q}dg(s)\right)^{\frac{q}{p+q}}$$

and

$$(h(a))^{\frac{p+q}{q}} \leq (g(a)-g(t_0))^{\frac{p}{q}} \int_{t_0}^{a} r(s)|y_g'(s)|^{p+q} dg(s).$$

Hence, applying (5.11), we obtain

$$\begin{aligned}
&\int_{t_0}^{a} r(t)|h(t)|^p |y_g'(t)| dg(t) \\
&\quad \leq \frac{q}{p+q}(g(a)-g(t_0))^{\frac{p(q-1)}{q}} (a-t_0)^{\frac{p}{q}} \int_{t_0}^{a} r(s)|y_g'(s)|^{p+q} dg(s) \\
&\quad = \frac{q}{p+q}(g(a)-g(t_0))^{p} \int_{t_0}^{a} r(s)|y_g'(s)|^{p+q} dg(s).
\end{aligned}$$

This completes the proof. □

Theorem 5.17. *Let $r \in \mathscr{C}(J)$ be a nonnegative nonincreasing function and $p \geq 0$, $q \geq 1$. Also, let $y \in \mathscr{C}_g^1(J)$ and $y(a) = 0$. Then,*

$$\int_{t_0}^{a} r(t)|y(t)|^p |y_g'(t)|^q dg(t) \leq \frac{q(g(a)-g(t_0))^p}{p+q} \int_{t_0}^{a} r(t)|y_g'(t)|^{p+q} dg(t).$$

Proof. Introduce the function

$$m(t) = \int_{t}^{a} (r(s))^{\frac{q}{p+q}} |y_g'(s)|^q dg(s), \quad t \in J.$$

Then,

$$m(t_0) = 0 \quad \text{and} \quad m_g'(t) = -(r(t))^{\frac{q}{p+q}} |y_g'(t)|^q \leq 0, \quad t \in J.$$

So, m is a nonincreasing function on J. Then, using the Stieltjes–Hölder inequality for $\frac{q}{q-q}$ and q, we find

$$\begin{aligned}
|y(t)| &= \left| \int_{t}^{a} y_g'(s) dg(s) \right| \\
&\leq \int_{t}^{a} |y_g'(s)| dg(s) \\
&= \int_{t}^{a} (r(s))^{-\frac{1}{p+q}} (r(s))^{\frac{1}{p+q}} |y_g'(s)| dg(s) \\
&\leq \left(\int_{t}^{a} (r(s))^{-\frac{1}{p+q}\frac{q}{q-1}} dg(s) \right)^{\frac{q-1}{q}} \left(\int_{t}^{a} (r(s))^{\frac{q}{p+q}} |y_g'(s)|^q dg(s) \right)^{\frac{1}{q}} \\
&\leq (r(t))^{-\frac{1}{p+q}} (g(a)-g(t))^{\frac{q-1}{q}} \left(\int_{t}^{a} (r(s))^{\frac{q}{p+q}} |y_g'(s)|^q dg(s) \right)^{\frac{1}{q}}, \quad t \in J,
\end{aligned}$$

whereupon

$$(r(t))^{\frac{1}{p+q}}|y(t)| \le (g(a)-g(t))^{\frac{q-1}{q}} \left(\int_t^a (r(s))^{\frac{q}{p+q}} |y_g'(s)|^q dg(s) \right)^{\frac{1}{q}}, \quad t \in J,$$

and

$$\begin{aligned}(r(t))^{\frac{p}{p+q}}|y(t)|^p &\le (g(a)-g(t))^{\frac{p(q-1)}{q}} \left(\int_t^a (r(s))^{\frac{q}{p+q}} |y_g'(s)|^q dg(s) \right)^{\frac{p}{q}} \\ &\le (g(a)-g(t_0))^{\frac{p(q-1)}{q}} (m(t))^{\frac{p}{q}}, \quad t \in J.\end{aligned}$$

From here,

$$\begin{aligned}\int_{t_0}^a r(t)|y(t)|^p|y_g'(t)|^q dg(t) &= \int_{t_0}^a (r(t))^{\frac{p}{p+q}}|y(t)|^p (r(t))^{\frac{q}{p+q}}|y_g'(t)|^q dg(t) \\ &= -(g(a)-g(t))^{\frac{p(q-1)}{q}} \int_{t_0}^a (g(t))^{\frac{p}{q}} m_g'(t)dg(t),\end{aligned}$$

i.e.,

$$\int_{t_0}^a r(t)|y(t)|^p|y_g'(t)|^q dg(t) \le -(g(a)-g(t_0))^{\frac{p(q-1)}{q}} \int_{t_0}^a (m(t))^{\frac{p}{q}} m_g'(t)dg(t). \tag{5.12}$$

By the chain rule, Theorem 1.42, and using the fact that m is nondecreasing, we get

$$\begin{aligned}-(m^{\frac{p+q}{q}}(t))_t' &= -\frac{p+q}{q} m_g'(t) \int_0^1 (m(ht+(1-h)t))^{\frac{p}{q}}\, dh \\ &\ge -\frac{p+q}{q} m_g'(t) \int_0^1 (m(ht+(1-h)t))^{\frac{p}{q}} dh \\ &= -\frac{p+q}{q} m_g'(t)(m(t))^{\frac{p}{q}}, \quad t \in J.\end{aligned}$$

From this and (5.12), we find

$$\begin{aligned}\int_{t_0}^a r(t)|m(t)|^p|y_g'(t)|dg(t) &\le -\frac{p}{p+q}(g(a)-g(t_0))^{\frac{p(q-1)}{q}} \int_{t_0}^a (m^{\frac{p+q}{q}}(t))_t' dg(t) \\ &= -\frac{q}{p+q}(g(a)-g(t_0))^{\frac{p(q-1)}{q}} (m(t))^{\frac{p+q}{q}} \Bigg|_{t=t_0}^{t=a} \\ &= \frac{q}{p+q}(g(a)-g(t_0))^{\frac{p(q-1)}{q}} (m(t_0))^{\frac{p+q}{q}},\end{aligned}$$

i.e.,

$$\int_{t_0}^{a} r(t)|m(t)|^p |y_g'(t)| dg(t) \leq \frac{q}{p+q}(g(a)-g(t_0))^{\frac{p(q-1)}{q}}(m(t_0))^{\frac{p+q}{q}}. \quad (5.13)$$

Now, applying the definition for m and the Stieltjes–Hölder inequality, we find

$$m(t) = \int_{t}^{a} (r(s))^{\frac{q}{p+q}} |y_g'(s)|^q dg(s)$$

$$\leq (g(a)-g(t))^{\frac{p}{p+q}} \left(\int_{t}^{a} r(s)|y_g'(s)|^{p+q} dg(s) \right)^{\frac{q}{p+q}}, \quad t \in J.$$

From here, we obtain

$$m(t_0) \leq (g(a)-g(t_0))^{\frac{p}{p+q}} \left(\int_{t_0}^{a} r(s)|y_g'(s)|^{p+q} dg(s) \right)^{\frac{q}{p+q}}$$

and

$$(m(t_0))^{\frac{p+q}{q}} \leq (g(a)-g(t_0))^{\frac{p}{q}} \int_{t_0}^{a} r(s)|y_g'(s)|^{p+q} dg(s).$$

Hence, applying (5.11), we obtain

$$\int_{t_0}^{a} r(t)|m(t)|^p |y_g'(t)| dg(t)$$

$$\leq \frac{q}{p+q}(g(a)-g(t_0))^{\frac{p(q-1)}{q}}(g(a)-g(t_0))^{\frac{p}{q}} \int_{t_0}^{a} r(s)|y_g'(s)|^{p+q} dg(s)$$

$$= \frac{q}{p+q}(g(a)-g(t_0))^p \int_{t_0}^{a} r(s)|y_g'(s)|^{p+q} dg(s).$$

This completes the proof. □

Theorem 5.18. *Let $r \in \mathscr{C}(J)$ be a nonnegative nonincreasing function and $p \geq 0$, $q \geq 1$. Also, let $y \in \mathscr{C}_g^1(J)$ and $y(t_0) = y(a) = 0$. Then,*

$$\int_{t_0}^{a} r(t)|y(t)|^p |y_g'(t)|^q dg(t) \leq \frac{q}{p+q} \left(\frac{g(a)-g(t_0)}{2} \right)^p \int_{t_0}^{a} r(t)|y_g'(t)|^{p+q} dg(t).$$

Proof. By Theorem 5.16, one gets

$$\int_{\frac{a+t_0}{2}}^{a} r(t)|y(t)|^p |y_g'(t)|^q dg(t) \leq \frac{q}{p+q} \left(\frac{g(a)-g(t_0)}{2} \right)^p \int_{\frac{a+t_0}{2}}^{a} r(t)|y_g'(t)|^{p+q} dg(t).$$

Applying Theorem 5.17, we find

$$\int_{t_0}^{\frac{a+t_0}{2}} r(t)|y(t)|^p|y_g'(t)|^q dg(t) \leq \frac{q}{p+q}\left(\frac{g(a)-g(t_0)}{2}\right)^p \int_{t_0}^{\frac{a+t_0}{2}} r(t)|y_g'(t)|^{p+q} dg(t).$$

Hence,

$$\begin{aligned}
&\int_{t_0}^{a} r(t)|y(t)|^p|y_g'(t)|^q dg(t)\\
&= \int_{t_0}^{\frac{a+t_0}{2}} r(t)|y(t)|^p|y_g'(t)|^q dg(t) + \int_{\frac{t_0+a}{2}}^{a} r(t)|y(t)|^p|y_g'(t)|^q dg(t)\\
&\leq \frac{q}{p+q}\left(\frac{g(a)-g(t_0)}{2}\right)^p \int_{t_0}^{\frac{a+t_0}{2}} r(t)|y_g'(t)|^{p+q} dg(t)\\
&\quad + \frac{q}{p+q}\left(\frac{g(a)-g(t_0)}{2}\right)^p \int_{\frac{a+t_0}{2}}^{a} r(t)|y_g'(t)|^{p+q} dg(t)\\
&= \frac{q}{p+q}\left(\frac{g(a)-g(t_0)}{2}\right)^p \int_{t_0}^{a} r(t)|y_g'(t)|^{p+q} dg(t).
\end{aligned}$$

This completes the proof. □

Theorem 5.19. *Let $p \in \mathscr{C}(J)$ be such that*

$$\int_{t_0}^{a} (p(t))^{1-\alpha} dg(t) < \infty, \quad \alpha > 1.$$

Also, let $q \in \mathscr{C}(J)$ be a positive, bounded, nonincreasing function and $y \in \mathscr{C}_g^1(J)$ be such that $y(t_0) = 0$. Then, for any $\gamma > 0$ and $\frac{1}{\alpha} + \frac{1}{\nu} = 1$, we have the inequality

$$\begin{aligned}
\int_{t_0}^{a} q(t)|y(t)|^\gamma |D_g y(t)| dg(t) &\leq \frac{1}{\gamma+1}\left(\int_{t_0}^{a} \frac{1}{(p(t))^{\alpha-1}} dg(t)\right)^{\frac{1+\gamma}{\alpha}}\\
&\quad \times \left(\int_{t_0}^{a} p(t)(q(t))^{\frac{\gamma}{1+\nu}} |y_g'(t)|^\nu dg(t)\right)^{\frac{1+\gamma}{\nu}}.
\end{aligned}$$

Proof. Let

$$m(t) = \int_{t_0}^{t} (q(s))^{\frac{1}{1+\gamma}} |y_g'(s)| dg(s), \quad t \in J.$$

Then,

$$m(t_0) = 0 \quad \text{and} \quad m_g'(t) = (q(t))^{\frac{1}{\gamma+1}} |y_g'(t)| \geq 0, \quad t \in J,$$

i.e., m is a nondecreasing function on J. Now, using the fact that m is a nonincreasing function, we obtain

$$\begin{aligned}
(m(t))^\gamma &= \left(\int_{t_0}^t (q(s))^{\frac{1}{1+\gamma}} |y_g'(s)| dg(s)\right)^\gamma \\
&\geq (q(t))^{\frac{\gamma}{\gamma+1}} \left(\int_{t_0}^t |y_g'(s)| dg(s)\right)^\gamma \\
&\geq (q(t))^{\frac{\gamma}{\gamma+1}} \left|\int_{t_0}^t y_g'(s) dg(s)\right|^\gamma \\
&= (q(t))^{\frac{\gamma}{\gamma+1}} \left| y(s)\Big|_{s=t_0}^{s=t} \right|^\gamma \\
&= (q(t))^{\frac{\gamma}{\gamma+1}} |y(t)|^\gamma, \quad t \in J.
\end{aligned}$$

Therefore,

$$\begin{aligned}
\int_{t_0}^a q(t)|y(t)|^\gamma |y_g'(t)| dg(t) &= \int_{t_0}^a (q(t))^{\frac{1}{\gamma+1}} (q(t))^{\frac{\gamma}{\gamma+1}} |y(t)|^\gamma |y_g'(t)| dg(t) \\
&\leq \int_{t_0}^a (m(t))^\gamma g_g'(t) dg(t). \qquad (5.14)
\end{aligned}$$

Applying the chain rule, Theorem 1.42, and using the fact that g is nondecreasing, we obtain

$$\begin{aligned}
(m^{\gamma+1}(t))_t' &= (\gamma+1) m_g'(t) \int_0^1 (m(hg(t) + (1-h)t))^\gamma dh \\
&\geq (\gamma+1) m_g'(t) \int_0^1 (m(ht + (1-h)t))^\gamma dh \\
&= (\gamma+1) m_g'(t) (m(t))^\gamma, \quad t \in J,
\end{aligned}$$

whereupon

$$m_g'(t)(m(t))^\gamma \leq \frac{1}{\gamma+1} (m^{\gamma+1}(t))_g', \quad t \in J.$$

Hence, using (5.14), we find

$$\begin{aligned}
\int_{t_0}^a q(t)|y(t)|^\gamma |y_g'(t)| dg(t) &\leq \frac{1}{\gamma+1} \int_{t_0}^a (m^{\gamma+1}(t))_g' dg(t) \\
&= \frac{1}{\gamma+1} (m(t))^{\gamma+1}\Big|_{t=t_0}^{t=a}
\end{aligned}$$

$$= \frac{1}{\gamma+1}(m(a))^{\gamma+1}$$

$$= \frac{1}{\gamma+1}\left(\int_{t_0}^{a}(q(t))^{\frac{1}{\gamma+1}}|y_g'(t)|dg(t)\right)^{\gamma+1},$$

i.e.,

$$\int_{t_0}^{a} q(t)|y(t)|^{\gamma}|y_g'(t)|dg(t) \le \frac{1}{\gamma+1}\left(\int_{t_0}^{a}(q(t))^{\frac{1}{\gamma+1}}|y_g'(t)|dg(t)\right)^{\gamma+1}. \quad (5.15)$$

Now, using the Stieltjes–Hölder inequality, we get

$$\int_{t_0}^{a}(q(t))^{\frac{1}{\gamma+1}}|y_g'(t)|dg(t) = \int_{t_0}^{a}(p(t))^{-\frac{1}{\gamma}}(q(t))^{\frac{1}{\gamma+1}}(p(t))^{\frac{1}{\gamma}}|y_g'(t)|dg(t)$$

$$\le \left(\int_{t_0}^{a}\frac{1}{(p(t))^{\alpha-1}}dg(t)\right)^{\frac{1}{\alpha}}$$

$$\times\left(\int_{t_0}^{a}p(t)(q(t))^{\frac{\gamma}{1+\nu}}|y_g'(t)|^{\nu}dg(t)\right)^{\frac{1}{\nu}}.$$

Now, employing (5.15), we find

$$\int_{t_0}^{a} q(t)|y(t)|^{\gamma}|y_g'(t)|dg(t) \le \frac{1}{\gamma+1}\left(\int_{t_0}^{a}\frac{1}{(p(t))^{\alpha-1}}dg(t)\right)^{\frac{1+\gamma}{\alpha}}$$

$$\times\left(\int_{t_0}^{a}p(t)(q(t))^{\frac{\gamma}{1+\nu}}|y_g'(t)|^{\nu}dg(t)\right)^{\frac{1+\gamma}{\nu}}.$$

This completes the proof. □

Theorem 5.20. *Let $p \in \mathscr{C}(J)$ be such that*

$$\int_{t_0}^{a}(p(t))^{1-\alpha}dg(t) < \infty, \quad \alpha > 1.$$

Also, let $q \in \mathscr{C}(J)$ be a positive, bounded, nonincreasing function and $y \in \mathscr{C}_g^1(J)$ be such that $y(a) = 0$. Then, for any $\gamma > 0$ and $\frac{1}{\alpha} + \frac{1}{\nu} = 1$, we have the inequality

$$\int_{t_0}^{a} q(t)|y(t)|^{\gamma}|y_g'(t)|dg(t) \le \frac{1}{\gamma+1}\left(\int_{t_0}^{a}\frac{1}{(p(t))^{\alpha-1}}dg(t)\right)^{\frac{1+\gamma}{\alpha}}$$

$$\times\left(\int_{t_0}^{a}p(t)(q(t))^{\frac{\gamma}{1+\nu}}|y_g'(t)|^{\nu}dg(t)\right)^{\frac{1+\gamma}{\nu}}.$$

Proof. Let

$$m(t)=\int_t^a (q(s))^{\frac{1}{1+\gamma}}|y_g'(s)|dg(s),\quad t\in J.$$

Then,

$$m(a)=0\quad\text{and}\quad m_g'(t)=-(q(t))^{\frac{1}{\gamma+1}}|y_g'(t)|\le 0,\quad t\in J,$$

so m is a nonincreasing function on J. Since q is a nonincreasing function, we obtain

$$\begin{aligned}(m(t))^\gamma&=\left(\int_t^a (q(s))^{\frac{1}{1+\gamma}}|y_g'(s)|dg(s)\right)^\gamma\\&\ge (q(t))^{\frac{\gamma}{\gamma+1}}\left(\int_t^a |y_g'(s)|dg(s)\right)^\gamma\\&\ge (q(t))^{\frac{\gamma}{\gamma+1}}\left|\int_t^a y_g'(s)dg(s)\right|^\gamma\\&=(q(t))^{\frac{\gamma}{\gamma+1}}\left|y(s)\Big|_{s=t}^{s=a}\right|\\&=(q(t))^{\frac{\gamma}{\gamma+1}}|y(t)|^\gamma,\quad t\in J.\end{aligned}$$

Therefore,

$$\begin{aligned}\int_{t_0}^a q(t)|y(t)|^\gamma|y_g'(t)|dg(t)&=\int_{t_0}^a (q(t))^{\frac{1}{\gamma+1}}(q(t))^{\frac{\gamma}{\gamma+1}}|y(t)|^\gamma|y_g'(t)|dg(t)\\&\le-\int_{t_0}^a (m(t))^\gamma m_g'(t)dg(t).\end{aligned}\tag{5.16}$$

Employing the chain rule, Theorem 1.42, and using the fact that g is nondecreasing, we arrive at

$$\begin{aligned}(m^{\gamma+1}(t))_g'&=-(\gamma+1)m_g'(t)\int_0^1 (m(hg(t)+(1-h)t))^\gamma dh\\&\ge-(\gamma+1)m_g'(t)\int_0^1 (m(ht+(1-h)t))^\gamma dh\\&=-(\gamma+1)m_g'(t)(m(t))^\gamma,\quad t\in J,\end{aligned}$$

whereupon

$$-m_g'(t)(m(t))^\gamma\le-\frac{1}{\gamma+1}(m^{\gamma+1}(t))_g',\quad t\in J.$$

From this and (5.16), we find

$$\begin{aligned}\int_{t_0}^{a} q(t)|y(t)|^{\gamma}|y_g'(t)|dg(t) &\leq -\frac{1}{\gamma+1}\int_{t_0}^{a}(m^{\gamma+1}(t))_g' dg(t)\\ &= -\frac{1}{\gamma+1}(m(t))^{\gamma+1}\Big|_{t=t_0}^{t=a}\\ &= \frac{1}{\gamma+1}(m(t_0))^{\gamma+1}\\ &= \frac{1}{\gamma+1}\left(\int_{t_0}^{a}(q(t))^{\frac{1}{\gamma+1}}|y_g'(t)|dg(t)\right)^{\gamma+1},\end{aligned}$$

i.e.,

$$\int_{t_0}^{a} q(t)|y(t)|^{\gamma}|y_g'(t)|dg(t) \leq \frac{1}{\gamma+1}\left(\int_{t_0}^{a}(q(t))^{\frac{1}{\gamma+1}}|y_g'(t)|dg(t)\right)^{\gamma+1}. \quad (5.17)$$

Applying the Stieltjes–Hölder inequality, we get

$$\begin{aligned}\int_{t_0}^{a}(q(t))^{\frac{1}{\gamma+1}}|y_g'(t)|dg(t) &= \int_{t_0}^{a}(p(t))^{-\frac{1}{\gamma}}(q(t))^{\frac{1}{\gamma+1}}(p(t))^{\frac{1}{\gamma}}|y_g'(t)|dg(t)\\ &\leq \left(\int_{t_0}^{a}\frac{1}{(p(t))^{\alpha-1}}dg(t)\right)^{\frac{1}{\alpha}}\\ &\quad\times\left(\int_{t_0}^{a}p(t)(q(t))^{\frac{\gamma}{1+\nu}}|y_g'(t)|^{\nu}dg(t)\right)^{\frac{1}{\nu}}.\end{aligned}$$

Now, employing (5.17), we find

$$\begin{aligned}\int_{t_0}^{a} q(t)|y(t)|^{\gamma}|y_g'(t)|dg(t) &\leq \frac{1}{\gamma+1}\left(\int_{t_0}^{a}\frac{1}{(p(t))^{\alpha-1}}dg(t)\right)^{\frac{1+\gamma}{\alpha}}\\ &\quad\times\left(\int_{t_0}^{a}p(t)(q(t))^{\frac{\gamma}{1+\nu}}|y_g'(t)|^{\nu}dg(t)\right)^{\frac{1+\gamma}{\nu}}.\end{aligned}$$

This completes the proof. □

Theorem 5.21. *Let $p \in \mathscr{C}(J)$ be such that*

$$\int_{t_0}^{a}(p(t))^{1-\alpha}dg(t) < \infty, \quad \alpha > 1.$$

Also, let $q \in \mathscr{C}(J)$ be a positive, bounded, nonincreasing function and $y \in \mathscr{C}_g^1(J)$ be such that $y(t_0) = y(a) = 0$. Then, for any $\gamma > 0$ and $\frac{1}{\alpha} + \frac{1}{\nu} = 1$, we have the inequality

$$\int_{t_0}^{a} q(t)|y(t)|^{\gamma}|y_g'(t)|dg(t) \le \max_{c\in J}\left\{\frac{1}{\gamma+1}\left(\int_{t_0}^{c}\frac{1}{(p(t))^{\alpha-1}}dg(t)\right)^{\frac{1+\gamma}{\alpha}}, \frac{1}{\gamma+1}\left(\int_{c}^{a}\frac{1}{(p(t))^{\alpha-1}}dg(t)\right)^{\frac{1+\gamma}{\alpha}}\right\} \times\left(\int_{t_0}^{a} p(t)(q(t))^{\frac{\gamma}{1+\nu}}|y_g'(t)|^{\nu}dg(t)\right)^{\frac{1+\gamma}{\nu}}.$$

Proof. Let $c \in J$ be arbitrarily chosen. Then, applying Theorem 5.19, we obtain

$$\begin{aligned}
&\int_{t_0}^{c} q(t)|y(t)|^{\gamma}|D_g y(t)|dg(t)\\
&\le \frac{1}{\gamma+1}\left(\int_{t_0}^{c}\frac{1}{(p(t))^{\alpha-1}}dg(t)\right)^{\frac{1+\gamma}{\alpha}} \times\left(\int_{t_0}^{c} p(t)(q(t))^{\frac{\gamma}{1+\nu}}|y_g'(t)|^{\nu}dg(t)\right)^{\frac{1+\gamma}{\nu}}\\
&\le \frac{1}{\gamma+1}\max_{c\in J}\left\{\left(\int_{t_0}^{c}\frac{1}{(p(t))^{\alpha-1}}dg(t)\right)^{\frac{1+\gamma}{\alpha}}, \left(\int_{c}^{a}\frac{1}{(p(t))^{\alpha-1}}dg(t)\right)^{\frac{1+\gamma}{\alpha}}\right\}\\
&\times\left(\int_{t_0}^{c} p(t)(q(t))^{\frac{\gamma}{1+\nu}}|y_g'(t)|^{\nu}dg(t)\right)^{\frac{1+\gamma}{\nu}}.
\end{aligned}$$

By Theorem 5.20, we obtain

$$\begin{aligned}
&\int_{c}^{a} q(t)|y(t)|^{\gamma}|D_g y(t)|dg(t)\\
&\le \frac{1}{\gamma+1}\left(\int_{c}^{a}\frac{1}{(p(t))^{\alpha-1}}dg(t)\right)^{\frac{1+\gamma}{\alpha}} \times\left(\int_{c}^{a} p(t)(q(t))^{\frac{\gamma}{1+\nu}}|y_g'(t)|^{\nu}dg(t)\right)^{\frac{1+\gamma}{\nu}}\\
&\le \frac{1}{\gamma+1}\max_{c\in J}\left\{\left(\int_{t_0}^{c}\frac{1}{(p(t))^{\alpha-1}}dg(t)\right)^{\frac{1+\gamma}{\alpha}}, \left(\int_{t}^{a}\frac{1}{(p(t))^{\alpha-1}}dg(t)\right)^{\frac{1+\gamma}{\alpha}}\right\}\\
&\times\left(\int_{c}^{a} p(t)(q(t))^{\frac{\gamma}{1+\nu}}|y_g'(t)|^{\nu}dg(t)\right)^{\frac{1+\gamma}{\nu}}.
\end{aligned}$$

Hence,

$$\begin{aligned}
&\int_{t_0}^{a} q(t)|y(t)|^{\gamma}|D_g y(t)|dg(t)\\
&= \int_{t_0}^{c} q(t)|y(t)|^{\gamma}|D_g y(t)|dg(t) + \int_{c}^{a} q(t)|y(t)|^{\gamma}|D_g y(t)|dg(t)\\
&\le \frac{1}{\gamma+1}\max_{c\in J}\left\{\left(\int_{t_0}^{c}\frac{1}{(p(t))^{\alpha-1}}dg(t)\right)^{\frac{1+\gamma}{\alpha}},\ \left(\int_{c}^{a}\frac{1}{(p(t))^{\alpha-1}}dg(t)\right)^{\frac{1+\gamma}{\alpha}}\right\}\\
&\quad\times\left(\int_{t_0}^{c} p(t)(q(t))^{\frac{\gamma}{1+\nu}}|y_g'(t)|^{\nu}dg(t)\right)^{\frac{1+\gamma}{\nu}} + \frac{1}{\gamma+1}\\
&\quad\times\max_{c\in J}\left\{\left(\int_{t_0}^{c}\frac{1}{(p(t))^{\alpha-1}}dg(t)\right)^{\frac{1+\gamma}{\alpha}},\ \left(\int_{c}^{a}\frac{1}{(p(t))^{\alpha-1}}dg(t)\right)^{\frac{1+\gamma}{\alpha}}\right\}\\
&\quad\times\left(\int_{c}^{a} p(t)(q(t))^{\frac{\gamma}{1+\nu}}|y_g'(t)|^{\nu}dg(t)\right)^{\frac{1+\gamma}{\nu}}\\
&= \frac{1}{\gamma+1}\max_{c\in J}\left\{\left(\int_{t_0}^{c}\frac{1}{(p(t))^{\alpha-1}}dg(t)\right)^{\frac{1+\gamma}{\alpha}},\ \left(\int_{c}^{a}\frac{1}{(p(t))^{\alpha-1}}dg(t)\right)^{\frac{1+\gamma}{\alpha}}\right\}\\
&\quad\times\left(\int_{t_0}^{a} p(t)(q(t))^{\frac{\gamma}{1+\nu}}|y_g'(t)|^{\nu}dg(t)\right)^{\frac{1+\gamma}{\nu}}.
\end{aligned}$$

This completes the proof. □

Theorem 5.22. *Let p and q be positive real numbers such that $p+q>1$. Also, let $r, s \in \mathscr{C}(J)$ be nonnegative functions such that*

$$\int_{t_0}^{a}\frac{1}{(r(t))^{\frac{1}{p+q-1}}}dg(t) < \infty.$$

If $y \in \mathscr{C}_g^1(J)$ is such that $y(t_0)=0$ and y_g' does not change its sign on J, then

$$\int_{t_0}^{a} s(t)|y(t)|^{p}|y_g'(t)|^{q}dg(t) \le K_1(t_0,a,p,q)\int_{t_0}^{a} r(t)|y_g'(t)|^{p+q}dg(t),$$

where

$$\begin{aligned}
&K_1(t_0,a,p,q)\\
&= \left(\frac{q}{p+q}\right)^{\frac{q}{p+q}}\left(\int_{t_0}^{a}\frac{(s(t))^{\frac{p+q}{q}}}{(r(t))^{\frac{q}{p}}}\left(\int_{t_0}^{t}\frac{1}{(r(\tau))^{\frac{1}{p+q-1}}}dg(\tau)\right)^{p+q-1}dg(t)\right)^{\frac{p}{p+q}}.
\end{aligned}$$

Proof. Since y_g' does not change its sign on J, we have

$$|y(t)| = \int_{t_0}^{t} |y_g'(\tau)| dg(\tau)$$
$$= \int_{t_0}^{t} \frac{1}{(r(\tau))^{\frac{1}{p+q}}} (r(\tau))^{\frac{1}{p+q}} |y_g'(\tau)| dg(\tau), \quad t \in J.$$

Applying the Stieltjes–Hölder inequality for $\gamma = \frac{p+q}{p+q-1}$ and $\nu = p+q$, we get

$$|y(t)| = \int_{t_0}^{t} \frac{1}{(r(\tau))^{\frac{1}{p+q}}} (r(\tau))^{\frac{1}{p+q}} |y_g'(\tau)| dg(\tau)$$
$$\leq \left(\int_{t_0}^{t} \frac{1}{(r(\tau))^{\frac{1}{p+q-1}}} dg(\tau) \right)^{\frac{p+q-1}{p+q}}$$
$$\times \left(\int_{t_0}^{t} r(\tau) |y_g'(\tau)|^{p+q} dg(\tau) \right)^{\frac{1}{p+q}}, \quad t \in J,$$

whereupon

$$|y(t)|^p \leq \left(\int_{t_0}^{t} \frac{1}{(r(\tau))^{\frac{1}{p+q-1}}} dg(\tau) \right)^{p\frac{p+q-1}{p+q}}$$
$$\times \left(\int_{t_0}^{t} r(\tau) |y_g'(\tau)|^{p+q} dg(\tau) \right)^{\frac{p}{p+q}}, \quad t \in J. \tag{5.18}$$

Define

$$z(t) = \int_{t_0}^{t} r(\tau) |y_g'(\tau)|^{p+q} dg(\tau), \quad t \in J.$$

Then,

$$z(t_0) = 0 \quad \text{and} \quad z_g'(t) = r(t) |y_g'(t)|^{p+q} \geq 0, \quad t \in J.$$

Thus, z is a nondecreasing function on J. Moreover, we have

$$(y_g'(t))^{p+q} = \frac{z_g'(t)}{r(t)}, \quad t \in J.$$

Now, using the fact that s is a nonnegative function on J and applying (5.18), we find

$$s(t)|y(t)|^p|y_g'(t)|^{p+q} \le s(t)\frac{1}{(r(t))^{p+q}}\left(\int_{t_0}^t \frac{1}{(r(\tau))^{\frac{1}{p+q-1}}}dg(\tau)\right)^{p\frac{p+q-1}{p+q}} \times (z(t))^{\frac{p}{p+q}}(z_g'(t))^{\frac{q}{p+q}}, \quad t \in J,$$

which we integrate in the sense of Stieltjes, from t_0 to a, and then applying the Stieltjes–Hölder inequality for $\frac{p+q}{p}$ and $\frac{p+q}{q}$, we arrive at

$$\begin{aligned}
&\int_{t_0}^a s(t)|y(t)|^p|y_g'(t)|^{p+q}dg(t) \\
&\le \int_{t_0}^a s(t)\frac{1}{(r(t))^{p+q}}\left(\int_{t_0}^t \frac{1}{(r(\tau))^{\frac{1}{p+q-1}}}dg(\tau)\right)^{p\frac{p+q-1}{p+q}} \\
&\times (z(t))^{\frac{p}{p+q}}(z_g'(t))^{\frac{q}{p+q}}dg(t) \\
&\le \left(\int_{t_0}^a (s(t))^{\frac{p+q}{p}}\frac{1}{(r(t))^{\frac{q}{p}}}\left(\int_{t_0}^t \frac{1}{(r(\tau))^{\frac{1}{p+q-1}}}dg(\tau)\right)^{\frac{p+q-1}{p}} dg(t)\right)^{\frac{p}{p+q}} \\
&\times \left(\int_{t_0}^a (z(t))^{\frac{p}{q}}z_g'(t)dg(t)\right)^{\frac{q}{p+q}}.
\end{aligned}$$

Now, applying the chain rule, Theorem 1.42, and using the fact that z is a nondecreasing function, we obtain

$$\begin{aligned}
\left(z^{\frac{p+q}{q}}\right)_g'(t) &= \frac{p+q}{q}z_g'(t)\int_0^1 (z(hg(t)+(1-h)t))^{\frac{p}{q}}dh \\
&\le \frac{p+q}{q}z_g'(t)\int_0^1 (z(ht+(1-h)t))^p q dh \\
&= \frac{p+q}{q}(z(t))^{\frac{q}{p}}z_g'(t), \quad t \in J,
\end{aligned}$$

whereupon

$$(z(t))^{\frac{p}{q}}z_g'(t) \le \frac{q}{p+q}(z_g')^{\frac{p+q}{q}}(t), \quad t \in J.$$

Therefore,

$$\int_{t_0}^{a} s(t)|y(t)|^p |y_g'(t)|^{p+q} dg(t)$$

$$\leq \left(\int_{t_0}^{a} (s(t))^{\frac{p+q}{p}} \frac{1}{(r(t))^{\frac{q}{p}}} \left(\int_{t_0}^{t} \frac{1}{(r(\tau))^{\frac{1}{p+q-1}}} dg(\tau) \right)^{\frac{p+q-1}{p}} dg(t) \right)^{\frac{p}{p+q}}$$

$$\times \left(\frac{q}{p+q} \int_{t_0}^{a} (z_g')^{\frac{p+q}{q}}(t) dg(t) \right)^{\frac{q}{p+q}}$$

$$= \left(\int_{t_0}^{a} (s(t))^{\frac{p+q}{p}} \frac{1}{(r(t))^{\frac{q}{p}}} \left(\int_{t_0}^{t} \frac{1}{(r(\tau))^{\frac{1}{p+q-1}}} dg(\tau) \right)^{\frac{p+q-1}{p}} dg(t) \right)^{\frac{p}{p+q}}$$

$$\times \left(\frac{q}{p+q} (z(t))^{\frac{p+q}{q}} \Big|_{t=t_0}^{t=a} \right)^{\frac{q}{p+q}}$$

$$= \left(\int_{t_0}^{a} (s(t))^{\frac{p+q}{p}} \frac{1}{(r(t))^{\frac{q}{p}}} \left(\int_{t_0}^{t} \frac{1}{(r(\tau))^{\frac{1}{p+q-1}}} dg(\tau) \right)^{\frac{p+q-1}{p}} dg(t) \right)^{\frac{p}{p+q}}$$

$$\times \left(\frac{q}{p+q} (z(a))^{\frac{p+q}{q}} \right)^{\frac{q}{p+q}}$$

$$= \left(\int_{t_0}^{a} (s(t))^{\frac{p+q}{p}} \frac{1}{(r(t))^{\frac{q}{p}}} \left(\int_{t_0}^{t} \frac{1}{(r(\tau))^{\frac{1}{p+q-1}}} dg(\tau) \right)^{\frac{p+q-1}{p}} dg(t) \right)^{\frac{p}{p+q}}$$

$$\times \left(\frac{q}{p+q} \right)^{\frac{q}{p+q}} z(a)$$

$$= \left(\int_{t_0}^{a} (s(t))^{\frac{p+q}{p}} \frac{1}{(r(t))^{\frac{q}{p}}} \left(\int_{t_0}^{t} \frac{1}{(r(\tau))^{\frac{1}{p+q-1}}} dg(\tau) \right)^{\frac{p+q-1}{p}} dg(t) \right)^{\frac{p}{p+q}}$$

$$\times \left(\frac{q}{p+q} \right)^{\frac{q}{p+q}} \int_{t_0}^{a} r(t) |y_g'(t)|^{p+q} dg(t)$$

$$= K_1(t_0, a, p, q) \int_{t_0}^{a} r(t) |y_g'(t)|^{p+q} dg(t).$$

This completes the proof. □

Theorem 5.23. *Let p and q be positive real numbers such that $p+q>1$. Also, let $r, s \in \mathscr{C}(J)$ be nonnegative functions such that*

$$\int_{t_0}^{a} \frac{1}{(r(t))^{\frac{1}{p+q-1}}} dg(t) < \infty.$$

If $y \in \mathscr{C}_g^1(J)$ is such that $y(a)=0$ and y'_g does not change its sign on J, then

$$\int_{t_0}^{a} s(t)|y(t)|^p |y'_g(t)|^q dg(t) \le K_2(t_0, a, p, q) \int_{t_0}^{a} r(t)|y'_g(t)|^{p+q} dg(t),$$

where

$$K_2(t_0, a, p, q)$$
$$= \left(\frac{q}{p+q}\right)^{\frac{q}{p+q}} \left(\int_{t_0}^{a} \frac{(s(t))^{\frac{p+q}{q}}}{(r(t))^{\frac{q}{p}}} \left(\int_{t}^{a} \frac{1}{(r(\tau))^{\frac{1}{p+q-1}}} dg(\tau)\right)^{p+q-1} dg(t)\right)^{\frac{p}{p+q}}.$$

Proof. Using the fact that y'_g does not change its sign on J, we obtain

$$\begin{aligned} |y(t)| &= \int_{t}^{a} |y'_g(\tau)| dg(\tau) \\ &= \int_{t}^{a} \frac{1}{(r(\tau))^{\frac{1}{p+q}}} (r(\tau))^{\frac{1}{p+q}} |y'_g(\tau)| dg(\tau), \quad t \in J. \end{aligned}$$

Employing the Stieltjes–Hölder inequality in the case when $\gamma = \frac{p+q}{p+q-1}$ and $\nu = p+q$, we find

$$\begin{aligned} |y(t)| &= \int_{t}^{a} \frac{1}{(r(\tau))^{\frac{1}{p+q}}} (r(\tau))^{\frac{1}{p+q}} |y'_g(\tau)| dg(\tau) \\ &\le \left(\int_{t}^{a} \frac{1}{(r(\tau))^{\frac{1}{p+q-1}}} dg(\tau)\right)^{\frac{p+q-1}{p+q}} \\ &\quad \times \left(\int_{t}^{a} r(\tau)|y'_g(\tau)|^{p+q} dg(\tau)\right)^{\frac{1}{p+q}}, \quad t \in J, \end{aligned}$$

from where

$$\begin{aligned} |y(t)|^p &\le \left(\int_{t}^{a} \frac{1}{(r(\tau))^{\frac{1}{p+q-1}}} dg(\tau)\right)^{p\frac{p+q-1}{p+q}} \\ &\quad \times \left(\int_{t}^{a} r(\tau)|y'_g(\tau)|^{p+q} dg(\tau)\right)^{\frac{p}{p+q}}, \quad t \in J. \end{aligned} \tag{5.19}$$

Define

$$z(t) = \int_t^a r(s)|y_g'(s)|^{p+q} dg(s), \quad t \in J.$$

Then,

$$z(a) = 0 \quad \text{and} \quad z_g'(t) = -r(t)|y_g'(t)|^{p+q} \leq 0, \quad t \in J,$$

i.e., z is a nonincreasing function on J. Next, we have

$$(y_g'(t))^{p+q} = -\frac{z_g'(t)}{r(t)}, \quad t \in J.$$

Now, using the fact that s is a nonnegative function on J and applying (5.19), we find

$$s(t)|y(t)|^p|y_g'(t)|^{p+q} \leq s(t)\frac{1}{(r(t))^{p+q}}\left(\int_t^a \frac{1}{(r(\tau))^{\frac{1}{p+q-1}}} dg(\tau)\right)^{p\frac{p+q-1}{p+q}} \times (z(t))^{\frac{p}{p+q}}(-z_g'(t))^{\frac{q}{p+q}}, \quad t \in J,$$

which we integrate in the sense of Stieltjes, from t_0 to a, and then applying the Stieltjes–Hölder inequality for $\frac{p+q}{p}$ and $\frac{p+q}{q}$, we arrive at

$$\begin{aligned}
&\int_{t_0}^a s(t)|y(t)|^p|y_g'(t)|^{p+q} dg(t) \\
&\leq \int_{t_0}^a s(t)\frac{1}{(r(t))^{p+q}}\left(\int_t^a \frac{1}{(r(\tau))^{\frac{1}{p+q-1}}} dg(\tau)\right)^{p\frac{p+q-1}{p+q}} (z(t))^{\frac{p}{p+q}} \\
&\quad \times (-z_g'(t))^{\frac{q}{p+q}} dg(t) \\
&\leq \left(\int_{t_0}^a (s(t))^{\frac{p+q}{p}} \frac{1}{(r(t))^{\frac{q}{p}}}\left(\int_t^a \frac{1}{(r(\tau))^{\frac{1}{p+q-1}}} dg(\tau)\right)^{\frac{p+q-1}{p}} dg(t)\right)^{\frac{p}{p+q}} \\
&\quad \times \left(-\int_{t_0}^a (z(t))^{\frac{p}{q}} z_g'(t) dg(t)\right)^{\frac{q}{p+q}}.
\end{aligned}$$

Employing the chain rule, Theorem 1.42, and using the fact that z is a nonincreasing function, we obtain

$$\begin{aligned}-(z_g')^{\frac{p+q}{q}}(t) &= -\frac{p+q}{q} z_g'(t) \int_0^1 (z(hg(t)+(1-h)t))^{\frac{p}{q}} dh \\ &\le -\frac{p+q}{q} z_g'(t) \int_0^1 (z(ht+(1-h)t))^p q dh \\ &= -\frac{p+q}{q} (z(t))^{\frac{q}{p}} z_g'(t), \quad t \in J,\end{aligned}$$

from where

$$-(z(t))^{\frac{p}{q}} z_g'(t) \le -\frac{q}{p+q} (z_g')^{\frac{p+q}{q}}(t), \quad t \in J.$$

Therefore,

$$\begin{aligned}&\int_{t_0}^a s(t)|y(t)|^p |y_g'(t)|^{p+q} dg(t) \\ &\le \left(\int_{t_0}^a (s(t))^{\frac{p+q}{p}} \frac{1}{(r(t))^{\frac{q}{p}}} \left(\int_t^a \frac{1}{(r(\tau))^{\frac{1}{p+q-1}}} dg(\tau)\right)^{\frac{p+q-1}{p}} dg(t)\right)^{\frac{p}{p+q}} \\ &\quad\times \left(-\frac{q}{p+q} \int_{t_0}^a (z_g')^{\frac{p+q}{q}}(t) dg(t)\right)^{\frac{q}{p+q}} \\ &= \left(\int_{t_0}^a (s(t))^{\frac{p+q}{p}} \frac{1}{(r(t))^{\frac{q}{p}}} \left(\int_t^a \frac{1}{(r(\tau))^{\frac{1}{p+q-1}}} dg(\tau)\right)^{\frac{p+q-1}{p}} dg(t)\right)^{\frac{p}{p+q}} \\ &\quad\times \left(-\frac{q}{p+q} (z(t))^{\frac{p+q}{q}} \Big|_{t=t_0}^{t=a}\right)^{\frac{q}{p+q}} \\ &= \left(\int_{t_0}^a (s(t))^{\frac{p+q}{p}} \frac{1}{(r(t))^{\frac{q}{p}}} \left(\int_t^a \frac{1}{(r(\tau))^{\frac{1}{p+q-1}}} dg(\tau)\right)^{\frac{p+q-1}{p}} dg(t)\right)^{\frac{p}{p+q}} \\ &\quad\times \left(\frac{q}{p+q} (z(t_0))^{\frac{p+q}{q}}\right)^{\frac{q}{p+q}} \\ &= \left(\int_{t_0}^a (s(t))^{\frac{p+q}{p}} \frac{1}{(r(t))^{\frac{q}{p}}} \left(\int_t^a \frac{1}{(r(\tau))^{\frac{1}{p+q-1}}} dg(\tau)\right)^{\frac{p+q-1}{p}} dg(t)\right)^{\frac{p}{p+q}}\end{aligned}$$

$$\times \left(\frac{q}{p+q} \right)^{\frac{q}{p+q}} z(t_0)$$

$$= \left(\int_{t_0}^{a} (s(t))^{\frac{p+q}{p}} \frac{1}{(r(t))^{\frac{q}{p}}} \left(\int_{t}^{a} \frac{1}{(r(\tau))^{\frac{1}{p+q-1}}} dg(\tau) \right)^{\frac{p+q-1}{p}} dg(t) \right)^{\frac{p}{p+q}}$$

$$\times \left(\frac{q}{p+q} \right)^{\frac{q}{p+q}} \int_{t_0}^{a} r(t)|y_g'(t)|^{p+q} dg(t)$$

$$= K_2(t_0, a, p, q) \int_{t_0}^{a} r(t)|y_g'(t)|^{p+q} dg(t).$$

This completes the proof. □

Theorem 5.24. *Let p and q be positive real numbers such that $p+q>1$. Also, let $r, s \in \mathscr{C}(J)$ be nonnegative functions such that*

$$\int_{t_0}^{a} \frac{1}{(r(t))^{\frac{1}{p+q-1}}} dg(t) < \infty.$$

If $y \in \mathscr{C}_g^1(J)$ is such that $y(t_0) = y(a) = 0$ and y_g' does not change its sign on J, then

$$\int_{t_0}^{a} s(t)|y(t)|^p |y_g'(t)|^q dg(t) \le K(t_0, a, p, q) \int_{t_0}^{a} r(t)|y_g'(t)|^{p+q} dg(t),$$

where

$$K(t_0, a, p, q) = \max\{K_1(t_0, a, p, q), K_2(t_0, a, p, q)\}.$$

Proof. Let $c \in J$ be arbitrarily chosen. By Theorem 5.22, we get

$$\int_{t_0}^{c} s(t)|y(t)|^p |y_g'(t)|^q dg(t) \le K_1(t_0, c, p, q) \int_{t_0}^{c} r(t)|y_g'(t)|^{p+q} dg(t)$$
$$\le K(t_0, a, p, q) \int_{t_0}^{c} r(t)|y_g'(t)|^{p+q} dg(t).$$

On the other hand, by Theorem 5.23, we get

$$\int_{c}^{a} s(t)|y(t)|^p |y_g'(t)|^q dg(t) \le K_2(c, a, p, q) \int_{c}^{a} r(t)|y_g'(t)|^{p+q} dg(t)$$
$$\le K(c, a, p, q) \int_{c}^{a} r(t)|y_g'(t)|^{p+q} dg(t).$$

Therefore,

$$\begin{aligned}
&\int_{t_0}^{a} s(t)|y(t)|^p|y_g'(t)|^q dg(t) \\
&= \int_{t_0}^{c} s(t)|y(t)|^p|y_g'(t)|^q dg(t) + \int_{c}^{a} s(t)|y(t)|^p|y_g'(t)|^q dg(t) \\
&\leq K(t_0,a,p,q) \int_{t_0}^{c} r(t)|y_g'(t)|^{p+q} dg(t) + K(t_0,a,p,q) \\
&\quad \times \int_{c}^{a} r(t)|y_g'(t)|^{p+q} dg(t) \\
&= K(t_0,a,p,q) \int_{t_0}^{a} r(t)|y_g'(t)|^{p+q} dg(t).
\end{aligned}$$

This completes the proof. □

Example 5.2. Suppose that p and q are positive real numbers such that $p+q>1$. First, note that by the chain rule, Theorem 1.42, we have

$$\begin{aligned}
((t-t_0)^{p+q})_g' &= (p+q)\int_0^1 (h(\Delta g(t_0) + (1-h)(t-t_0))^{p+q-1} dh \\
&\geq (p+q)\int_0^1 (h(t-t_0) + (1-h)(t-t_0))^{p+q-1} dh \\
&= (p+q)(t-t_0)^{p+q-1}, \quad t \in J.
\end{aligned}$$

Hence,

$$(t-t_0)^{p+q-1} \leq \frac{1}{p+q}((t-t_0)^{p+q})_g', \quad t \in J.$$

Also,

$$\begin{aligned}
\int_{t_0}^{a} (t-t_0)^{p+q-1} dg(t) &\leq \frac{1}{p+q}\int_{t_0}^{a} ((t-t_0)^{p+q})_g' dg(t) \\
&= \frac{1}{p+q}(t-t_0)^{p+q}\Big|_{t=t_0}^{t=a} \\
&= \frac{1}{p+q}(a-t_0)^{p+q}.
\end{aligned}$$

Let $r(t) = s(t) = 1$, $t \in J$. Then,

$$\begin{aligned} K_1(t_0, a, p, q) &= \left(\frac{q}{p+q}\right)^{\frac{q}{p+q}} \left(\int_{t_0}^{a} \left(\int_{t_0}^{t} dg(s)\right)^{p+q-1} dg(t)\right)^{\frac{p}{p+q}} \\ &= \left(\frac{q}{p+q}\right)^{\frac{q}{p+q}} \left(\int_{t_0}^{a} (t-t_0)^{p+q-1} dg(t)\right)^{\frac{p}{p+q}} \\ &\le \left(\frac{q}{p+q}\right)^{\frac{q}{p+q}} \left(\frac{1}{p+q}(a-t_0)^{p+q}\right)^{\frac{p}{p+q}} \\ &= \frac{q^{\frac{p}{p+q}}}{(p+q)^{\frac{p}{p+q}}(p+q)^{\frac{q}{p+q}}}(a-t_0)^p \\ &= \frac{q^{\frac{p}{p+q}}}{p+q}(a-t_0)^p. \end{aligned}$$

Now, applying Theorem 5.22, we get

$$\int_{t_0}^{a} |y(t)|^p |y_g'(t)|^q dg(t) \le \frac{q^{\frac{p}{p+q}}}{p+q}(a-t_0)^p \int_{t_0}^{a} |y_g'(t)|^{p+q} dg(t).$$

Problem 5.3. Let p and q be positive real numbers such that $p+q>1$. Suppose that $y \in \mathscr{C}_g^1(J)$ is such that $y(t_0) = 0$ and y_g' does not change its sign on J. Then, prove that

$$\int_{t_0}^{a} (t-t_0)|y(t)|^p |y_g'(t)|^q dg(t) \le \frac{q^{\frac{q}{p+q}} p^{\frac{p}{p+q}}}{(p+q)(1+p)^{\frac{p}{p+q}}}(a-t_0)^{1+p}.$$

5.5 An Application

In this section, we establish a Stieltjes–Opial inequality for the smallest eigenvalue of the following Stieltjes boundary value problem (SBVP):

$$\begin{aligned} \left(r(t)(y_g'(t))^p\right)_g' &= \gamma s_g'(t)(y(t))^p, \quad t \in J, \\ y(t_0) = 0, \quad y(a)(y_g'(a))^p &- \gamma s(a)(y(a))^p = 0, \end{aligned} \tag{5.20}$$

where $p > 0$, $y \in \mathscr{C}_g^1(J)$ is a nonnegative solution such that $y_g'(t) \ge 0$, $t \in J$, and γ is a positive eigenvalue that corresponds to y, $p > 0$, $r, s \in \mathscr{C}(J)$ are nonnegative functions. Set

$$z(t) = y_g'(t), \quad Z(t) = \int_{t_0}^{t} z(\tau) d_g\tau, \quad w(t) = \left(\frac{y_g'(t)}{y(t)}\right)^p, \quad t \in J, \ \lambda = \frac{p+1}{p}.$$

In the inequality (5.2), we take $a = z$, $b = w^{\frac{1}{p}} Z$, then we get

$$z^{p+1} + pw^{\frac{p+1}{p}} Z^{p+1} - (p+1)zwZ^p \geq 0 \quad \text{on} \quad J$$

or

$$z^{p+1} + pw^{\lambda} Z^{p+1} - (p+1)zwZ^p \geq 0 \quad \text{on} \quad J.$$

We multiply the last inequality by $r(t)$, $t \in J$, and find

$$\begin{aligned} &r(t)(z(t))^{p+1} + pr(t)(w(t))^{\lambda} (Z(t))^{p+1} \\ &\quad - (p+1)r(t)z(t)w(t) (Z(t))^p \geq 0, \quad t \in J, \end{aligned}$$

or

$$r(t)(z(t))^{p+1} + pr(t)(w(t))^{\lambda} (Z(t))^{p+1} \geq (p+1)r(t)z(t)w(t) (Z(t))^p, \quad t \in J.$$

Now, integrating the last inequality in the sense of Stieltjes, from t_0 to a, we obtain

$$\begin{aligned} &\int_{t_0}^{a} r(t)(z(t))^{p+1} dg(t) + p \int_{t_0}^{a} r(t)(w(t))^{\lambda} (Z(t))^{p+1} dg(t) \\ &\geq (p+1) \int_{t_0}^{a} r(t)z(t)w(t) (Z(t))^p \, dg(t) \\ &= (p+1) \int_{t_0}^{a} r(t)w(t)(Z(t))^{p+1} Z_g'(t) dg(t). \end{aligned} \tag{5.21}$$

By the definition of Z, we have $Z_g'(t) = z(t) = y_g'(t) \geq 0$, $t \in J$. Thus, Z is a nondecreasing function on J. From this and the chain rule, Theorem 1.42, we find

$$\begin{aligned} (Z^{p+1}(t))_g' &= (p+1)Z_g'(t) \int_0^1 (Z(ht + (1-h)t))^p dh \\ &\geq (p+1)Z_g'(t) \int_0^1 (Z(ht + (1-h)t))^p dh \\ &= (p+1)Z_g'(t)(Z(t))^p, \quad t \in J, \end{aligned}$$

and

$$\begin{aligned} (Z^{p+1}(t))_g' &= (p+1)Z_g'(t) \int_0^1 (Z(ht + (1-h)t))^p dh \\ &\leq (p+1)Z_g'(t) \int_0^1 (Z(htt + (1-h)t))^p dh \\ &= (p+1)Z_g'(t)(Z(t))^p, \quad t \in J. \end{aligned}$$

Hence, applying (5.21) and integration by parts, we arrive at

$$\begin{aligned}
&\int_{t_0}^{a} r(t)(z(t))^{p+1}dg(t) + p\int_{t_0}^{a} r(t)(w(t))^{\lambda}\left(Z(t)\right)^{p+1}dg(t)\\
&\quad\geq \int_{t_0}^{a} r(t)w(t)(Z^{p+1}(t))'_g dg(t)\\
&\quad= r(t)w(t)(Z(t))^{p+1}\Big|_{t=t_0}^{t=a} - \int_{t_0}^{a}(r(t)w(t))'_g Z(t)dg(t)\\
&\quad= r(a)w(a)(Z(a))^{p+1} - \int_{t_0}^{a}(r(t)w(t))'_g Z(t)dg(t),
\end{aligned}$$

i.e.,

$$\begin{aligned}
&\int_{t_0}^{a} r(t)(z(t))^{p+1}dg(t) + p\int_{t_0}^{a} r(t)(w(t))^{\lambda}\left(Z(t)\right)^{p+1}dg(t)\\
&\quad\geq r(a)w(a)(Z(a))^{p+1} - \int_{t_0}^{a}(r(t)w(t))'_g Z(t)dg(t). \qquad (5.22)
\end{aligned}$$

Now, we return to the definition of w and observe that

$$r(t)w(t) = \frac{r(t)(y'_g(t))^p}{(y(t))^p}, \quad t \in J,$$

from where, and with (5.20), we arrive at

$$\begin{aligned}
(r(t)w(t))'_g &= \frac{1}{(y(t))^p}\left(r(t)(y'_g(t))^p\right)'_g\\
&\quad + \left(r(t)(y'_g(t))^p\right)'_g\left(-\frac{(y'_g)^p(t)}{(y(t))^p(y(t))^p}\right)\\
&= \gamma s'_g(t) - \frac{\left(r(t)(y'_g(t))^p\right)^g (y'_g)^p(t)}{(y(t))^p(y(t))^p}, \quad t \in J. \qquad (5.23)
\end{aligned}$$

Applying the chain rule, Theorem 1.42, and using the fact that $y'_g(t) \geq 0$, $t \in J$, we get

$$\begin{aligned}
(y^p)'_g(t) &= py'_g(t)\int_{t_0}^{a}\left(y(ht+(1-h)t)\right)^{p-1}dh\\
&\geq py'_g(t)\int_{t_0}^{a}\left(y(ht+(1-h)t)\right)^{p-1}dh\\
&= p(y(t))^{p-1}y'_g(t), \quad t \in J.
\end{aligned}$$

From this and (5.23), we find

$$\begin{aligned}(r(t)w(t))'_g &\le \gamma s'_g(t) - \frac{\left(r(t)(y'_g(t))^p\right)^g py'_g(t)(y(t))^{p-1}}{(y(t))^p(y(t))^p}\\ &= \gamma s'_g(t) - \frac{\left(r(t)(y'_g(t))^p\right)^g py'_g(t)}{(y(t))(y(t))^p}\\ &= \gamma s'_g(t) - \frac{pr(t)((y'_g(t))^p)^g y'_g(t)}{y(t)((y(t))^p)^g}\\ &= \gamma s'_g(t) - pr(t)w(t)(w(t))^{\frac{1}{p}}, \quad t \in J.\end{aligned}$$

Hence, applying (5.22), we get

$$\begin{aligned}&\int_{t_0}^a r(t)(z(t))^{p+1}dg(t) + p\int_{t_0}^a r(t)(w(t))^{\lambda}\,(Z(t))^{p+1}\,dg(t)\\ &\quad\ge r(a)w(a)(Z(a))^{p+1} - \gamma\int_{t_0}^a s'_g(t)(Z(t))^{p+1}dg(t)\\ &\qquad + p\int_{t_0}^a r(t)w(t)(w(t))^{\frac{1}{p}}(Z(t))^{p+1}dg(t),\end{aligned}$$

whereupon

$$\begin{aligned}&\int_{t_0}^a r(t)(z(t))^{p+1}dg(t)\\ &\qquad + p\int_{t_0}^a \left(r(t)(w(t))^{\lambda} - r(t)w(t)(w(t))^{\frac{1}{p}}\right)(Z(t))^{p+1}\,dg(t)\\ &\quad\ge r(a)w(a)(Z(a))^{p+1} - \gamma\int_{t_0}^a s'_g(t)(Z(t))^{p+1}dg(t).\end{aligned} \tag{5.24}$$

Using integration by parts, we find

$$\begin{aligned}&-\gamma\int_{t_0}^a s'_g(t)(Z(t))^{p+1}dg(t)\\ &\quad= -\gamma s(t)(Z(t))^{p+1}\Big|_{t=t_0}^{t=a} + \gamma\int_{t_0}^a s(t)(Z^{p+1}(t))'_g dg(t)\\ &\quad= -\gamma s(a)(Z(a))^{p+1} + \gamma\int_{t_0}^a s(t)(Z^{p+1}(t))'_g dg(t)\\ &\quad\ge -\gamma s(a)(Z(a))^{p-1} + \gamma(p+1)\int_{t_0}^a s(t)(Z(t))^p Z'_g(t)dg(t).\end{aligned}$$

From this and (5.24), we find

$$\begin{aligned}\int_{t_0}^{a} & r(t)(z(t))^{p+1}dg(t)\\ &+p\int_{t_0}^{a}\left(r(t)(w(t))^{\lambda}-r(t)w(t)(w(t))^{\frac{1}{p}}\right)(Z(t))^{p+1}\,dg(t)\\ &\geq (r(a)w(a)-\gamma s(a))(Z(a))^{p+1}+\gamma(p+1)\int_{t_0}^{a}s(t)(Z(t))^{p}Z_g'(t)dg(t)\\ &=\gamma(p+1)\int_{t_0}^{a}s(t)(Z(t))^{p}Z_g'(t)dg(t),\end{aligned}$$

whereupon

$$\begin{aligned}\int_{t_0}^{a} & s(t)(Z(t))^{p}Z_g'(t)dg(t)\\ &\leq \frac{1}{\gamma(p+1)}\int_{t_0}^{a}r(t)(z(t))^{p+1}dg(t)\\ &\quad+\frac{p}{\gamma(p+1)}\int_{t_0}^{a}\left(r(t)(w(t))^{\lambda}-r(t)w(t)(w(t))^{\frac{1}{p}}\right)(Z(t))^{p+1}\,dg(t).\end{aligned}$$

In the last inequality, we replace Z with y and z with y_g', and obtain

$$\begin{aligned}\int_{t_0}^{a} & s(t)(y(t))^{p}y_g'(t)dg(t)\\ &\leq \frac{1}{\gamma(p+1)}\int_{t_0}^{a}r(t)(y_g'(t))^{p+1}dg(t)\\ &\quad+\frac{p}{\gamma(p+1)}\int_{t_0}^{a}\left(r(t)(w(t))^{\lambda}-r(t)w(t)(w(t))^{\frac{1}{p}}\right)(y(t))^{p+1}\,dg(t).\end{aligned}$$

5.6 Stieltjes–Opial-Type Inequalities III

In this section, we derive some other kinds of Stieltjes–Opial-type inequalities. We start with the following inequality.

Theorem 5.25. *Let p and q be positive real numbers such that $p \geq 1$ and $r, s \in \mathscr{C}(J)$ be nonnegative functions such that*

$$\int_{t_0}^{a}\frac{1}{(r(t))^{\frac{1}{p+q-1}}}dg(t)<\infty.$$

Also, let $y \in \mathscr{C}_q^1(J)$ be such that $y(t_0) = 0$ and y'_g does not change its sign on J. Then, we have the inequality

$$\int_{t_0}^{a} s(t)|y(t) + y(t+)|^p |y'_g(t)|^q dg(t) \leq K_1(t_0, a, p, q) \int_{t_0}^{a} r(t)|y'_g(t)|^{p+q} dg(t),$$

where

$$K_1(t_0, a, p, q) = 2^{2p-1} \left(\frac{q}{p+q}\right)^{\frac{q}{p+q}} \times \left(\int_{t_0}^{a} \frac{(s(t))^{\frac{p+q}{q}}}{(r(t))^{\frac{q}{p}}} \left(\int_{t_0}^{t} \frac{1}{(r(\tau))^{\frac{1}{p+q-1}}} dg(\tau)\right)^{p+q-1} dg(t)\right)^{\frac{p}{p+q}} + 2^{p-1} \sup_{t \in J} \left((\Delta g(t))^p \frac{s(t)}{r(t)}\right).$$

Proof. Since y'_g does not change its sign on J and $y(t_0) = 0$, we have

$$|y(t)| = \int_{t_0}^{t} |y'_g(\tau)| dg(\tau), \quad t \in J.$$

Then, applying the Stieltjes–Hölder inequality for $\frac{p+q}{p+q-1}$ and $p+q$, we find

$$\begin{aligned} |y(t)| &= \int_{t_0}^{t} \frac{1}{(r(\tau))^{\frac{1}{p+q}}} (r(\tau))^{\frac{1}{p+q}} |y'_g(\tau)| dg(\tau) \\ &\leq \left(\int_{t_0}^{t} \frac{1}{(r(\tau))^{\frac{1}{p+q-1}}} d_g \tau\right)^{\frac{p+q-1}{p+q}} \\ &\quad \times \left(\int_{t_0}^{t} r(\tau)|y'_g(\tau)|^{p+q} dg(\tau)\right)^{\frac{1}{p+q}}, \quad t \in J, \end{aligned}$$

and

$$|y(t)|^p \leq \left(\int_{t_0}^{t} \frac{1}{(r(\tau))^{\frac{1}{p+q-1}}} d_g \tau\right)^{p\frac{p+q-1}{p+q}} \times \left(\int_{t_0}^{t} r(\tau)|y'_g(\tau)|^{p+q} dg(\tau)\right)^{\frac{p}{p+q}},$$

$t \in J$. Note that

$$\begin{aligned} y(t) + y(t+) &= y(t) + y(t) + (\Delta g(t)) y'_g(t) \\ &= 2y(t) + (\Delta g(t)) y'_g(t), \quad t \in J. \end{aligned}$$

Now, applying Lemma 4.1, we find

$$\begin{aligned}|y(t)+y(t+)|^p &= |2y(t)+(\Delta g(t))y_g'(t)|^p\\ &\le 2^{2p-1}|y(t)|^p + 2^{p-1}(\Delta g(t))^p|y_g'(t)|^p, \quad t\in J.\end{aligned}$$

Set

$$z(t) = \int_{t_0}^t r(\tau)|y_g'(\tau)|^{p+q}dg(\tau), \quad t\in J.$$

Then,

$$z(t_0)=0 \quad \text{and} \quad z_g'(t) = r(t)|y_g'(t)|^{p+q}\ge 0, \quad t\in J.$$

Thus, z is a nondecreasing function on J and

$$|y_g'(t)|^{p+q} = \frac{z_g'(t)}{r(t)}, \quad t\in J,$$

and

$$|y_g'(t)|^q = \left(\frac{z_g'(t)}{r(t)}\right)^{\frac{q}{p+q}}, \quad t\in J.$$

Therefore,

$$\begin{aligned}&s(t)|y(t)+y(t+)|^p|y_g'(t)|^q\\ &\quad\le 2^{2p-1}s(t)|y(t)|^p|y_g'(t)|^p + 2^{p-1}(\Delta g(t))^p s(t)|y_g'(t)|^{p+q}\\ &\quad\le 2^{2p-1}s(t)\left(\frac{1}{r(t)}\right)^{\frac{q}{p+q}}\left(\int_{t_0}^t \frac{1}{(r(\tau))^{\frac{1}{p+q-1}}}dg(\tau)\right)^{p\frac{p+q-1}{p+q}}\\ &\qquad\times (z(t))^{\frac{p}{p+q}}(z_g'(t))^{\frac{q}{p+q}} + 2^{p-1}(\Delta g(t))^p s(t)\frac{z_g'(t)}{r(t)}, \quad t\in J.\end{aligned}$$

Integrating the last inequality in the sense of Stieltjes, from t_0 to a, we find

$$\begin{aligned}&\int_{t_0}^a s(t)|y(t)+y(t+)|^p|y_g'(t)|^q dg(t)\\ &\quad\le 2^{2p-1}\int_{t_0}^a s(t)\left(\frac{1}{r(t)}\right)^{\frac{q}{p+q}}\left(\int_{t_0}^t \frac{1}{(r(\tau))^{\frac{1}{p+q-1}}}dg(\tau)\right)^{p\frac{p+q-1}{p+q}} dg(t)\\ &\qquad\times (z(t))^{\frac{p}{p+q}}(z_g'(t))^{\frac{q}{p+q}}dg(t) + 2^{p-1}\int_{t_0}^a (\Delta g(t))^p s(t)\frac{z_g'(t)}{r(t)}dg(t)\end{aligned}$$

$$\le 2^{2p-1}\int_{t_0}^{a} s(t)\left(\frac{1}{r(t)}\right)^{\frac{q}{p+q}}\left(\int_{t_0}^{t}\frac{1}{(r(\tau))^{\frac{1}{p+q-1}}}dg(\tau)\right)^{p\frac{p+q-1}{p+q}}$$
$$\times (z(t))^{\frac{p}{p+q}}(z_g'(t))^{\frac{q}{p+q}}dg(t)+2^{p-1}\sup_{t\in J}\left((\Delta g(t))^p\frac{s(t)}{r(t)}\right)\int_{t_0}^{a} z_g'(t)dg(t)$$
$$= 2^{2p-1}\int_{t_0}^{a} s(t)\left(\frac{1}{r(t)}\right)^{\frac{q}{p+q}}\left(\int_{t_0}^{t}\frac{1}{(r(\tau))^{\frac{1}{p+q-1}}}dg(\tau)\right)^{p\frac{p+q-1}{p+q}}$$
$$\times (z(t))^{\frac{p}{p+q}}(z_g'(t))^{\frac{q}{p+q}}dg(t)+2^{p-1}\sup_{t\in J}\left((\Delta g(t))^p\frac{s(t)}{r(t)}\right)z(t)\Big|_{t=t_0}^{t=a}$$
$$= 2^{2p-1}\int_{t_0}^{a} s(t)\left(\frac{1}{r(t)}\right)^{\frac{q}{p+q}}\left(\int_{t_0}^{t}\frac{1}{(r(\tau))^{\frac{1}{p+q-1}}}dg(\tau)\right)^{p\frac{p+q-1}{p+q}}$$
$$\times (z(t))^{\frac{p}{p+q}}(z_g'(t))^{\frac{q}{p+q}}dg(t)+2^{p-1}\sup_{t\in J}\left((\Delta g(t))^p\frac{s(t)}{r(t)}\right)z(a).$$

Now, applying the Stieltjes–Hölder inequality for $\frac{p+q}{p}$ and $\frac{p+q}{q}$, we arrive at

$$\int_{t_0}^{a} s(t)|y(t)+y(t+)|^p|y_g'(t)|^q dg(t)$$
$$\le 2^{2p-1}\left(\int_{t_0}^{a}(s(t))^{\frac{p+q}{p}}\left(\frac{1}{r(t)}\right)^{\frac{q}{p}}\left(\int_{t_0}^{t}\frac{1}{(r(\tau))^{\frac{1}{p+q-1}}}dg(\tau)\right)^{p+q-1}dg(t)\right)^{\frac{p}{p+q}}$$
$$\times\left(\int_{t_0}^{a}(z(t))^{\frac{p}{q}}z_g'(t)dg(t)\right)^{\frac{q}{p+q}}+2^{p-1}\sup_{t\in J}\left((\Delta g(t))^p\frac{s(t)}{r(t)}\right)z(a).$$

Note that

$$\begin{aligned}(z^{\frac{p+q}{q}})_g'(t)&=\frac{p+q}{q}z_g'(t)\int_0^1(z(ht+(1-h)t))^{\frac{p}{q}}dh\\&\ge\frac{p+q}{q}z_g'(t)\int_0^1(z(ht+(1-h)t))^{\frac{p}{q}}dh\\&=\frac{p+q}{q}(z(t))^{\frac{p}{q}}z_g'(t),\quad t\in J,\end{aligned}$$

whereupon

$$(z(t))^{\frac{p}{q}} z_g'(t) \le \frac{q}{p+q} (z_g')^{\frac{p+q}{q}}(t), \quad t \in J.$$

Therefore,

$$\begin{aligned}
&\int_{t_0}^{a} s(t)|y(t)+y(t+)|^p |y_g'(t)|^q dg(t) \\
&\le 2^{2p-1}\left(\int_{t_0}^{a} (s(t))^{\frac{p+q}{p}} \left(\frac{1}{r(t)}\right)^{\frac{q}{p}} \left(\int_{t_0}^{t} \frac{1}{(r(\tau))^{\frac{1}{p+q-1}}} dg(\tau)\right)^{p+q-1} dg(t)\right)^{\frac{p}{p+q}} \\
&\quad \times \left(\frac{q}{p+q}\right)^{\frac{q}{p+q}} \left(\int_{t_0}^{a} (z_g')^{\frac{p+q}{q}}(t) dg(t)\right)^{\frac{q}{p+q}} + 2^{p-1} \sup_{t\in J} \left((\Delta g(t))^p \frac{s(t)}{r(t)}\right) z(a) \\
&= 2^{2p-1}\left(\int_{t_0}^{a} (s(t))^{\frac{p+q}{p}} \left(\frac{1}{r(t)}\right)^{\frac{q}{p}} \left(\int_{t_0}^{t} \frac{1}{(r(\tau))^{\frac{1}{p+q-1}}} dg(\tau)\right)^{p+q-1} dg(t)\right)^{\frac{p}{p+q}} \\
&\quad \times \left(\frac{q}{p+q}\right)^{\frac{q}{p+q}} \left((z(t))^{\frac{p+q}{q}} \Big|_{t=t_0}^{t=a} \right)^{\frac{q}{p+q}} + 2^{p-1} \sup_{t\in J} \left((\Delta g(t))^p \frac{s(t)}{r(t)}\right) z(a) \\
&= 2^{2p-1}\left(\int_{t_0}^{a} (s(t))^{\frac{p+q}{p}} \left(\frac{1}{r(t)}\right)^{\frac{q}{p}} \left(\int_{t_0}^{t} \frac{1}{(r(\tau))^{\frac{1}{p+q-1}}} dg(t)\right)^{p+q-1} dg(\tau)\right)^{\frac{p}{p+q}} \\
&\quad \times \left(\frac{q}{p+q}\right)^{\frac{q}{p+q}} z(a) + 2^{p-1} \sup_{t\in J} \left((\Delta g(t))^p \frac{s(t)}{r(t)}\right) z(a) \\
&= 2^{2p-1}\left(\int_{t_0}^{a} (s(t))^{\frac{p+q}{p}} \left(\frac{1}{r(t)}\right)^{\frac{q}{p}} \left(\int_{t_0}^{t} \frac{1}{(r(\tau))^{\frac{1}{p+q-1}}} dg(\tau)\right)^{p+q-1} dg(t)\right)^{\frac{p}{p+q}} \\
&\quad \times \left(\frac{q}{p+q}\right)^{\frac{q}{p+q}} \left(\int_{t_0}^{a} r(t)|y_g'(t)|^{p+q} dg(t)\right) \\
&\quad + 2^{p-1} \sup_{t\in J} \left((\Delta g(t))^p \frac{s(t)}{r(t)}\right) \left(\int_{t_0}^{a} r(t)|y_g'(t)|^{p+q} dg(t)\right) \\
&= \left(2^{2p-1}\left(\int_{t_0}^{a} (s(t))^{\frac{p+q}{p}} \left(\frac{1}{r(t)}\right)^{\frac{q}{p}} \left(\int_{t_0}^{t} \frac{1}{(r(\tau))^{\frac{1}{p+q-1}}} dg(\tau)\right)^{p+q-1} dg(t)\right)^{\frac{p}{p+q}}\right.
\end{aligned}$$

$$\times \left(\frac{q}{p+q}\right)^{\frac{q}{p+q}} + 2^{p-1}\sup_{t\in J}\left((\Delta g(t))^p \frac{s(t)}{r(t)}\right)\Bigg) \left(\int_{t_0}^a r(t)|y_g'(t)|^{p+q} dg(t)\right)$$

$$= K_1(t_0, a, p, q) \int_{t_0}^a r(t)|y_g'(t)|^{p+q} dg(t).$$

This completes the proof. □

Theorem 5.26. *Let p and q be positive real numbers such that $p \geq 1$ and $r, s \in \mathscr{C}(J)$ be nonnegative functions such that*

$$\int_{t_0}^a \frac{1}{(r(t))^{\frac{1}{p+q-1}}} dg(t) < \infty.$$

Also, let $y \in \mathscr{C}_g^1(J)$ be such that $y(a) = 0$ and y_g' does not change its sign on J. Then, we have

$$\int_{t_0}^a s(t)|y(t)+y(t+)|^p |y_g'(t)|^q dg(t) \leq K_2(t_0, a, p, q) \int_{t_0}^a r(t)|y_g'(t)|^{p+q} dg(t),$$

where

$$K_2(t_0, a, p, q) = 2^{2p-1}\left(\frac{q}{p+q}\right)^{\frac{q}{p+q}}$$

$$\times \left(\int_{t_0}^a \frac{(s(t))^{\frac{p+q}{q}}}{(r(t))^{\frac{q}{p}}} \left(\int_t^a \frac{1}{(r(\tau))^{\frac{1}{p+q-1}}} dg(\tau)\right)^{p+q-1} dg(t)\right)^{\frac{p}{p+q}}$$

$$+ 2^{p-1}\sup_{t\in J}\left((\Delta g(t))^p \frac{s(t)}{r(t)}\right).$$

Proof. Since y_g' does not change its sign on J and $y(a) = 0$, we have

$$|y(t)| = \int_t^a |y_g'(\tau)| dg(\tau), \quad t \in J.$$

Then, applying the Stieltjes–Hölder inequality for $\frac{p+q}{p+q-1}$ and $p+q$, we find

$$|y(t)| = \int_t^a \frac{1}{(r(\tau))^{\frac{1}{p+q}}} (r(\tau))^{\frac{1}{p+q}} |y_g'(\tau)| dg(\tau)$$

$$\leq \left(\int_t^a \frac{1}{(r(\tau))^{\frac{1}{p+q-1}}} d_g\tau\right)^{\frac{p+q-1}{p+q}} \left(\int_t^a r(\tau)|y_g'(\tau)|^{p+q} dg(\tau)\right)^{\frac{1}{p+q}},$$

$t \in J$, and

$$|y(t)|^p \leq \left(\int_t^a \frac{1}{(r(\tau))^{\frac{1}{p+q-1}}} d_g\tau\right)^{p\frac{p+q-1}{p+q}} \left(\int_t^a r(\tau)|y_g'(\tau)|^{p+q} dg(\tau)\right)^{\frac{p}{p+q}},$$

$t \in J$. Observe that

$$\begin{aligned} y(t) + y(t+) &= y(t) + y(t) + (\Delta g(t))y_g'(t) \\ &= 2y(t) + (\Delta g(t))y_g'(t), \quad t \in J. \end{aligned}$$

Now, using Lemma 4.1, we find

$$\begin{aligned} |y(t) + y(t+)|^p &= |2y(t) + (\Delta g(t))y_g'(t)|^p \\ &\leq 2^{2p-1}|y(t)|^p + 2^{p-1}(\Delta g(t))^p |y_g'(t)|^p, \quad t \in J. \end{aligned}$$

Set

$$z(t) = \int_t^a r(\tau)|y_g'(\tau)|^{p+q} dg(\tau), \quad t \in J.$$

Then,

$$z(a) = 0 \quad \text{and} \quad z_g'(t) = -r(t)|y_g'(t)|^{p+q} \leq 0, \quad t \in J.$$

So, z is a nonincreasing function on J and

$$|y_g'(t)|^{p+q} = -\frac{z_g'(t)}{r(t)}, \quad t \in J,$$

and

$$|y_g'(t)|^q = \left(-\frac{z_g'(t)}{r(t)}\right)^{\frac{q}{p+q}}, \quad t \in J.$$

Therefore,

$$\begin{aligned} & s(t)|y(t) + y(t+)|^p |y_g'(t)|^q \\ & \leq 2^{2p-1}s(t)|y(t)|^p|y_g'(t)|^p + 2^{p-1}(\Delta g(t))^p s(t)|y_g'(t)|^{p+q} \\ & \leq 2^{2p-1}s(t)\left(\frac{1}{r(t)}\right)^{\frac{q}{p+q}} \left(\int_t^a \frac{1}{(r(\tau))^{\frac{1}{p+q-1}}} dg(\tau)\right)^{p\frac{p+q-1}{p+q}} \\ & \quad \times (z(t))^{\frac{p}{p+q}} (z_g'(t))^{\frac{q}{p+q}} \\ & \quad - 2^{p-1}(\Delta g(t))^p s(t)\frac{z_g'(t)}{r(t)}, \quad t \in J. \end{aligned}$$

Integrating the last inequality in the sense of Stieltjes, from t_0 to a, we find

$$\int_{t_0}^{a} s(t)|y(t)+y(t+)|^p|y_g'(t)|^q dg(t)$$

$$\le 2^{2p-1}\int_{t_0}^{a} s(t)\left(\frac{1}{r(t)}\right)^{\frac{q}{p+q}}\left(\int_t^a \frac{1}{(r(\tau))^{\frac{1}{p+q-1}}}dg(\tau)\right)^{p\frac{p+q-1}{p+q}}$$

$$\times (z(t))^{\frac{p}{p+q}}(z_g'(t))^{\frac{q}{p+q}}dg(t) - 2^{p-1}\int_{t_0}^{a}(\Delta g(t))^p s(t)\frac{z_g'(t)}{r(t)}dg(t)$$

$$\le 2^{2p-1}\int_{t_0}^{a} s(t)\left(\frac{1}{r(t)}\right)^{\frac{q}{p+q}}\left(\int_t^a \frac{1}{(r(\tau))^{\frac{1}{p+q-1}}}dg(\tau)\right)^{p\frac{p+q-1}{p+q}}$$

$$\times (z(t))^{\frac{p}{p+q}}(z_g'(t))^{\frac{q}{p+q}}dg(t) - 2^{p-1}\sup_{t\in J}\left((\Delta g(t))^p\frac{s(t)}{r(t)}\right)\int_{t_0}^{a} z_g'(t)dg(t)$$

$$= 2^{2p-1}\int_{t_0}^{a} s(t)\left(\frac{1}{r(t)}\right)^{\frac{q}{p+q}}\left(\int_t^a \frac{1}{(r(\tau))^{\frac{1}{p+q-1}}}dg(\tau)\right)^{p\frac{p+q-1}{p+q}}$$

$$\times (z(t))^{\frac{p}{p+q}}(z_g'(t))^{\frac{q}{p+q}}dg(t) - 2^{p-1}\sup_{t\in J}\left((\Delta g(t))^p\frac{s(t)}{r(t)}\right)z(t)\Big|_{t=t_0}^{t=a}$$

$$= 2^{2p-1}\int_{t_0}^{a} s(t)\left(\frac{1}{r(t)}\right)^{\frac{q}{p+q}}\left(\int_t^a \frac{1}{(r(\tau))^{\frac{1}{p+q-1}}}dg(\tau)\right)^{p\frac{p+q-1}{p+q}}$$

$$\times (z(t))^{\frac{p}{p+q}}(z_g'(t))^{\frac{q}{p+q}}dg(t) + 2^{p-1}\sup_{t\in J}\left((\Delta g(t))^p\frac{s(t)}{r(t)}\right)z(t_0).$$

Now, applying the Stieltjes–Hölder inequality for $\frac{p+q}{p}$ and $\frac{p+q}{q}$, we arrive at

$$\int_{t_0}^{a} s(t)|y(t)+y(t+)|^p|y_g'(t)|^q dg(t)$$

$$\le 2^{2p-1}\left(\int_{t_0}^{a}(s(t))^{\frac{p+q}{p}}\left(\frac{1}{r(t)}\right)^{\frac{q}{p}}\left(\int_t^a \frac{1}{(r(\tau))^{\frac{1}{p+q-1}}}dg(\tau)\right)^{p+q-1}dg(t)\right)^{\frac{p}{p+q}}$$

$$\times\left(-\int_{t_0}^{a}(z(t))^{\frac{p}{q}}z_g'(t)dg(t)\right)^{\frac{q}{p+q}} + 2^{p-1}\sup_{t\in J}\left((\Delta g(t))^p\frac{s(t)}{r(t)}\right)z(t_0).$$

Note by the chain rule, Theorem 1.42, that

$$\begin{aligned}-(z^{\frac{p+q}{q}})'_g(t) &= -\frac{p+q}{q} z'_g(t)\int_0^1 (z(ht+(1-h)t))^{\frac{p}{q}}dh\\ &\geq -\frac{p+q}{q} z'_g(t)\int_0^1 (z(ht+(1-h)t))^{\frac{p}{q}}dh\\ &= -\frac{p+q}{q}(z(t))^{\frac{p}{q}} z'_g(t), \quad t\in J,\end{aligned}$$

whereupon

$$-(z(t))^{\frac{p}{q}} z'_g(t) \leq -\frac{q}{p+q}(z'_g)^{\frac{p+q}{q}}(t), \quad t\in J.$$

Therefore,

$$\begin{aligned}&\int_{t_0}^a s(t)|y(t)+y(t+)|^p|y'_g(t)|^q dg(t)\\ &\leq 2^{2p-1}\left(\int_{t_0}^a (s(t))^{\frac{p+q}{p}}\left(\frac{1}{r(t)}\right)^{\frac{q}{p}}\left(\int_t^a \frac{1}{(r(\tau))^{\frac{1}{p+q-1}}}dg(\tau)\right)^{p+q-1} dg(t)\right)^{\frac{p}{p+q}}\\ &\quad\times\left(\frac{q}{p+q}\right)^{\frac{q}{p+q}}\left(-\int_{t_0}^a (z'_g)^{\frac{p+q}{q}}(t)dg(t)\right)^{\frac{q}{p+q}}\\ &\quad+2^{p-1}\sup_{t\in J}\left((\Delta g(t))^p\frac{s(t)}{r(t)}\right)z(t_0)\\ &= 2^{2p-1}\left(\int_{t_0}^a (s(t))^{\frac{p+q}{p}}\left(\frac{1}{r(t)}\right)^{\frac{q}{p}}\left(\int_t^a \frac{1}{(r(\tau))^{\frac{1}{p+q-1}}}dg(\tau)\right)^{p+q-1} dg(t)\right)^{\frac{p}{p+q}}\\ &\quad\times\left(\frac{q}{p+q}\right)^{\frac{q}{p+q}}\left(-(z(t))^{\frac{p+q}{q}}\Big|_{t=t_0}^{t=a}\right)^{\frac{q}{p+q}}+2^{p-1}\sup_{t\in J}\left((\Delta g(t))^p\frac{s(t)}{r(t)}\right)z(t_0)\\ &= 2^{2p-1}\left(\int_{t_0}^a (s(t))^{\frac{p+q}{p}}\left(\frac{1}{r(t)}\right)^{\frac{q}{p}}\left(\int_t^a \frac{1}{(r(\tau))^{\frac{1}{p+q-1}}}dg(\tau)\right)^{p+q-1} dg(t)\right)^{\frac{p}{p+q}}\\ &\quad\times\left(\frac{q}{p+q}\right)^{\frac{q}{p+q}} z(t_0)+2^{p-1}\sup_{t\in J}\left((\Delta g(t))^p\frac{s(t)}{r(t)}\right)z(t_0)\end{aligned}$$

$$
\begin{aligned}
&= 2^{2p-1}\left(\int_{t_0}^{a}(s(t))^{\frac{p+q}{p}}\left(\frac{1}{r(t)}\right)^{\frac{q}{p}}\left(\int_t^a \frac{1}{(r(\tau))^{\frac{1}{p+q-1}}}dg(\tau)\right)^{p+q-1}dg(t)\right)^{\frac{p}{p+q}}\\
&\quad\times\left(\frac{q}{p+q}\right)^{\frac{q}{p+q}}\left(\int_{t_0}^{a} r(t)|y_g'(t)|^{p+q}dg(t)\right)\\
&\quad+2^{p-1}\sup_{t\in J}\left((\Delta g(t))^p\frac{s(t)}{r(t)}\right)\left(\int_{t_0}^{a} r(t)|y_g'(t)|^{p+q}dg(t)\right)\\
&=\left(2^{2p-1}\left(\int_{t_0}^{a}(s(t))^{\frac{p+q}{p}}\left(\frac{1}{r(t)}\right)^{\frac{q}{p}}\left(\int_t^a \frac{1}{(r(\tau))^{\frac{1}{p+q-1}}}dg(\tau)\right)^{p+q-1}dg(t)\right)^{\frac{p}{p+q}}\right.\\
&\quad\left.\times\left(\frac{q}{p+q}\right)^{\frac{q}{p+q}}+2^{p-1}\sup_{t\in J}\left((\Delta g(t))^p\frac{s(t)}{r(t)}\right)\right)\left(\int_{t_0}^{a} r(t)|y_g'(t)|^{p+q}dg(t)\right)\\
&=K_2(t_0,a,p,q)\int_{t_0}^{a} r(t)|y_g'(t)|^{p+q}dg(t).
\end{aligned}
$$

This completes the proof. □

Theorem 5.27. *Let p and q be positive real numbers such that $p \geq 1$ and $r, s \in \mathscr{C}(J)$ be nonnegative functions such that*

$$\int_{t_0}^{a}\frac{1}{(r(t))^{\frac{1}{p+q-1}}}dg(t) < \infty.$$

Also, let $y \in \mathscr{C}_g^1(J)$ be such that $y(t_0) = y(a) = 0$ and y_g' does not change its sign on J. Then, we have

$$\int_{t_0}^{a} s(t)|y(t)+y(t+)|^p|y_g'(t)|^q dg(t) \leq K(t_0,a,p,q)\int_{t_0}^{a} r(t)|y_g'(t)|^{p+q}dg(t),$$

where

$$K(t_0,a,p,q)=\max_{c\in J}\{K_1(t_0,c,p,q),K_2(c,a,p,q)\}.$$

Proof. Let $c \in J$ be arbitrarily chosen. Then, applying Theorem 5.25, we find

$$
\begin{aligned}
\int_{t_0}^{c} s(t)|y(t)+y(t+)|^p|y_g'(t)|^q dg(t) &\leq K_1(t_0,c,p,q)\int_{t_0}^{c} r(t)|y_g'(t)|^{p+q}dg(t)\\
&\leq K(t_0,a,p,q)\int_{t_0}^{c} r(t)|y_g'(t)|^{p+q}dg(t).
\end{aligned}
$$

Now, by Theorem 5.26, we find

$$\int_c^a s(t)|y(t)+y(t+)|^p|y_g'(t)|^q dg(t) \le K_2(c,a,p,q)\int_c^a r(t)|y_g'(t)|^{p+q}dg(t)$$
$$\le K(t_0,a,p,q)\int_c^a r(t)|y_g'(t)|^{p+q}dg(t).$$

Therefore,

$$\begin{aligned}
&\int_{t_0}^a s(t)|y(t)+y(t+)|^p|y_g'(t)|^q dg(t)\\
&= \int_{t_0}^c s(t)|y(t)+y(t+)|^p|y_g'(t)|^q dg(t)\\
&\quad + \int_c^a s(t)|y(t)+y(t+)|^p|y_g'(t)|^q dg(t)\\
&\le K(t_0,a,p,q)\int_{t_0}^c r(t)|y_g'(t)|^{p+q}dg(t)\\
&\quad + K(t_0,a,p,q)\int_c^a r(t)|y_g'(t)|^{p+q}dg(t)\\
&= K(t_0,a,p,q)\int_{t_0}^a r(t)|y_g'(t)|^{p+q}dg(t).
\end{aligned}$$

This completes the proof. □

Example 5.3. Let p and q be positive real numbers such that $p \ge 1$. Take

$$r(t) = 1 \quad \text{and} \quad s(t) = 1, \quad t \in J.$$

By the chain rule, Theorem 1.42, we find

$$\begin{aligned}
((t-t_0)^{p+q})_g' &= (p+q)\int_0^1 (h(\Delta g(t_0)) + (1-h)(t-t_0))^{p+q-1}dh\\
&\ge (p+q)\int_0^1 (h(t-t_0) + (1-h)(t-t_0))^{p+q-1}dh\\
&= (p+q)(t-t_0)^{p+q-1}, \quad t \in J,
\end{aligned}$$

whereupon

$$(t-t_0)^{p+q-1} \le \frac{1}{p+q}((t-t_0)^{p+q})_g', \quad t \in J.$$

Hence,

$$\begin{aligned}
&K_1(t_0,a,p,q)\\
&= 2^{2p-1}\left(\frac{q}{p+q}\right)^{\frac{q}{p+q}}\left(\int_{t_0}^{a}(t-t_0)^{p+q-1}dg(t)\right)^{\frac{p}{p+q}} + 2^{p-1}\sup_{t\in J}\left((\Delta g(t))^p\right)\\
&\leq 2^{2p-1}\left(\frac{q}{p+q}\right)^{\frac{q}{p+q}}\left(\frac{1}{p+q}\int_{t_0}^{a}((t-t_0)^{p+q})'_g dg(t)\right)^{\frac{p}{p+q}}\\
&\quad + 2^{p-1}\sup_{t\in J}\left((\Delta g(t))^p\right)\\
&= 2^{2p-1}\left(\frac{q}{p+q}\right)^{\frac{q}{p+q}}\left(\frac{1}{p+q}(t-t_0)^{p+q}\Big|_{t=t_0}^{t=a}\right)^{\frac{p}{p+q}} + 2^{p-1}\sup_{t\in J}\left((\Delta g(t))^p\right)\\
&= 2^{2p-1}\left(\frac{q}{p+q}\right)^{\frac{q}{p+q}}\left(\frac{1}{p+q}\right)^{\frac{p}{p+q}}(a-t_0)^{p+q} + 2^{p-1}\sup_{t\in J}\left((\Delta g(t))^p\right)\\
&= 2^{2p-1}\frac{q^{\frac{q}{p+q}}}{p+q}(a-t_0)^p + 2^{p-1}\sup_{t\in J}\left((\Delta g(t))^p\right).
\end{aligned}$$

Let $y \in \mathscr{C}_g^1(J)$ be such that $y(t_0) = 0$ and y'_g does not change its sign on J. Then, applying Theorem 5.25, we find the inequality

$$\begin{aligned}
&\int_{t_0}^{a} |y(t)+y(t)|^p |y'_g(t)|^q dg(t)\\
&\leq \left(2^{2p-1}\frac{q^{\frac{q}{p+q}}}{p+q}(a-t_0)^p + 2^{p-1}\sup_{t\in J}\left((\Delta g(t))^p\right)\right) \times \int_{t_0}^{a} |y'_g(t)|^{p+q} dg(t).
\end{aligned}$$

Problem 5.4. Let p and q be positive real numbers such that $p \geq 1$ and $y \in \mathscr{C}_g^1(J)$ be such that $y(t_0) = 0$ and y'_g does not change its sign on J. Then, prove the inequality

$$\begin{aligned}
&\int_{t_0}^{a} (t-t_0)|y(t)+y(t)|^p |y'_g(t)|^q dg(t)\\
&\leq \left(2^{2p-1}\frac{q}{(p+q)(1+q)^{\frac{p}{p+q}}}(a-t_0)^{\frac{p(1+q)}{q}}\right.\\
&\quad \left. +2^{p-1}\sup_{t\in J}\left((t-t_0)(\Delta g(t))^p\right)\right)\int_{t_0}^{a} |y'_g(t)|^{p+q} dg(t).
\end{aligned}$$

Theorem 5.28. *Let p and q be positive real numbers such that $p \le 1$, $p+q > 1$, and $r, s \in \mathscr{C}(J)$ be nonnegative functions such that*

$$\int_{t_0}^{a} \frac{1}{(r(t))^{\frac{1}{p+q-1}}} dg(t) < \infty.$$

Also, let $y \in \mathscr{C}_g^1(J)$ be such that $y(t_0) = 0$ and y'_g does not change its sign on J. Then, we have the inequality

$$\int_{t_0}^{a} s(t)|y(t) + y(t+)|^p |y'_g(t)|^q dg(t) \le L_1(t_0, a, p, q) \int_{t_0}^{a} r(t)|y'_g(t)|^{p+q} dg(t),$$

where

$$L_1(t_0, a, p, q) = 2^p \left(\frac{q}{p+q}\right)^{\frac{q}{p+q}} \times \left(\int_{t_0}^{a} \frac{(s(t))^{\frac{p+q}{q}}}{(r(t))^{\frac{q}{p}}} \right.$$
$$\left. \times \left(\int_{t_0}^{t} \frac{1}{(r(\tau))^{\frac{1}{p+q-1}}} dg(\tau)\right)^{p+q-1} dg(t)\right)^{\frac{p}{p+q}}$$
$$+ \sup_{t \in J} \left((\Delta g(t))^p \frac{s(t)}{r(t)}\right).$$

Proof. Since y'_g does not change its sign on J and $y(t_0) = 0$, we have

$$|y(t)| = \int_{t_0}^{t} |y'_g(\tau)| dg(\tau), \quad t \in J.$$

Then, applying the Stieltjes–Hölder inequality for $\frac{p+q}{p+q-1}$ and $p+q$, we find

$$|y(t)| = \int_{t_0}^{t} \frac{1}{(r(\tau))^{\frac{1}{p+q}}} (r(\tau))^{\frac{1}{p+q}} |y'_g(\tau)| dg(\tau)$$
$$\le \left(\int_{t_0}^{t} \frac{1}{(r(\tau))^{\frac{1}{p+q-1}}} d_g\tau\right)^{\frac{p+q-1}{p+q}}$$
$$\times \left(\int_{t_0}^{t} r(\tau)|y'_g(\tau)|^{p+q} dg(\tau)\right)^{\frac{1}{p+q}}, \quad t \in J,$$

$$|y(t)|^p \le \left(\int_{t_0}^{t} \frac{1}{(r(\tau))^{\frac{1}{p+q-1}}} d_g\tau\right)^{p\frac{p+q-1}{p+q}}$$
$$\times \left(\int_{t_0}^{t} r(\tau)|y'_g(\tau)|^{p+q} dg(\tau)\right)^{\frac{p}{p+q}}, \quad t \in J.$$

Note that

$$\begin{aligned} y(t) + y(t+) &= y(t) + y(t) + (\Delta g(t)) y_g'(t) \\ &= 2y(t) + (\Delta g(t)) y_g'(t), \quad t \in J. \end{aligned}$$

Now, applying Lemma 4.1, we find

$$\begin{aligned} |y(t) + y(t+)|^p &= |2y(t) + (\Delta g(t)) y_g'(t)|^p \\ &\leq 2^p |y(t)|^p + (\Delta g(t))^p |y_g'(t)|^p, \quad t \in J. \end{aligned}$$

Set

$$z(t) = \int_{t_0}^{t} r(\tau) |y_g'(\tau)|^{p+q} dg(\tau), \quad t \in J.$$

Then,

$$z(t_0) = 0 \quad \text{and} \quad z_g'(t) = r(t) |y_g'(t)|^{p+q} \geq 0, \quad t \in J.$$

Thus, z is a nondecreasing function on J and

$$|y_g'(t)|^{p+q} = \frac{z_g'(t)}{r(t)}, \quad t \in J,$$

and

$$|y_g'(t)|^{q} = \left(\frac{z_g'(t)}{r(t)} \right)^{\frac{q}{p+q}}, \quad t \in J.$$

Therefore,

$$\begin{aligned} & s(t) |y(t) + y(t+)|^p |y_g'(t)|^q \\ & \quad \leq 2^p s(t) |y(t)|^p |y_g'(t)|^p + (\Delta g(t))^p s(t) |y_g'(t)|^{p+q} \\ & \quad \leq 2^p s(t) \left(\frac{1}{r(t)} \right)^{\frac{q}{p+q}} \left(\int_{t_0}^{t} \frac{1}{(r(\tau))^{\frac{1}{p+q-1}}} dg(\tau) \right)^{p \frac{p+q-1}{p+q}} (z(t))^{\frac{p}{p+q}} (z_g'(t))^{\frac{q}{p+q}} \\ & \qquad + (\Delta g(t))^p s(t) \frac{z_g'(t)}{r(t)}, \quad t \in J. \end{aligned}$$

Integrating the last inequality in the sense of Stieltjes, from t_0 to a, we find

$$\int_{t_0}^{a} s(t)|y(t)+y(t+)|^p|y_g'(t)|^q dg(t)$$

$$\leq 2^p \int_{t_0}^{a} s(t)\left(\frac{1}{r(t)}\right)^{\frac{q}{p+q}}\left(\int_{t_0}^{t}\frac{1}{(r(\tau))^{\frac{1}{p+q-1}}}dg(\tau)\right)^{p\frac{p+q-1}{p+q}}$$

$$\times (z(t))^{\frac{p}{p+q}}(z_g'(t))^{\frac{q}{p+q}}dg(t)+\int_{t_0}^{a}(\Delta g(t))^p s(t)\frac{z_g'(t)}{r(t)}dg(t)$$

$$\leq 2^p \int_{t_0}^{a} s(t)\left(\frac{1}{r(t)}\right)^{\frac{q}{p+q}}\left(\int_{t_0}^{t}\frac{1}{(r(\tau))^{\frac{1}{p+q-1}}}dg(\tau)\right)^{p\frac{p+q-1}{p+q}}$$

$$\times (z(t))^{\frac{p}{p+q}}(z_g'(t))^{\frac{q}{p+q}}dg(t)+\sup_{t\in J}\left((\Delta g(t))^p\frac{s(t)}{r(t)}\right)\int_{t_0}^{a} z_g'(t)dg(t)$$

$$= 2^p \int_{t_0}^{a} s(t)\left(\frac{1}{r(t)}\right)^{\frac{q}{p+q}}\left(\int_{t_0}^{t}\frac{1}{(r(\tau))^{\frac{1}{p+q-1}}}dg(\tau)\right)^{p\frac{p+q-1}{p+q}}$$

$$\times (z(t))^{\frac{p}{p+q}}(z_g'(t))^{\frac{q}{p+q}}dg(t)+\sup_{t\in J}\left((\Delta g(t))^p\frac{s(t)}{r(t)}\right) z(t)\Big|_{t=t_0}^{t=a}$$

$$= 2^p \int_{t_0}^{a} s(t)\left(\frac{1}{r(t)}\right)^{\frac{q}{p+q}}\left(\int_{t_0}^{t}\frac{1}{(r(\tau))^{\frac{1}{p+q-1}}}dg(\tau)\right)^{p\frac{p+q-1}{p+q}}$$

$$\times (z(t))^{\frac{p}{p+q}}(z_g'(t))^{\frac{q}{p+q}}dg(t)+\sup_{t\in J}\left((\Delta g(t))^p\frac{s(t)}{r(t)}\right) z(a).$$

Now, applying the Stieltjes–Hölder inequality for $\frac{p+q}{p}$ and $\frac{p+q}{q}$, we arrive at

$$\int_{t_0}^{a} s(t)|y(t)+y(t+)|^p|y_g'(t)|^q dg(t)$$

$$\leq 2^p\left(\int_{t_0}^{a}(s(t))^{\frac{p+q}{p}}\left(\frac{1}{r(t)}\right)^{\frac{q}{p}}\left(\int_{t_0}^{t}\frac{1}{(r(\tau))^{\frac{1}{p+q-1}}}dg(\tau)\right)^{p+q-1}dg(\tau)\right)^{\frac{p}{p+q}}$$

$$\times\left(\int_{t_0}^{a}(z(t))^{\frac{p}{q}}z_g'(t)dg(t)\right)^{\frac{q}{p+q}}+\sup_{t\in J}\left((\Delta g(t))^p\frac{s(t)}{r(t)}\right) z(a).$$

Note by the chain rule, Theorem 1.42, that

$$\begin{aligned}
(z^{\frac{p+q}{q}})'_g(t) &= \frac{p+q}{q} z'_g(t) \int_0^1 (z(ht+(1-h)t))^{\frac{p}{q}} dh \\
&\geq \frac{p+q}{q} z'_g(t) \int_0^1 (z(ht+(1-h)t))^{\frac{p}{q}} dh \\
&= \frac{p+q}{q} (z(t))^{\frac{p}{q}} z'_g(t), \quad t \in J,
\end{aligned}$$

whereupon

$$(z(t))^{\frac{p}{q}} z'_g(t) \leq \frac{q}{p+q} (z^{\frac{p+q}{q}})'_g(t), \quad t \in J.$$

Therefore,

$$\begin{aligned}
&\int_{t_0}^a s(t)|y(t)+y(t+)|^p |y'_g(t)|^q dg(t) \\
&\quad \leq 2^p \left(\int_{t_0}^a (s(t))^{\frac{p+q}{p}} \left(\frac{1}{r(t)}\right)^{\frac{q}{p}} \left(\int_{t_0}^t \frac{1}{(r(\tau))^{\frac{1}{p+q-1}}} dg(\tau) \right)^{p+q-1} dg(t) \right)^{\frac{p}{p+q}} \\
&\qquad \times \left(\frac{q}{p+q}\right)^{\frac{q}{p+q}} \left(\int_{t_0}^a (z^{\frac{p+q}{q}})'_g(t) dg(t) \right)^{\frac{q}{p+q}} + \sup_{t\in J} \left((\Delta g(t))^p \frac{s(t)}{r(t)} \right) z(a) \\
&\quad = 2^p \left(\int_{t_0}^a (s(t))^{\frac{p+q}{p}} \left(\frac{1}{r(t)}\right)^{\frac{q}{p}} \left(\int_{t_0}^t \frac{1}{(r(\tau))^{\frac{1}{p+q-1}}} dg(\tau) \right)^{p+q-1} dg(t) \right)^{\frac{p}{p+q}} \\
&\qquad \times \left(\frac{q}{p+q}\right)^{\frac{q}{p+q}} \left((z(t))^{\frac{p+q}{q}} \Big|_{t=t_0}^{t=a} \right)^{\frac{q}{p+q}} + \sup_{t\in J} \left((\Delta g(t))^p \frac{s(t)}{r(t)} \right) z(a) \\
&\quad = 2^p \left(\int_{t_0}^a (s(t))^{\frac{p+q}{p}} \left(\frac{1}{r(t)}\right)^{\frac{q}{p}} \left(\int_{t_0}^t \frac{1}{(r(\tau))^{\frac{1}{p+q-1}}} dg(\tau) \right)^{p+q-1} dg(t) \right)^{\frac{p}{p+q}} \\
&\qquad \times \left(\frac{q}{p+q}\right)^{\frac{q}{p+q}} z(a) + \sup_{t\in J} \left((\Delta g(t))^p \frac{s(t)}{r(t)} \right) z(a) \\
&\quad = 2^p \left(\int_{t_0}^a (s(t))^{\frac{p+q}{p}} \left(\frac{1}{r(t)}\right)^{\frac{q}{p}} \left(\int_{t_0}^t \frac{1}{(r(\tau))^{\frac{1}{p+q-1}}} dg(\tau) \right)^{p+q-1} dg(t) \right)^{\frac{p}{p+q}}
\end{aligned}$$

$$\times \left(\frac{q}{p+q}\right)^{\frac{q}{p+q}} \left(\int_{t_0}^{a} r(t)|y_g'(t)|^{p+q} dg(t)\right)$$

$$+ \sup_{t\in J} \left((\Delta g(t))^p \frac{s(t)}{r(t)}\right) \left(\int_{t_0}^{a} r(t)|y_g'(t)|^{p+q} dg(t)\right)$$

$$= \left(2^p \left(\int_{t_0}^{a} (s(t))^{\frac{p+q}{p}} \left(\frac{1}{r(t)}\right)^{\frac{q}{p}} \left(\int_{t_0}^{t} \frac{1}{(r(\tau))^{\frac{1}{p+q-1}}} dg(\tau)\right)^{p+q-1} dg(t)\right)^{\frac{p}{p+q}}\right.$$

$$\left. \times \left(\frac{q}{p+q}\right)^{\frac{q}{p+q}} + \sup_{t\in J} \left((\Delta g(t))^p \frac{s(t)}{r(t)}\right)\right) \left(\int_{t_0}^{a} r(t)|y_g'(t)|^{p+q} dg(t)\right)$$

$$= K_1(t_0, a, p, q) \int_{t_0}^{a} r(t)|y_g'(t)|^{p+q} dg(t).$$

This completes the proof. □

Theorem 5.29. *Let p and q be positive real numbers such that $p \leq 1$, $p+q > 1$, and $r, s \in \mathscr{C}(J)$ be nonnegative functions such that*

$$\int_{t_0}^{a} \frac{1}{(r(t))^{\frac{1}{p+q-1}}} dg(t) < \infty.$$

Also, let $y \in \mathscr{C}_g^1(J)$ be such that $y(a) = 0$ and y_g' does not change its sign on J. Then, we have

$$\int_{t_0}^{a} s(t)|y(t) + y(t+)|^p |y_g'(t)|^q dg(t) \leq L_2(t_0, a, p, q) \int_{t_0}^{a} r(t)|y_g'(t)|^{p+q} dg(t),$$

where

$$L_2(t_0, a, p, q) = 2^p \left(\frac{q}{p+q}\right)^{\frac{q}{p+q}} \times \left(\int_{t_0}^{a} \frac{(s(t))^{\frac{p+q}{q}}}{(r(t))^{\frac{q}{p}}}\right.$$

$$\left. \times \left(\int_{t}^{a} \frac{1}{(r(\tau))^{\frac{1}{p+q-1}}} dg(\tau)\right)^{p+q-1} dg(t)\right)^{\frac{p}{p+q}}$$

$$+ \sup_{t\in J} \left((\Delta g(t))^p \frac{s(t)}{r(t)}\right).$$

Proof. Since y_g' does not change its sign on J and $y(a) = 0$, we have

$$|y(t)| = \int_{t}^{a} |y_g'(\tau)| dg(\tau), \quad t \in J.$$

Then, applying the Stieltjes–Hölder inequality for $\frac{p+q}{p+q-1}$ and $p+q$, we find

$$
\begin{aligned}
|y(t)| &= \int_t^a \frac{1}{(r(\tau))^{\frac{1}{p+q}}} (r(\tau))^{\frac{1}{p+q}} |y_g'(\tau)| dg(\tau) \\
&\leq \left(\int_t^a \frac{1}{(r(\tau))^{\frac{1}{p+q-1}}} d_g \tau \right)^{\frac{p+q-1}{p+q}} \\
&\quad \times \left(\int_t^a r(\tau) |y_g'(\tau)|^{p+q} dg(\tau) \right)^{\frac{1}{p+q}}, \quad t \in J, \\
|y(t)|^p &< \left(\int_t^a \frac{1}{(r(\tau))^{\frac{1}{p+q-1}}} d_g \tau \right)^{p \frac{p+q-1}{p+q}} \\
&\quad \times \left(\int_t^a r(\tau) |y_g'(\tau)|^{p+q} dg(\tau) \right)^{\frac{p}{p+q}}, \quad t \in J.
\end{aligned}
$$

Observe that

$$
\begin{aligned}
y(t) + y(t+) &= y(t) + y(t) + (\Delta g(t)) y_g'(t) \\
&= 2y(t) + (\Delta g(t)) y_g'(t), \quad t \in J.
\end{aligned}
$$

Now, using Lemma 4.1, we find

$$
\begin{aligned}
|y(t) + y(t+)|^p &= |2y(t) + (\Delta g(t)) y_g'(t)|^p \\
&\leq 2^p |y(t)|^p + (\Delta g(t))^p |y_g'(t)|^p, \quad t \in J.
\end{aligned}
$$

Set

$$
z(t) = \int_t^a r(\tau) |y_g'(\tau)|^{p+q} dg(\tau), \quad t \in J.
$$

Then,

$$
z(a) = 0 \quad \text{and} \quad z_g'(t) = -r(t) |y_g'(t)|^{p+q} \leq 0, \quad t \in J.
$$

So, z is a nonincreasing function on J and

$$
|y_g'(t)|^{p+q} = -\frac{z_g'(t)}{r(t)}, \quad t \in J.
$$

Then,

$$
|y_g'(t)|^q = \left(-\frac{z_g'(t)}{r(t)} \right)^{\frac{q}{p+q}}, \quad t \in J.
$$

Therefore,

$$\begin{aligned}
&s(t)|y(t)+y(t+)|^p|y_g'(t)|^q \\
&\quad\le 2^p s(t)|y(t)|^p|y_g'(t)|^p+(\Delta g(t))^p s(t)|y_g'(t)|^{p+q} \\
&\quad\le 2^p s(t)\left(\frac{1}{r(t)}\right)^{\frac{q}{p+q}}\left(\int_t^a \frac{1}{(r(\tau))^{\frac{1}{p+q-1}}}dg(\tau)\right)^{p\frac{p+q-1}{p+q}} \\
&\qquad\times (z(t))^{\frac{p}{p+q}}(z_g'(t))^{\frac{q}{p+q}}-(\Delta g(t))^p s(t)\frac{z_g'(t)}{r(t)}, \quad t\in J.
\end{aligned}$$

Integrating the last inequality in the sense of Stieltjes, from t_0 to a, we find

$$\begin{aligned}
&\int_{t_0}^a s(t)|y(t)+y(t+)|^p|y_g'(t)|^q dg(t) \\
&\quad\le 2^p\int_{t_0}^a s(t)\left(\frac{1}{r(t)}\right)^{\frac{q}{p+q}}\left(\int_t^a \frac{1}{(r(\tau))^{\frac{1}{p+q-1}}}dg(\tau)\right)^{p\frac{p+q-1}{p+q}} \\
&\qquad\times (z(t))^{\frac{p}{p+q}}(z_g'(t))^{\frac{q}{p+q}}dg(t)-\int_{t_0}^a(\Delta g(t))^p s(t)\frac{z_g'(t)}{r(t)}dg(t) \\
&\quad\le 2^p\int_{t_0}^a s(t)\left(\frac{1}{r(t)}\right)^{\frac{q}{p+q}}\left(\int_t^a \frac{1}{(r(\tau))^{\frac{1}{p+q-1}}}dg(\tau)\right)^{p\frac{p+q-1}{p+q}} \\
&\qquad\times (z(t))^{\frac{p}{p+q}}(z_g'(t))^{\frac{q}{p+q}}dg(t)-\sup_{t\in J}\left((\Delta g(t))^p\frac{s(t)}{r(t)}\right)\int_{t_0}^a z_g'(t)dg(t) \\
&\quad= 2^p\int_{t_0}^a s(t)\left(\frac{1}{r(t)}\right)^{\frac{q}{p+q}}\left(\int_t^a \frac{1}{(r(\tau))^{\frac{1}{p+q-1}}}dg(\tau)\right)^{p\frac{p+q-1}{p+q}} \\
&\qquad\times (z(t))^{\frac{p}{p+q}}(z_g'(t))^{\frac{q}{p+q}}dg(t)-\sup_{t\in J}\left((\Delta g(t))^p\frac{s(t)}{r(t)}\right)z(t)\Big|_{t=t_0}^{t=a} \\
&\quad= 2^p\int_{t_0}^a s(t)\left(\frac{1}{r(t)}\right)^{\frac{q}{p+q}}\left(\int_t^a \frac{1}{(r(\tau))^{\frac{1}{p+q-1}}}dg(\tau)\right)^{p\frac{p+q-1}{p+q}} \\
&\qquad\times (z(t))^{\frac{p}{p+q}}(z_g'(t))^{\frac{q}{p+q}}dg(t)+\sup_{t\in J}\left((\Delta g(t))^p\frac{s(t)}{r(t)}\right)z(t_0).
\end{aligned}$$

Now, applying the Stieltjes–Hölder inequality for $\frac{p+q}{p}$ and $\frac{p+q}{q}$, we arrive at

$$\int_{t_0}^{a} s(t)|y(t)+y(t+)|^p|y_g'(t)|^q dg(t)$$
$$\leq 2^p \left(\int_{t_0}^{a} (s(t))^{\frac{p+q}{p}} \left(\frac{1}{r(t)}\right)^{\frac{q}{p}} \left(\int_t^a \frac{1}{(r(\tau))^{\frac{1}{p+q-1}}} dg(\tau)\right)^{p+q-1} dg(t)\right)^{\frac{p}{p+q}}$$
$$\times \left(-\int_{t_0}^{a} (z(t))^{\frac{p}{q}} z_g'(t) dg(t)\right)^{\frac{q}{p+q}} + \sup_{t\in J}\left((\Delta g(t))^p \frac{s(t)}{r(t)}\right) z(t_0).$$

Note by the chain rule, Theorem 1.42, that

$$\begin{aligned} -(z^{\frac{p+q}{q}})_g'(t) &= -\frac{p+q}{q} z_g'(t) \int_0^1 (z(ht+(1-h)t))^{\frac{p}{q}} dh \\ &\geq -\frac{p+q}{q} z_g'(t) \int_0^1 (z(ht+(1-h)t))^{\frac{p}{q}} dh \\ &= -\frac{p+q}{q} (z(t))^{\frac{p}{q}} z_g'(t), \quad t\in J, \end{aligned}$$

whereupon

$$-(z(t))^{\frac{p}{q}} z_g'(t) \leq -\frac{q}{p+q} (z^{\frac{p+q}{q}})_g'(t), \quad t \in J.$$

Therefore,

$$\int_{t_0}^{a} s(t)|y(t)+y(t+)|^p|y_g'(t)|^q dg(t)$$
$$\leq 2^p \left(\int_{t_0}^{a} (s(t))^{\frac{p+q}{p}} \left(\frac{1}{r(t)}\right)^{\frac{q}{p}} \left(\int_t^a \frac{1}{(r(\tau))^{\frac{1}{p+q-1}}} dg(\tau)\right)^{p+q-1} dg(t)\right)^{\frac{p}{p+q}}$$
$$\times \left(\frac{q}{p+q}\right)^{\frac{q}{p+q}} \left(-\int_{t_0}^{a} (z^{\frac{p+q}{q}})_g'(t) dg(t)\right)^{\frac{q}{p+q}} + \sup_{t\in J}\left((\Delta g(t))^p \frac{s(t)}{r(t)}\right) z(t_0)$$
$$= 2^p \left(\int_{t_0}^{a} (s(t))^{\frac{p+q}{p}} \left(\frac{1}{r(t)}\right)^{\frac{q}{p}} \left(\int_t^a \frac{1}{(r(\tau))^{\frac{1}{p+q-1}}} dg(\tau)\right)^{p+q-1} dg(t)\right)^{\frac{p}{p+q}}$$
$$\times \left(\frac{q}{p+q}\right)^{\frac{q}{p+q}} \left(-(z(t))^{\frac{p+q}{q}}\Big|_{t=t_0}^{t=a}\right)^{\frac{q}{p+q}} + \sup_{t\in J}\left((\Delta g(t))^p \frac{s(t)}{r(t)}\right) z(t_0)$$

$$= 2^p \left(\int_{t_0}^{a} (s(t))^{\frac{p+q}{p}} \left(\frac{1}{r(t)} \right)^{\frac{q}{p}} \left(\int_t^a \frac{1}{(r(\tau))^{\frac{1}{p+q-1}}} dg(\tau) \right)^{p+q-1} dg(t) \right)^{\frac{p}{p+q}}$$

$$\times \left(\frac{q}{p+q} \right)^{\frac{q}{p+q}} z(t_0) + \sup_{t \in J} \left((\Delta g(t))^p \frac{s(t)}{r(t)} \right) z(t_0)$$

$$= 2^p \left(\int_{t_0}^{a} (s(t))^{\frac{p+q}{p}} \left(\frac{1}{r(t)} \right)^{\frac{q}{p}} \left(\int_t^a \frac{1}{(r(\tau))^{\frac{1}{p+q-1}}} dg(\tau) \right)^{p+q-1} dg(t) \right)^{\frac{p}{p+q}}$$

$$\times \left(\frac{q}{p+q} \right)^{\frac{q}{p+q}} \left(\int_{t_0}^{a} r(t) |y_g'(t)|^{p+q} dg(t) \right)$$

$$+ \sup_{t \in J} \left((\Delta g(t))^p \frac{s(t)}{r(t)} \right) \left(\int_{t_0}^{a} r(t) |y_g'(t)|^{p+q} dg(t) \right)$$

$$= \left(2^p \left(\int_{t_0}^{a} (s(t))^{\frac{p+q}{p}} \left(\frac{1}{r(t)} \right)^{\frac{q}{p}} \left(\int_t^a \frac{1}{(r(\tau))^{\frac{1}{p+q-1}}} dg(\tau) \right)^{p+q-1} dg(t) \right)^{\frac{p}{p+q}} \right.$$

$$\left. \times \left(\frac{q}{p+q} \right)^{\frac{q}{p+q}} + \sup_{t \in J} \left((\Delta g(t))^p \frac{s(t)}{r(t)} \right) \right) \left(\int_{t_0}^{a} r(t) |y_g'(t)|^{p+q} dg(t) \right)$$

$$= K_2(t_0, a, p, q) \int_{t_0}^{a} r(t) |y_g'(t)|^{p+q} dg(t).$$

This completes the proof. □

Theorem 5.30. *Let p and q be positive real numbers such that $p \leq 1$, $p + q > 1$, and $r, s \in \mathscr{C}(J)$ be nonnegative functions such that*

$$\int_{t_0}^{a} \frac{1}{(r(t))^{\frac{1}{p+q-1}}} dg(t) < \infty.$$

Also, let $y \in \mathscr{C}_g^1(J)$ be such that $y(t_0) = y(a) = 0$ and y_g' does not change its sign on J. Then, we have

$$\int_{t_0}^{a} s(t) |y(t) + y(t+)|^p |y_g'(t)|^q dg(t) \leq L(t_0, a, p, q) \int_{t_0}^{a} r(t) |y_g'(t)|^{p+q} dg(t),$$

where

$$L(t_0, a, p, q) = \max_{c \in J} \{ L_1(t_0, c, p, q), L_2(c, a, p, q) \}.$$

Proof. Let $c \in J$ be arbitrarily chosen. Then, applying Theorem 5.28, we find

$$\int_{t_0}^{c} s(t)|y(t)+y(t+)|^p|y_g'(t)|^q dg(t) \le L_1(t_0,c,p,q)\int_{t_0}^{c} r(t)|y_g'(t)|^{p+q}dg(t)$$
$$\le L(t_0,a,p,q)\int_{t_0}^{c} r(t)|y_g'(t)|^{p+q}dg(t).$$

Now, by Theorem 5.29, we find

$$\int_{c}^{a} s(t)|y(t)+y(t+)|^p|y_g'(t)|^q dg(t) \le L_2(c,a,p,q)\int_{c}^{a} r(t)|y_g'(t)|^{p+q}dg(t)$$
$$\le L(t_0,a,p,q)\int_{c}^{a} r(t)|y_g'(t)|^{p+q}dg(t).$$

Therefore,

$$\begin{aligned}
&\int_{t_0}^{a} s(t)|y(t)+y(t+)|^p|y_g'(t)|^q dg(t)\\
&\quad= \int_{t_0}^{c} s(t)|y(t)+y(t+)|^p|y_g'(t)|^q dg(t) + \int_{c}^{a} s(t)|y(t)+y(t+)|^p|y_g'(t)|^q dg(t)\\
&\quad\le L(t_0,a,p,q)\int_{t_0}^{c} r(t)|y_g'(t)|^{p+q}dg(t) + L(t_0,a,p,q)\int_{c}^{a} r(t)|y_g'(t)|^{p+q}dg(t)\\
&\quad= L(t_0,a,p,q)\int_{t_0}^{a} r(t)|y_g'(t)|^{p+q}dg(t).
\end{aligned}$$

This completes the proof. □

Theorem 5.31. *Let $n \in \mathbb{N}$ and $f_j \in \mathscr{C}_g^1(J)$, $j \in \{1,\ldots,n\}$. Then,*

$$\left(\prod_{j=1}^{n} f_j\right)_g'(t) = \sum_{j=1}^{n}\left(\left(\prod_{k=1}^{j-1} f_k(t)\right)(f_j)_g'(t)\left(\prod_{k=j+1}^{n} f_k(t)\right)\right), \quad t \in J.$$

Proof. We have

$$\begin{aligned}(f_1f_2f_3\dots f_n)'_g(t) &= (f_1(f_2f_3\dots f_n))'_g(t)\\ &= ((f_1)'_g(t))(f_2(t)f_3(t)\dots f_n(t)) + f_1(t)(f_2f_3\dots f_n)'_g(t)\\ &= ((f_1)'_g(t))(f_2(t)f_3(t)\dots f_n(t))\\ &\quad + f_1(t)\left((f_2)'_g(t)\right)(f_3(t)\dots f_n(t))\\ &\quad + f_1(t)f_2(t)(f_3\dots f_n)'_g(t)\\ &= \sum_{j=1}^{n}\left(\left(\prod_{k=1}^{j-1} f_k(t)\right)(f_j)'_g(t)\left(\prod_{k=j+1}^{n} f_k(t)\right)\right), \quad t\in J.\end{aligned}$$

This completes the proof. □

Theorem 5.32. *Let $n\in\mathbb{N}$ and $f_j\in\mathscr{C}^1_g(J)$, $j\in\{1,\dots,n\}$, be such that $f_j(t_0)=0$, $j\in\{1,\dots,n\}$. Then,*

$$\begin{aligned}&\int_{t_0}^{a}\sum_{j=1}^{n-1}\left|\left(\prod_{k=1}^{j-1} f_k(t)\right)(f_j)'_g\left(\prod_{k=j+1}^{n} f_k(t)\right)\right| dg(t)\\ &\qquad \le \frac{2^{n^2}(a-t_0)^{\frac{n}{n+1}}}{(n+1)^{n+1}}\int_{t_0}^{a}\sum_{j=1}^{n+1}\left((f_j)'_g(t)\right)^{n+1} dg(t).\end{aligned}$$

Proof. Let

$$F_j(t) = \int_{t_0}^{t} |(f_j)'_g(\tau)| dg(\tau), \quad j\in\{1,\dots,n\}, \quad t\in J.$$

Then, $F_j(t_0)=0$ and $(F_j)'_g(t) = |(f_j)'_g(t)|$, $t\in J$, $j\in\{1,\dots,n+1\}$, and

$$\begin{aligned}F_j(t) &= \int_{t_0}^{t} |(f_j)'_g(\tau)| dg(\tau)\\ &\ge \left|\int_{t_0}^{t} (f_j)'_g(\tau) dg(\tau)\right|\\ &= \left| f_j(\tau)\Big|_{\tau=t_0}^{\tau=t}\right|\\ &= |f_j(t)|, \quad t\in J, \quad j\in\{1,\dots,n\}.\end{aligned}$$

Set

$$F(t) = \prod_{j=1}^{n} F_j(t), \quad t \in J.$$

Then, applying the arithmetic mean and geometric mean inequalities and the Stieltjes–Hölder inequality, we get

$$\begin{aligned}
&\int_{t_0}^{a} \sum_{j=1}^{n-1} \left| \left(\prod_{k=1}^{j-1} f_k(t) \right) (f_j)'_g(t) \left(\prod_{k=j+1}^{n} f_k(t) \right) \right| dg(t) \\
&\le \int_{t_0}^{a} \sum_{j=1}^{n-1} \left(\left(\prod_{k=1}^{j-1} |f_k(t)| \right) |(f_j)'_g(t)| \left(\prod_{k=j+1}^{n} |f_k(t)| \right) \right) dg(t) \\
&\le \int_{t_0}^{a} \sum_{j=1}^{n-1} \left(\left(\prod_{k=1}^{j-1} F_k(t) \right) (F_j)'_g(t) \left(\prod_{k=j+1}^{n} F_k(t) \right) \right) dg(t) \\
&= \int_{t_0}^{a} \left(\prod_{j=1}^{n+1} F_j \right)'_g (t) dg(t) \\
&= \int_{t_0}^{a} F'_g(t) dg(t) \\
&= F(t) \Big|_{t=t_0}^{t=a} \\
&= F(a) \\
&= \prod_{j=1}^{n+1} F_j(a) \\
&\le \frac{1}{(n+1)^{n+1}} \left(\sum_{j=1}^{n+1} F_j(a) \right)^{n+1} \\
&\le \frac{2^{n^2}}{(n+1)^{n+1}} \sum_{j=1}^{n+1} (F_j(a))^{n+1} \\
&= \frac{2^{n^2}}{(n+1)^{n+1}} \sum_{j=1}^{n+1} \left(\int_{t_0}^{a} |(f_j)'_g(t)| dg(t) \right)^{n+1}
\end{aligned}$$

$$\leq \frac{2^{n^2}}{(n+1)^{n+1}} \sum_{j=1}^{n+1} (a-t_0)^{\frac{n}{n+1}} \int_{t_0}^{a} |(f_j)'_g(t)|^{n+1} dg(t)$$

$$= \frac{2^{n^2}(a-t_0)^{\frac{n}{n+1}}}{(n+1)^{n+1}} \int_{t_0}^{a} \sum_{j=1}^{n+1} \left((f_j)'_g(t)\right)^{n+1} dg(t).$$

This completes the proof. □

Theorem 5.33. *Let $m, n \in \mathbb{N}$ and $f_j \in \mathscr{C}_g^1(J)$, $j \in \{1, \ldots, n+1\}$, be such that $f_j(t_0) = 0$, $j \in \{1, \ldots, n+1\}$. Then,*

$$\int_{t_0}^{a} \sum_{j=0}^{m} \left| \left(\prod_{k=1}^{n+1} f_k(t+) \right)^{j} \left(\prod_{k=1}^{n+1} f_k(t) \right)^{m-j} \right|$$

$$\times \sum_{j=1}^{n+1} \left| \left(\prod_{k=1}^{j-1} f_k(t+) \right) (f_j)'_g(t) \left(\prod_{k=j+1}^{n+1} f_k(t) \right) \right| dg(t)$$

$$\leq 2^{n^2} \frac{(a-t_0)^{\frac{(n+1)(m+1)-1}{(n+1)(m+1)}}}{(n+1)^{n+1}} \sum_{j=1}^{n+1} \int_{t_0}^{a} |(f_j)'_g(t)|^{(n+1)(m+1)} dg(t).$$

Proof. Let

$$F_j(t) = \int_{t_0}^{t} |(f_j)'_g(\tau)| dg(\tau), \quad j \in \{1, \ldots, n+1\}, \quad t \in J.$$

Then, $F_j(t_0) = 0$, $(F_j)'_g(t) = |(f_j)'_g(t)|$, $t \in J$, $j \in \{1, \ldots, n+1\}$, and

$$\begin{aligned} F_j(t) &= \int_{t_0}^{t} |(f_j)'_g(\tau)| dg(\tau) \\ &\geq \left| \int_{t_0}^{t} (f_j)'_g(\tau) dg(\tau) \right| \\ &= \left| f_j(\tau) \Big|_{\tau=t_0}^{\tau=t} \right| \\ &= |f_j(t)|, \quad t \in J, \quad j \in \{1, \ldots, n+1\}. \end{aligned}$$

Set

$$F(t) = \prod_{j=1}^{n+1} F_j(t), \quad t \in J.$$

Then, applying the arithmetic mean and geometric mean inequalities and the Stieltjes–Hölder inequality, we get

$$\begin{aligned}
&\int_{t_0}^{a} \sum_{j=0}^{m} \left| \left(\prod_{k=1}^{n+1} f_k(t+) \right)^{j} \left(\prod_{k=1}^{n+1} f_k(t) \right)^{m-j} \right| \\
&\quad \times \sum_{j=1}^{n+1} \left| \left(\prod_{k=1}^{j-1} f_k(t+) \right) (f_j)'_g(t) \left(\prod_{k=j+1}^{n+1} f_k(t) \right) \right| dg(t) \\
&\le \int_{t_0}^{a} \sum_{j=0}^{m} \left(\left(\prod_{k=1}^{n+1} |f_k(t+)| \right)^{j} \left(\prod_{k=1}^{n+1} |f_k(t)| \right)^{m-j} \right) \\
&\quad \times \sum_{j=1}^{n+1} \left(\left(\prod_{k=1}^{j-1} |f_k(t+)| \right) |(f_j)'_g(t)| \left(\prod_{k=j+1}^{n+1} |f_k(t)| \right) \right) dg(t) \\
&= \int_{t_0}^{a} \sum_{j=0}^{m} \left(\left(\prod_{k=1}^{n+1} F_k(t+) \right)^{j} \left(\prod_{k=1}^{n+1} F_k(t) \right)^{m-j} \right) \\
&\quad \times \sum_{j=1}^{n+1} \left(\left(\prod_{k=1}^{j-1} F_k(t+) F \right) (F_j)'_g(t) \left(\prod_{k=j+1}^{n+1} F_k(t) \right) \right) dg(t) \\
&= \int_{t_0}^{a} \sum_{j=0}^{m} (F(t+))^{j} (F(t))^{m-j} F'_g(t) dg(t) \\
&= \int_{t_0}^{a} (F^{m+1})'_g(\tau) dg(\tau) \\
&= (F(\tau))^{m+1} \Big|_{\tau=t_0}^{\tau=a} \\
&= (F(a))^{m+1} \\
&= \prod_{j=1}^{n+1} (F_j(a))^{m+1} \\
&\le \frac{1}{(n+1)^{n+1}} \left(\sum_{j=1}^{n+1} (F_j(a))^{(m+1)} \right)^{n+1}
\end{aligned}$$

$$
\begin{aligned}
&\leq \frac{2^{n^2}}{(n+1)^{n+1}} \sum_{j=1}^{n+1} (F_j(a))^{(n+1)(m+1)} \\
&= \frac{2^{n^2}}{(n+1)^{n+1}} \sum_{j=1}^{n+1} \left(\int_{t_0}^{a} |(f_j)'_g(t)| dg(t) \right)^{(n+1)(m+1)} \\
&\leq \sum_{j=1}^{n+1} \frac{2^{n^2} (a-t_0)^{\frac{(n+1)(m+1)-1}{(n+1)(m+1)}}}{(n+1)^{n+1}} \int_{t_0}^{a} |(f_j)'_g(t)|^{(n+1)(m+1)} dg(t) \\
&= \frac{2^{n^2} (a-t_0)^{\frac{(n+1)(m+1)-1}{(n+1)(m+1)}}}{(n+1)^{n+1}} \sum_{j=1}^{n+1} \int_{t_0}^{a} |(f_j)'_g(t)|^{(n+1)(m+1)} dg(t).
\end{aligned}
$$

This completes the proof. □

Theorem 5.34. *Let $n \in \mathbb{N}$ and $f_j \in \mathscr{C}_g^1(J)$, $j \in \{1, \ldots, n\}$, be such that $f_j(t_0) = 0$, $j \in \{1, \ldots, n\}$. Also, let $p \in \mathscr{C}(J)$ be a nonnegative function such that*

$$
\int_{t_0}^{a} \frac{1}{(p(\tau))^{\frac{1}{n}}} dg(\tau) < \infty
$$

and $q \in \mathscr{C}(J)$ be a nonnegative nonincreasing function. Then,

$$
\begin{aligned}
&\int_{t_0}^{a} q(\tau+) \sum_{j=1}^{n-1} \left| \left(\prod_{k=1}^{j-1} f_k(t) \right) (f_j)'_g(t) \left(\prod_{k=j+1}^{n} f_k(t) \right) \right| dg(t) \\
&\quad \leq \frac{2^{n^2}}{(n+1)^{n+1}} \left(\int_{t_0}^{a} \frac{1}{(p(\tau))^{\frac{1}{n}}} dg(\tau) \right)^{n} \\
&\qquad \times \left(\int_{t_0}^{a} p(\tau) q(\tau+) \sum_{j=1}^{n+1} \left((f_j)'_g(t) \right)^{n+1} dg(t) \right).
\end{aligned}
$$

Proof. Let

$$
F_j(t) = \int_{t_0}^{t} (q(\tau+))^{\frac{1}{n+1}} |(f_j)'_g(\tau)| dg(\tau), \quad j \in \{1, \ldots, n\}, \quad t \in J.
$$

Then, $F_j(t_0) = 0$ and $(F_j)'_g(t) = (q(t))^{\frac{1}{n+1}} |(f_j)'_g(t)| \geq 0$, $t \in J$,

$j \in \{1, \ldots, n+1\}$, and using the fact that q is nonincreasing, we find

$$\begin{aligned}
F_j(t) &\geq (q(t))^{\frac{1}{n+1}} \int_{t_0}^{t} |(f_j)'_g(\tau)| dg(\tau) \\
&\geq (q(t))^{\frac{1}{n+1}} \left| \int_{t_0}^{t} (f_j)'_g(\tau) dg(\tau) \right| \\
&= (q(t))^{\frac{1}{n+1}} \left| f_j(\tau) \Big|_{\tau=t_0}^{\tau=t} \right| \\
&= (q(t))^{\frac{1}{n+1}} |f_j(t)| \\
&\geq (q(t))^{\frac{1}{n+1}} |f_j(t)|, \quad t \in J, \quad j \in \{1, \ldots, n\}.
\end{aligned}$$

Define

$$F(t) = \prod_{j=1}^{n} F_j(t), \quad t \in J.$$

Then, applying the arithmetic mean and geometric mean inequalities and the Stieltjes–Hölder inequality, we get

$$\begin{aligned}
&\int_{t_0}^{a} q(t+) \sum_{j=1}^{n-1} \left| \left(\prod_{k=1}^{j-1} f_k(t) \right) (f_j)'_g(t) \left(\prod_{k=j+1}^{n} f_k(t) \right) \right| dg(t) \\
&\leq \int_{t_0}^{a} q(t+) \sum_{j=1}^{n-1} \left(\left(\prod_{k=1}^{j-1} |f_k(t)| \right) |(f_j)'_g(t)| \left(\prod_{k=j+1}^{n} |f_k(t)| \right) \right) dg(t) \\
&\leq \int_{t_0}^{a} \sum_{j=1}^{n-1} \left(\left(\prod_{k=1}^{j-1} F_k(t) \right) (F_j)'_g(t) \left(\prod_{k=j+1}^{n} F_k(t) \right) \right) dg(t) \\
&= \int_{t_0}^{a} \left(\prod_{j=1}^{n+1} F_j \right)'_g (t) dg(t) \\
&= \int_{t_0}^{a} F'_g(t) dg(t) \\
&= F(t) \Big|_{t=t_0}^{t=a} \\
&= F(a)
\end{aligned}$$

$$
\begin{aligned}
&= \prod_{j=1}^{n+1} F_j(a) \\
&\le \frac{1}{(n+1)^{n+1}} \left(\sum_{j=1}^{n+1} F_j(a) \right)^{n+1} \\
&\le \frac{2^{n^2}}{(n+1)^{n+1}} \sum_{j=1}^{n+1} (F_j(a))^{n+1} \\
&= \frac{2^{n^2}}{(n+1)^{n+1}} \sum_{j=1}^{n+1} \left(\int_{t_0}^{a} |(f_j)'_g(t)| dg(t) \right)^{n+1} \\
&\le \frac{2^{n^2}}{(n+1)^{n+1}} \sum_{j=1}^{n+1} \left(\int_{t_0}^{a} (q(t+))^{\frac{1}{n+1}} |(f_j)'_g(t)| dg(t) \right)^{n+1} \\
&= \frac{2^{n^2}}{(n+1)^{n+1}} \sum_{j=1}^{n+1} \left(\int_{t_0}^{a} \frac{1}{(p(t))^{\frac{1}{n+1}}} (p(t)q(t+))^{\frac{1}{n+1}} |(f_j)'_g(t)| dg(t) \right)^{n+1} \\
&\le \frac{2^{n^2}}{(n+1)^{n+1}} \left(\int_{t_0}^{a} \frac{1}{(p(t))^{\frac{1}{n}}} dg(t) \right)^{n} \left(\int_{t_0}^{a} p(t)q(t)|f'_g(t)|^{n+1} dg(t) \right).
\end{aligned}
$$

This completes the proof. □

Theorem 5.35. *Let $m, n \in \mathbb{N}$ and $f_j \in \mathscr{C}_g^1(J)$, $j \in \{1, \ldots, n+1\}$, be such that $f_j(t_0) = 0$, $j \in \{1, \ldots, n+1\}$. Also, let $p \in \mathscr{C}(J)$ be a positive function such that*

$$
\int_{t_0}^{a} \frac{1}{(p(t))^{\frac{1}{m+n}}} dg(t) < \infty
$$

and $q \in \mathscr{C}(J)$ be a nonnegative function. Then,

$$
\begin{aligned}
\int_{t_0}^{a} \sum_{j=0}^{m} q(t+) &\left| \left(\prod_{k=1}^{n+1} f_k(t+) \right)^{j} \left(\prod_{k=1}^{n+1} f_k(t) \right)^{m-j} \right| \\
&\times \sum_{j=1}^{n+1} \left| \left(\prod_{k=1}^{j-1} f_k(t+) \right) (f_j)'_g(t) \left(\prod_{k=j+1}^{n+1} f_k(t) \right) \right| dg(t)
\end{aligned}
$$

$$\leq \frac{2^{n^2}}{(n+1)^{n+1}} \sum_{j=1}^{n+1} \left(\int_{t_0}^{a} \frac{1}{(p(t))^{\frac{1}{m+n}}} dg(t) \right)^{\frac{(m+1)(n+1)(m+n)}{m+n+1}}$$

$$\times \left(\int_{t_0}^{a} p(t)q(t)|f_g'(t)|^{m+n+1} dg(t) \right)^{\frac{(n+1)(m+1)}{m+n+1}}.$$

Proof. Let

$$F_j(t) = \int_{t_0}^{t} (q(\tau+))^{\frac{1}{m+n+1}} |(f_j)_g'(\tau)| d_g\xi, \quad j \in \{1, \ldots, n+1\}, \quad t \in J.$$

Then, $F_j(t_0) = 0$ and $(F_j)_g'(t) = (q(t))^{\frac{1}{m+n+1}} |(f_j)_g'(t)|$, $t \in J$, $j \in \{1, \ldots, n+1\}$, and

$$\begin{aligned} F_j(t) &= \int_{t_0}^{t} (q(\tau+))^{\frac{1}{m+n+1}} |(f_j)_g'(\tau)| dg(\tau) \\ &\geq (q(t))^{\frac{1}{m+n+1}} \left| \int_{t_0}^{t} (f_j)_g'(\tau) dg(\tau) \right| \\ &= (q(t))^{\frac{1}{m+n+1}} \left| f_j(\tau) \Big|_{\tau=t_0}^{\tau=t} \right| \\ &= (q(t))^{\frac{1}{m+n+1}} |f_j(t)| \\ &\geq (q(t))^{\frac{1}{m+n+1}} |f_j(t)|, \quad t \in J, \quad j \in \{1, \ldots, n+1\}. \end{aligned}$$

Set

$$F(t) = \prod_{j=1}^{n+1} F_j(t), \quad t \in J.$$

Then, applying the arithmetic mean and geometric mean inequalities and the Stieltjes–Hölder inequality, we get

$$\int_{t_0}^{a} (q(t+))^{\frac{1}{m+n+1}} \sum_{j=0}^{m} \left| \left(\prod_{k=1}^{n+1} f_k(t+) \right)^j \left(\prod_{k=1}^{n+1} f_k(t) \right)^{m-j} \right|$$

$$\times \sum_{j=1}^{n+1} \left| \left(\prod_{k=1}^{j-1} f_k(\tau+) \right) (f_j)_g'(t) \left(\prod_{k=j+1}^{n+1} f_k(t) \right) \right| dg(t)$$

$$\leq \int_{t_0}^{a} (q(t+))^{\frac{1}{m+n+1}} \sum_{j=0}^{m} \left(\left(\prod_{k=1}^{n+1} |f_k(t+)| \right)^{j} \left(\prod_{k=1}^{n+1} |f_k(t)| \right)^{m-j} \right|$$

$$\times \sum_{j=1}^{n+1} \left(\left(\prod_{k=1}^{j-1} |f_k(t+)| \right) |(f_j)_g'(t)| \left(\prod_{k=j+1}^{n+1} |f_k(t)| \right) \right) dg(t)$$

$$= \int_{t_0}^{a} \sum_{j=0}^{m} \left(\left(\prod_{k=1}^{n+1} F_k(t+) \right)^{j} \left(\prod_{k=1}^{n+1} F_k(t) \right)^{m-j} \right|$$

$$\times \sum_{j=1}^{n+1} \left(\left(\prod_{k=1}^{j-1} F_k(t+)F \right) (F_j)_g'(t) \left(\prod_{k=j+1}^{n+1} F_k(t) \right) \right) dg(t)$$

$$= \int_{t_0}^{a} \sum_{j=0}^{m} (F(t+))^{j} (F(t))^{m-j} F_t' + dg(t)$$

$$= \int_{t_0}^{a} (F^{m+1})_g'(t) dg(t)$$

$$= (F(t))^{m+1} \Big|_{t=t_0}^{t=a}$$

$$= (F(a))^{m+1}$$

$$= \prod_{j=1}^{n+1} (F_j(a))^{m+1}$$

$$\leq \frac{1}{(n+1)^{n+1}} \left(\sum_{j=1}^{n+1} (F_j(a))^{m+1} \right)^{n+1}$$

$$\leq \frac{2^{n^2}}{(n+1)^{n+1}} \sum_{j=1}^{n+1} (F_j(a))^{(n+1)(m+1)}$$

$$= \frac{2^{n^2}}{(n+1)^{n+1}} \sum_{j=1}^{n+1} \left(\int_{t_0}^{a} (q(t))^{\frac{1}{m+n+1}} |(f_j)_g'(t)| dg(t) \right)^{(n+1)(m+1)}$$

$$= \frac{2^{n^2}}{(n+1)^{n+1}} \sum_{j=1}^{n+1} \left(\int_{t_0}^{a} \frac{1}{(p(t))^{\frac{1}{m+n+1}}} \right.$$

$$\times \left.(p(t))q(t))^{\frac{1}{m+n+1}}|(f_j)'_g(t)|dg(t)\right)^{(n+1)(m+1)}$$

$$\leq \frac{2^{n^2}}{(n+1)^{n+1}}\sum_{j=1}^{n+1}\left(\int_{t_0}^{a}\frac{1}{(p(t))^{\frac{1}{m+n}}}dg(t)\right)^{\frac{(m+1)(n+1)(m+n)}{m+n+1}}$$

$$\times\left(\int_{t_0}^{a}p(t)q(t)|f'_g(t)|^{m+n+1}dg(t)\right)^{\frac{(n+1)(m+1)}{m+n+1}}.$$

This completes the proof. □

5.7 Higher-Order Stieltjes–Opial Inequalities

Theorem 5.36. *Let $y \in \mathscr{C}_g^n(J)$ and*

$$y_g^{(k)}(t_0) = 0, \quad k \in \{1, \ldots, n-1\}. \tag{5.25}$$

Then, we have the inequality

$$\int_{t_0}^{a}\left|\left(y\left(g^{-(n-1)}(t)\right)_g^{(n)}\right)\right|\left|y\left(g^{-(n-1)}(t)\right)\right|dg(t)$$

$$\leq \frac{1}{\sqrt{2}}\left(\int_{t_0}^{a}\left(\left(y\left(g^{-(n-1)}(t)\right)_g^{(n)}\right)\right)^2 dg(t)\right)^{\frac{1}{2}}$$

$$\times\left(\int_{t_0}^{a}\left(\int_{t_0}^{t}\left|h_{n-1}\left(s, g^{-(n-1)}(t)\right)\right|^2 dg(s)\right)dg(t)\right)^{\frac{1}{2}}.$$

Proof. By the Stieltjes–Taylor formula and (5.25), we get

$$y(t) = (-1)^{n-1}\int_{t_0}^{g^{n-1}(t)} y_g^{(n)}\left(g^{-(n-1)}(s)\right)h_{n-1}(s,t)dg(s), \quad t \in J,$$

whereupon

$$y\left(g^{-(n-1)}(t)\right) = (-1)^{n-1}\int_{t_0}^{t} y_g^{(n)}\left(g^{-(n-1)}(s)\right)$$

$$\times h_{n-1}\left(s, g^{-(n-1)}(t)\right)dg(s), \quad t \in J,$$

and

$$\begin{aligned}
&\left|y\left(g^{-(n-1)}(t)\right)\right| \\
&\quad = \left|(-1)^{n-1}\int_{t_0}^{t} y_g^{(n)}\left(g^{-(n-1)}(s)\right) h_{n-1}\left(s, g^{-(n-1)}(t)\right) dg(s)\right| \\
&\quad \le \int_{t_0}^{t} \left|y_g^{(n)}\left(g^{-(n-1)}(s)\right)\right| \left|h_{n-1}\left(s, g^{-(n-1)}(t)\right)\right| dg(s), \quad t \in J.
\end{aligned}$$

Therefore,

$$\begin{aligned}
&\left|\left(y\left(g^{-(n-1)}(t)\right)\right)_g^{(n)}\right| \left|y\left(g^{-(n-1)}(t)\right)\right| \\
&\quad \le \left|\left(y\left(g^{-(n-1)}(t)\right)\right)_g^{(n)}\right| \times \int_{t_0}^{t} \left|y_g^{(n)}\left(g^{-(n-1)}(s)\right)\right| \\
&\qquad \times \left|h_{n-1}\left(s, g^{-(n-1)}(t)\right)\right| dg(s) \\
&\quad \le \left|\left(y\left(g^{-(n-1)}(t)\right)\right)_g^{(n)}\right| \left(\int_{t_0}^{t} \left|y_g^{(n)}\left(g^{-(n-1)}(s)\right)\right|^2 dg(s)\right)^{\frac{1}{2}} \\
&\qquad \times \left(\int_{t_0}^{t} \left|h_{n-1}\left(s, g^{-(n-1)}(t)\right)\right|^2 dg(s)\right)^{\frac{1}{2}}, \quad t \in J,
\end{aligned}$$

$$\begin{aligned}
&\int_{t_0}^{a} \left|\left(y\left(g^{-(n-1)}(t)\right)\right)_g^{(n)}\right| \left|y\left(g^{-(n-1)}(t)\right)\right| dg(t) \\
&\quad \le \int_{t_0}^{a} \left|\left(y\left(g^{-(n-1)}(t)\right)\right)_g^{(n)}\right| \times \left(\int_{t_0}^{t} \left|y_g^{(n)}\left(g^{-(n-1)}(s)\right)\right|^2 dg(s)\right)^{\frac{1}{2}} \\
&\qquad \times \left(\int_{t_0}^{t} \left|h_{n-1}\left(s, g^{-(n-1)}(t)\right)\right|^2 dg(s)\right)^{\frac{1}{2}} dg(t).
\end{aligned}$$

Let

$$z(t) = \int_{t_0}^{t} \left(\left(y\left(g^{-(n-1)}(s)\right)\right)_g^{(n)}\right)^2 dg(s), \quad t \in J.$$

Then, $z(t_0) = 0$ and

$$z_g'(t) = \left(\left(y\left(g^{-(n-1)}(t)\right)\right)_g^{(n)}\right)^2 \ge 0, \quad t \in J.$$

Therefore, z is a nondecreasing function on J and

$$\int_{t_0}^{a} \left| \left(y \left(g^{-(n-1)}(t) \right) \right)_g^{(n)} \right| \left| y \left(g^{-(n-1)}(t) \right) \right| dg(t)$$
$$\leq \int_{t_0}^{a} \left(\int_{t_0}^{t} \left| h_{n-1} \left(s, g^{-(n-1)}(t) \right) \right|^2 dg(s) \right)^{\frac{1}{2}} (z(t) z_g'(t))^{\frac{1}{2}} dg(t). \quad (5.26)$$

By the chain rule, Theorem 1.42, and the fact that z is a nondecreasing function on J, we find

$$\begin{aligned}(z^2)_g'(t) &= 2z_g'(t) \int_0^1 (hz(t) + (1-h)z(t))dh \\ &\geq 2z_g'(t) \int_0^1 (hz(t) + (1-h)z(t))dh \\ &= 2z(t)z_g'(t), \quad t \in J.\end{aligned}$$

Then,

$$z(t)z_g'(t) \leq \frac{1}{2}(z^2)_g'(t), \quad t \in J,$$

and

$$\begin{aligned}\int_{t_0}^{a} z(t)z_g'(t)dg(t) &\leq \frac{1}{2} \int_{t_0}^{a} (z^2)_g'(t)dg(t) \\ &= \frac{1}{2}(z(t))^2 \Big|_{t=t_0}^{t=a} \\ &= \frac{1}{2}(z(a))^2 \\ &= \frac{1}{2} \int_{t_0}^{a} \left(\left(y \left(g^{-(n-1)}(s) \right) \right)_g^{(n)} \right)^2 dg(s).\end{aligned}$$

From this and (5.26), applying the Stieltjes–Hölder inequality, we arrive at

$$\int_{t_0}^{a} \left| \left(y \left(g^{-(n-1)}(t) \right) \right)_g^{(n)} \right| \left| y \left(g^{-(n-1)}(t) \right) \right| dg(t)$$
$$\leq \int_{t_0}^{a} \left(\int_{t_0}^{t} \left| h_{n-1} \left(s, g^{-(n-1)}(t) \right) \right|^2 dg(s) \right)^{\frac{1}{2}} (z(t) z_g'(t))^{\frac{1}{2}} dg(t)$$

$$\le \left(\int_{t_0}^{a} z(t)z_g'(t)dg(t)\right)^{\frac{1}{2}}$$

$$\times\left(\int_{t_0}^{a}\left(\int_{t_0}^{t}\left(h_{n-1}\left(s,g^{-(n-1)}(t)\right)\right)^2 dg(s)\right)dg(t)\right)^{\frac{1}{2}}$$

$$\le \frac{1}{\sqrt{2}}\left(\int_{t_0}^{a}\left(\left(y\left(g^{-(n-1)}(s)\right)\right)_g^{(n)}\right)^2 dg(s)\right)^{\frac{1}{2}}$$

$$\times\left(\int_{t_0}^{a}\left(\int_{t_0}^{t}\left(h_{n-1}\left(s,g^{-(n-1)}(t)\right)\right)^2 dg(s)\right)dg(t)\right)^{\frac{1}{2}}.$$

This completes the proof. □

Theorem 5.37. *Let $y \in \mathscr{C}_g^n(J)$ and*

$$y_g^{(k)}(a) = 0, \quad k \in \{1, \ldots, n-1\}. \tag{5.27}$$

Then, we have the inequality

$$\int_{t_0}^{a}\left|\left(y\left(g^{-(n-1)}(t)\right)\right)_g^{(n)}\right|\left|y\left(g^{-(n-1)}(t)\right)\right|dg(t)$$

$$\le \frac{1}{\sqrt{2}}\left(\int_{t_0}^{a}\left(\left(y\left(g^{-(n-1)}(s)\right)\right)_g^{(n)}\right)^2 dg(s)\right)^{\frac{1}{2}}$$

$$\times\left(\int_{t_0}^{a}\left(\int_{t}^{a}\left|h_{n-1}\left(s,g^{-(n-1)}(t)\right)\right|^2 dg(s)\right)dg(t)\right)^{\frac{1}{2}}.$$

Proof. By the Stieltjes–Taylor formula and (5.27), we get

$$y(t) = (-1)^{n-1}\int_{a}^{g^{n-1}(t)} y_g^{(n)}\left(g^{-(n-1)}(s)\right)h_{n-1}(s,t)dg(s), \quad t \in J,$$

whereupon

$$y\left(g^{-(n-1)}(t)\right)$$

$$= (-1)^{n-1}\int_{a}^{t} y_g^{(n)}\left(g^{-(n-1)}(s)\right)h_{n-1}\left(s,g^{-(n-1)}(t)\right)dg(s)$$

$$= (-1)^{n}\int_{t}^{a} y_g^{(n)}\left(g^{-(n-1)}(s)\right)h_{n-1}\left(s,g^{-(n-1)}(t)\right)dg(s), \quad t \in J,$$

and

$$\begin{aligned}
&\left|y\left(g^{-(n-1)}(t)\right)\right| \\
&\quad = \left|(-1)^n \int_t^a y_g^{(n)}\left(g^{-(n-1)}(s)\right) h_{n-1}\left(s, g^{-(n-1)}(t)\right) dg(s)\right| s \\
&\quad \leq \int_t^a \left|y_g^{(n)}\left(g^{-(n-1)}(s)\right)\right| \left|h_{n-1}\left(s, g^{-(n-1)}(t)\right)\right| dg(s), \quad t \in J.
\end{aligned}$$

Therefore,

$$\begin{aligned}
&\left|\left(y\left(g^{-(n-1)}(t)\right)\right)_g^{(n)}\right| \left|y\left(g^{-(n-1)}(t)\right)\right| \\
&\quad \leq \left|\left(y\left(g^{-(n-1)}(t)\right)\right)_g^{(n)}\right| \times \int_t^a \left|y_g^{(n)}\left(g^{-(n-1)}(s)\right)\right| \\
&\qquad \times \left|h_{n-1}\left(s, g^{-(n-1)}(t)\right)\right| dg(s) \\
&\quad \leq \left|\left(y\left(g^{-(n-1)}(t)\right)\right)_g^{(n)}\right| \\
&\qquad \times \left(\int_t^a \left|y_g^{(n)}\left(g^{-(n-1)}(s)\right)\right|^2 dg(s)\right)^{\frac{1}{2}} \\
&\qquad \times \left(\int_t^a \left|h_{n-1}\left(s, g^{-(n-1)}(t)\right)\right|^2 dg(s)\right)^{\frac{1}{2}},
\end{aligned}$$

$t \in J$, and

$$\begin{aligned}
&\int_{t_0}^a \left|\left(y\left(g^{-(n-1)}(t)\right)\right)_g^{(n)}\right| \left|y\left(g^{-(n-1)}(t)\right)\right| dg(t) \\
&\quad \leq \int_{t_0}^a \left|\left(y\left(g^{-(n-1)}(t)\right)\right)_g^{(n)}\right| \left(\int_t^a \left|y_g^{(n)}\left(g^{-(n-1)}(s)\right)\right|^2 dg(s)\right)^{\frac{1}{2}} \\
&\quad \times \left(\int_t^a \left|h_{n-1}\left(s, g^{-(n-1)}(t)\right)\right|^2 dg(s)\right)^{\frac{1}{2}} dg(t).
\end{aligned}$$

Let

$$z(t) = \int_t^a \left(\left(y\left(g^{-(n-1)}(s)\right)\right)_g^{(n)}\right)^2 dg(s), \quad t \in J.$$

Then, $z(a) = 0$ and

$$z_g'(t) = -\left(\left(y\left(g^{-(n-1)}(t)\right)\right)_g^{(n)}\right)^2 \leq 0, \quad t \in J.$$

Therefore, z is a nondecreasing function on J and

$$\int_{t_0}^{a}\left|\left(y\left(g^{-(n-1)}(t)\right)\right)_g^{(n)}\right|\left|y\left(g^{-(n-1)}(t)\right)\right|dg(t)$$
$$\leq\int_{t_0}^{a}\left(\int_{t_0}^{t}\left|h_{n-1}\left(s,g^{-(n-1)}(t)\right)\right|^2dg(s)\right)^{\frac{1}{2}}(-z(t)z_g'(t))^{\frac{1}{2}}dg(t). \tag{5.28}$$

By the chain rule, Theorem 1.42, and the fact that z is a nondecreasing function on J, we find

$$\begin{aligned}-(z^2)_g'(t)&=-2z_g'(t)\int_0^1(hz(t)+(1-h)z(t))dh\\&\geq-2z_g'(t)\int_0^1(hz(t)+(1-h)z(t))dh\\&=-2z(t)z_g'(t),\quad t\in J.\end{aligned}$$

Then,

$$-z(t)z_g'(t)\leq-\frac{1}{2}(z^2)_g'(t),\quad t\in J,$$

and

$$\begin{aligned}-\int_{t_0}^{a}z(t)z_g'(t)dg(t)&\leq-\frac{1}{2}\int_{t_0}^{a}(z^2)_g'(t)dg(t)\\&=-\frac{1}{2}(z(t))^2\Big|_{t=t_0}^{t=a}\\&=\frac{1}{2}(z(t_0))^2\\&=\frac{1}{2}\int_{t_0}^{a}\left(\left(y\left(g^{-(n-1)}(s)\right)\right)_g^{(n)}\right)^2dg(s).\end{aligned}$$

From this and (5.28), applying the Stieltjes–Hölder inequality, we arrive at

$$\int_{t_0}^{a}\left|\left(y\left(g^{-(n-1)}(t)\right)\right)_g^{(n)}\right|\left|y\left(g^{-(n-1)}(t)\right)\right|dg(t)$$
$$\leq\int_{t_0}^{a}\left(\int_{t}^{a}\left|h_{n-1}\left(s,g^{-(n-1)}(t)\right)\right|^2dg(s)\right)^{\frac{1}{2}}(-z(t)z_g'(t))^{\frac{1}{2}}dg(t)$$

$$\leq \left(-\int_{t_0}^{a} z(t)z_g'(t)dg(t)\right)^{\frac{1}{2}}$$

$$\times \left(\int_{t_0}^{a}\left(\int_{t}^{a}\left(h_{n-1}\left(s, g^{-(n-1)}(t)\right)\right)^2 dg(s)\right)dg(t)\right)^{\frac{1}{2}}$$

$$\leq \frac{1}{\sqrt{2}}\left(\int_{t_0}^{a}\left(\left(y\left(g^{-(n-1)}(s)\right)\right)_g^{(n)}\right)^2 dg(s)\right)^{\frac{1}{2}}$$

$$\times \left(\int_{t_0}^{a}\left(\int_{t_0}^{t}\left(h_{n-1}\left(s, g^{-(n-1)}(t)\right)\right)^2 dg(s)\right)dg(t)\right)^{\frac{1}{2}}.$$

This completes the proof. □

Theorem 5.38. *Let $y \in \mathscr{C}_g^n(J)$ and (5.25), (5.27) hold. Then, we have the inequality*

$$\int_{t_0}^{a}\left|\left(y\left(g^{-(n-1)}(t)\right)\right)_g^{(n)}\right|\left|y\left(g^{-(n-1)}(t)\right)\right|dg(t)$$

$$\leq M\sqrt{2}\left(\int_{t_0}^{a}\left(\left(y\left(g^{-(n-1)}(s)\right)\right)_g^{(n)}\right)^2 dg(s)\right)^{\frac{1}{2}},$$

where

$$M = \max\left\{\left(\int_{t_0}^{\frac{t_0+a}{2}}\left(\int_{t_0}^{t}\left|h_{n-1}\left(s, g^{-(n-1)}(t)\right)\right|^2 dg(s)\right)dg(t)\right)^{\frac{1}{2}},\right.$$

$$\left.\times\left(\int_{a}^{\frac{a+t_0}{2}}\left(\int_{t}^{a}\left|h_{n-1}\left(s, g^{-(n-1)}(t)\right)\right|^2 dg(s)\right)dg(t)\right)^{\frac{1}{2}}\right\}.$$

Proof. By Theorem 5.36, it follows that

$$\int_{t_0}^{\frac{t_0+a}{2}}\left|\left(y\left(g^{-(n-1)}(t)\right)\right)_g^{(n)}\right|\left|y\left(g^{-(n-1)}(t)\right)\right|dg(t)$$

$$\leq \frac{1}{\sqrt{2}}\left(\int_{t_0}^{\frac{t_0+a}{2}}\left(\left(y\left(g^{-(n-1)}(s)\right)\right)_g^{(n)}\right)^2 dg(s)\right)^{\frac{1}{2}}$$

$$\times \left(\int_{t_0}^{\frac{t_0+a}{2}} \left(\int_{t_0}^{t} \left| h_{n-1}\left(s, g^{-(n-1)}(t)\right) \right|^2 dg(s) \right) dg(t) \right)^{\frac{1}{2}}$$

$$\leq \frac{M}{\sqrt{2}} \left(\int_{t_0}^{a} \left(\left(y\left(g^{-(n-1)}(s)\right)\right)_g^{(n)} \right)^2 dg(s) \right)^{\frac{1}{2}}.$$

By Theorem 5.37, we get

$$\int_{\frac{t_0+a}{2}}^{a} \left| \left(y\left(g^{-(n-1)}(t)\right)\right)_g^{(n)} \right| \left| y\left(g^{-(n-1)}(t)\right) \right| dg(t)$$

$$\leq \frac{1}{\sqrt{2}} \left(\int_{\frac{t_0+a}{2}}^{a} \left(\left(y\left(g^{-(n-1)}(s)\right)\right)_g^{(n)} \right)^2 dg(s) \right)^{\frac{1}{2}}$$

$$\times \left(\int_{\frac{t_0+a}{2}}^{a} \left(\int_{t_0}^{t} \left| h_{n-1}\left(s, g^{-(n-1)}(t)\right) \right|^2 dg(s) \right) dg(t) \right)^{\frac{1}{2}}$$

$$\leq \frac{M}{\sqrt{2}} \left(\int_{t_0}^{a} \left(\left(y\left(g^{-(n-1)}(s)\right)\right)_g^{(n)} \right)^2 dg(s) \right)^{\frac{1}{2}}.$$

Therefore,

$$\int_{t_0}^{a} \left| \left(y\left(g^{-(n-1)}(t)\right)\right)_g^{(n)} \right| \left| y\left(g^{-(n-1)}(t)\right) \right| dg(t)$$

$$= \int_{t_0}^{\frac{t_0+a}{2}} \left| \left(y\left(g^{-(n-1)}(t)\right)\right)_g^{(n)} \right| \left| y\left(g^{-(n-1)}(t)\right) \right| dg(t)$$

$$+ \int_{\frac{t_0+a}{2}}^{a} \left| \left(y\left(g^{-(n-1)}(t)\right)\right)_g^{(n)} \right| \left| y\left(g^{-(n-1)}(t)\right) \right| dg(t)$$

$$\leq \frac{M}{\sqrt{2}} \left(\int_{t_0}^{a} \left(\left(y\left(g^{-(n-1)}(s)\right)\right)_g^{(n)} \right)^2 dg(s) \right)^{\frac{1}{2}}$$

$$+ \frac{M}{\sqrt{2}} \left(\int_{t_0}^{a} \left(\left(y\left(g^{-(n-1)}(s)\right)\right)_g^{(n)} \right)^2 dg(s) \right)^{\frac{1}{2}}$$

$$= \frac{2M}{\sqrt{2}} \left(\int_{t_0}^{a} \left(\left(y \left(g^{-(n-1)}(s) \right) \right)_g^{(n)} \right)^2 dg(s) \right)^{\frac{1}{2}}$$

$$= M\sqrt{2} \left(\int_{t_0}^{a} \left(\left(y \left(g^{-(n-1)}(s) \right) \right)_g^{(n)} \right)^2 dg(s) \right)^{\frac{1}{2}}.$$

This completes the proof. □

Theorem 5.39. *Let $y \in \mathscr{C}_g^n(J)$ and (5.25) hold. Also, let $l, m > 0$ be such that $l + m > 1$. Then, we have the inequality*

$$\int_{t_0}^{a} \left| y_g^{(n)} \left(g^{-(n-1)}(t) \right) \right|^m \left| y \left(g^{-(n-1)}(t) \right) \right|^l dg(t)$$

$$\leq \left(\int_{t_0}^{a} \left(\int_{t_0}^{t} \left| h_{n-1} \left(s, g^{-(n-1)}(t) \right) \right|^{\frac{l+m}{l+m-1}} dg(s) \right)^{l+m-1} dg(t) \right)^{\frac{l}{l+m}}$$

$$\times \left(\frac{m}{m+l} \right)^{\frac{m}{m+l}} \int_{t_0}^{a} \left| y_g^{(n)} \left(g^{-(n-1)}(t) \right) \right|^{m+l} dg(t).$$

Proof. By the Stieltjes–Taylor formula and (5.25), we find

$$y(t) = (-1)^{n-1} \int_{t_0}^{g^{n-1}(t)} y_g^{(n)} \left(g^{-(n-1)}(s) \right) h_{n-1}(s, t) dg(s), \quad t \in J,$$

whereupon

$$y \left(g^{-(n-1)}(t) \right) = (-1)^{n-1} \int_{t_0}^{t} y_g^{(n)} \left(g^{-(n-1)}(s) \right)$$
$$\times h_{n-1} \left(s, g^{-(n-1)}(t) \right) dg(s), \quad t \in J,$$

and

$$\left| y \left(g^{-(n-1)}(t) \right) \right|$$
$$= \left| (-1)^{n-1} \int_{t_0}^{t} y_g^{(n)} \left(g^{-(n-1)}(s) \right) h_{n-1} \left(s, g^{-(n-1)}(t) \right) dg(s) \right|$$
$$\leq \int_{t_0}^{t} \left| y_g^{(n)} \left(g^{-(n-1)}(s) \right) \right| \left| h_{n-1} \left(s, g^{-(n-1)}(t) \right) \right| dg(s), \quad t \in J.$$

Now, applying the Stieltjes–Hölder inequality for $l+m$ and $\frac{l+m}{l+m-1}$, we obtain

$$\begin{aligned}
&\left|y_g^{(n)}\left(g^{-(n-1)}(t)\right)\right|^m \left|y\left(g^{-(n-1)}(t)\right)\right|^l \\
&\quad\le \left|y_g^{(n)}\left(g^{-(n-1)}(t)\right)\right|^m \\
&\qquad\times \left(\int_{t_0}^t \left|y_g^{(n)}\left(g^{-(n-1)}(s)\right)\right| \left|h_{n-1}\left(s, g^{-(n-1)}(t)\right)\right| dg(s)\right)^l \\
&\quad\le \left|y_g^{(n)}\left(g^{-(n-1)}(t)\right)\right|^m \left(\int_{t_0}^t \left|y_g^{(n)}\left(g^{-(n-1)}(s)\right)\right|^{l+m} dg(s)\right)^{\frac{l}{l+m}} \\
&\qquad\times \left(\int_{t_0}^t \left|h_{n-1}\left(s, g^{-(n-1)}(t)\right)\right|^{\frac{l+m}{l+m-1}} dg(s)\right)^{\frac{l(l+m-1)}{l+m}}, \quad t \in J.
\end{aligned}$$

Thus,

$$\begin{aligned}
&\int_{t_0}^a \left|y_g^{(n)}\left(g^{-(n-1)}(t)\right)\right|^m \left|y\left(g^{-(n-1)}(t)\right)\right|^l dg(t) \\
&\quad\le \int_{t_0}^a \left|y_g^{(n)}\left(g^{-(n-1)}(t)\right)\right|^m \left(\int_{t_0}^t \left|y_g^{(n)}\left(g^{-(n-1)}(s)\right)\right|^{l+m} dg(s)\right)^{\frac{l}{l+m}} \\
&\qquad\times \left(\int_{t_0}^t \left|h_{n-1}\left(s, g^{-(n-1)}(t)\right)\right|^{\frac{l+m}{l+m-1}} dg(s)\right)^{\frac{l(l+m-1)}{l+m}} dg(t), \quad t \in J.
\end{aligned} \tag{5.29}$$

Set

$$z(t) = \int_{t_0}^t \left|y_g^{(n)}\left(g^{-(n-1)}(s)\right)\right|^{m+l} dg(s), \quad t \in J.$$

Then, $z(t_0) = 0$ and

$$z_g'(t) = \left|y_g'\left(g^{-(n-1)}(t)\right)\right|^{m+l} \ge 0, \quad t \in J.$$

Therefore, z is a nondecreasing function on J, and applying the chain rule, Theorem 1.42, we arrive at

$$\begin{aligned}
\left(z^{\frac{l+m}{m}}\right)_g'(t) &= \frac{l+m}{m} z_g'(t) \int_0^1 (hz(t) + (1-h)z(t))^{\frac{l}{m}}\, dh \\
&\ge \frac{l+m}{m} z_g'(t) \int_0^1 (hz(t) + (1-h)z(t))^{\frac{l}{m}}\, dh \\
&= \frac{l+m}{m} (z(t))^{\frac{l}{m}} z_g'(t), \quad t \in J.
\end{aligned}$$

Hence,

$$(z(t))^{\frac{l}{m}} z_g'(t) \leq \frac{m}{l+m}\left(z^{\frac{l+m}{m}}\right)_g'(t), \quad t \in J,$$

and

$$\begin{aligned}
\int_{t_0}^{a} (z(t))^{\frac{l}{m}} z_g'(t) &\leq \frac{m}{l+m} \int_{t_0}^{a} \left(z^{\frac{l+m}{m}}\right)_g'(t) dg(t) \\
&= \frac{m}{l+m} (z(t))^{\frac{l+m}{m}} \Big|_{t=t_0}^{t=a} \\
&= \frac{m}{l+m} (z(a))^{\frac{l+m}{m}} \\
&= \frac{m}{l+m} \left(\int_{t_0}^{a} \left| y_g^{(n)}\left(g^{-(n-1)}(t)\right)\right|^{m+l} dg(t)\right)^{\frac{l+m}{m}}.
\end{aligned}$$

From this and (5.29), we find

$$\begin{aligned}
&\int_{t_0}^{a} \left| y_g^{(n)}\left(g^{-(n-1)}(t)\right)\right|^{m} \left| y\left(g^{-(n-1)}(t)\right)\right|^{l} dg(t) \\
&\leq \int_{t_0}^{a} (z(t))^{\frac{l}{l+m}} \left(z_g'(t)\right)^{\frac{m}{l+m}} \\
&\quad \times \left(\int_{t_0}^{t} \left| h_{n-1}\left(s, g^{-(n-1)}(t)\right)\right|^{\frac{l+m}{l+m-1}} dg(s)\right)^{\frac{l(l+m-1)}{l+m}} dg(t), \quad t \in J.
\end{aligned}$$

Now, applying the Stieltjes–Hölder inequality for $\frac{l+m}{l}$ and $\frac{l+m}{m}$, we find

$$\begin{aligned}
&\int_{t_0}^{a} \left| y_g^{(n)}\left(g^{-(n-1)}(t)\right)\right|^{m} \left| y\left(g^{-(n-1)}(t)\right)\right|^{l} dg(t) \\
&\leq \left(\int_{t_0}^{a} \left(\int_{t_0}^{t} \left| h_{n-1}\left(s, g^{-(n-1)}(t)\right)\right|^{\frac{l+m}{l+m-1}} dg(s)\right)^{\frac{l+m}{l+m-1}} dg(t)\right)^{\frac{l}{l+m}} \\
&\quad \times \left(\int_{t_0}^{a} (z(t))^{\frac{l}{m}} z_g'(t) dg(t)\right)^{\frac{m}{l+m}} \\
&\leq \left(\int_{t_0}^{a} \left(\int_{t_0}^{t} \left| h_{n-1}\left(s, g^{-(n-1)}(t)\right)\right|^{\frac{l+m}{l+m-1}} dg(s)\right)^{\frac{l+m}{l+m-1}} dg(t)\right)^{\frac{l}{l+m}} \\
&\quad \times \left(\frac{m}{l+m}\right)^{\frac{m}{l+m}} \int_{t_0}^{a} \left| y_g^{(n)}\left(g^{-(n-1)}(t)\right)\right|^{m+l} dg(t).
\end{aligned}$$

This completes the proof. □

Theorem 5.40. *Let $y \in \mathscr{C}_g^n(J)$ and (5.27) hold. Also, let $l, m > 0$ be such that $l + m > 1$. Then, we have the inequality*

$$\int_{t_0}^{a} \left| y_g^{(n)} \left(g^{-(n-1)}(t) \right) \right|^m \left| y \left(g^{-(n-1)}(t) \right) \right|^l dg(t)$$

$$\leq \left(\int_{t_0}^{a} \left(\int_{t}^{a} \left| h_{n-1} \left(s, g^{-(n-1)}(t) \right) \right|^{\frac{l+m}{l+m-1}} dg(s) \right)^{l+m-1} dg(t) \right)^{\frac{l}{l+m}}$$

$$\times \left(\frac{m}{m+l} \right)^{\frac{m}{m+l}} \int_{t_0}^{a} \left| y_g^{(n)} \left(g^{-(n-1)}(t) \right) \right|^{m+l} dg(t).$$

Proof. By the Stieltjes–Taylor formula and (5.27), we find

$$y(t) = (-1)^{n-1} \int_{a}^{g^{n-1}(t)} y_g^{(n)} \left(g^{-(n-1)}(s) \right) h_{n-1}(s,t) dg(s), \quad t \in J,$$

whereupon

$$y \left(g^{-(n-1)}(t) \right)$$

$$= (-1)^{n-1} \int_{a}^{t} y_g^{(n)} \left(g^{-(n-1)}(s) \right) h_{n-1} \left(s, g^{-(n-1)}(t) \right) dg(s)$$

$$= (-1)^{n} \int_{t}^{a} y_g^{(n)} \left(g^{-(n-1)}(s) \right) h_{n-1} \left(s, g^{-(n-1)}(t) \right) dg(s), \quad t \in J,$$

and

$$\left| y \left(g^{-(n-1)}(t) \right) \right|$$

$$= \left| (-1)^{n} \int_{t}^{a} y_g^{(n)} \left(g^{-(n-1)}(s) \right) h_{n-1} \left(s, g^{-(n-1)}(t) \right) dg(s) \right| s$$

$$\leq \int_{t}^{a} \left| y_g^{(n)} \left(g^{-(n-1)}(s) \right) \right| \left| h_{n-1} \left(s, g^{-(n-1)}(t) \right) \right| dg(s), \quad t \in J.$$

Now, applying the Stieltjes–Hölder inequality for $l + m$ and $\frac{l+m}{l+m-1}$, we obtain

$$\begin{aligned}
&\left|y_g^{(n)}\left(g^{-(n-1)}(t)\right)\right|^m \left|y\left(g^{-(n-1)}(t)\right)\right|^l \\
&\quad \leq \left|y_g^{(n)}\left(g^{-(n-1)}(t)\right)\right|^m \\
&\qquad \times \left(\int_t^a \left|y_g^{(n)}\left(g^{-(n-1)}(s)\right)\right| \left|h_{n-1}\left(s, g^{-(n-1)}(t)\right)\right| dg(s)\right)^l \\
&\quad \leq \left|y_g^{(n)}\left(g^{-(n-1)}(t)\right)\right|^m \left(\int_t^a \left|y_g^{(n)}\left(g^{-(n-1)}(s)\right)\right|^{l+m} dg(s)\right)^{\frac{l}{l+m}} \\
&\qquad \times \left(\int_t^a \left|h_{n-1}\left(s, g^{-(n-1)}(t)\right)\right|^{\frac{l+m}{l+m-1}} dg(s)\right)^{\frac{l(l+m-1)}{l+m}}, \quad t \in J.
\end{aligned}$$

Thus,

$$\begin{aligned}
&\int_{t_0}^a \left|y_g^{(n)}\left(g^{-(n-1)}(t)\right)\right|^m \left|y\left(g^{-(n-1)}(t)\right)\right|^l dg(t) \\
&\quad \leq \int_{t_0}^a \left|y_g^{(n)}\left(g^{-(n-1)}(t)\right)\right|^m \left(\int_t^a \left|y_g^{(n)}\left(g^{-(n-1)}(s)\right)\right|^{l+m} dg(s)\right)^{\frac{l}{l+m}} \\
&\qquad \times \left(\int_t^a \left|h_{n-1}\left(s, g^{-(n-1)}(t)\right)\right|^{\frac{l+m}{l+m-1}} dg(s)\right)^{\frac{l(l+m-1)}{l+m}} dg(t), \quad t \in J.
\end{aligned} \tag{5.30}$$

Set

$$z(t) = \int_t^a \left|y_g^{(n)}\left(g^{-(n-1)}(s)\right)\right|^{m+l} dg(s), \quad t \in J.$$

Then, $z(a) = 0$ and

$$z_g'(t) = -\left|y_g'\left(g^{-(n-1)}(t)\right)\right|^{m+l} \leq 0, \quad t \in J.$$

Therefore, z is a nonincreasing function on J and applying the chain rule, Theorem 1.42, we arrive at

$$\begin{aligned}
\left(-(z^{\frac{l+m}{m}})_g'\right)(t) &= -\frac{l+m}{m} z_g'(t) \int_0^1 (hz(t) + (1-h)z(t))^{\frac{l}{m}} dh \\
&\geq -\frac{l+m}{m} z_g'(t) \int_0^1 (hz(t) + (1-h)z(t))^{\frac{l}{m}} dh \\
&= -\frac{l+m}{m} (z(t))^{\frac{l}{m}} z_g'(t), \quad t \in J.
\end{aligned}$$

Hence,

$$-(z(t))^{\frac{l}{m}} z'_g(t) \le -\frac{m}{l+m}\left((z^{\frac{l+m}{m}})'_g\right)(t), \quad t \in J,$$

and

$$\begin{aligned}
-\int_{t_0}^{a} (z(t))^{\frac{l}{m}} z'_g(t) &\le -\frac{m}{l+m}\int_{t_0}^{a}\left((z^{\frac{l+m}{m}})'_g\right)(t)dg(t)\\
&= -\frac{m}{l+m}(z(t))^{\frac{l+m}{m}}\Big|_{t=t_0}^{t=a}\\
&= \frac{m}{l+m}(z(t_0))^{\frac{l+m}{m}}\\
&= \frac{m}{l+m}\left(\int_{t_0}^{a}\left|y_g^{(n)}\left(g^{-(n-1)}(t)\right)\right|^{m+l} dg(t)\right)^{\frac{l+m}{m}}.
\end{aligned}$$

From this and (5.30), we find

$$\begin{aligned}
&\int_{t_0}^{a}\left|y_g^{(n)}\left(g^{-(n-1)}(t)\right)\right|^{m}\left|y\left(g^{-(n-1)}(t)\right)\right|^{l} dg(t)\\
&\quad\le \int_{t_0}^{a} (z(t))^{\frac{l}{l+m}}\left(z'_g(t)\right)^{\frac{m}{l+m}}\\
&\qquad\times\left(\int_{t}^{a}\left|h_{n-1}\left(s, g^{-(n-1)}(t)\right)\right|^{\frac{l+m}{l+m-1}} dg(s)\right)^{\frac{l(l+m-1)}{l+m}} dg(t), \quad t \in J.
\end{aligned}$$

Now, applying the Stieltjes–Hölder inequality for $\frac{l+m}{l}$ and $\frac{l+m}{m}$, we find

$$\begin{aligned}
&\int_{t_0}^{a}\left|y_g^{(n)}\left(g^{-(n-1)}(t)\right)\right|^{m}\left|y\left(g^{-(n-1)}(t)\right)\right|^{l} dg(t)\\
&\quad\le\left(\int_{t_0}^{a}\left(\int_{t}^{a}\left|h_{n-1}\left(s, g^{-(n-1)}(t)\right)\right|^{\frac{l+m}{l+m-1}} dg(s)\right)^{\frac{l+m}{l+m-1}} dg(t)\right)^{\frac{l}{l+m}}\\
&\qquad\times\left(\int_{t_0}^{a}(z(t))^{\frac{l}{m}} z'_g(t)dg(t)\right)^{\frac{m}{l+m}}\\
&\quad\le\left(\int_{t_0}^{a}\left(\int_{t_0}^{t}\left|h_{n-1}\left(s, g^{-(n-1)}(t)\right)\right|^{\frac{l+m}{l+m-1}} dg(s)\right)^{\frac{l+m}{l+m-1}} dg(t)\right)^{\frac{l}{l+m}}\\
&\qquad\times\left(\frac{m}{l+m}\right)^{\frac{m}{l+m}}\int_{t_0}^{a}\left|y_g^{(n)}\left(g^{-(n-1)}(t)\right)\right|^{m+l} dg(t).
\end{aligned}$$

This completes the proof. □

Theorem 5.41. *Let $y \in \mathscr{C}_g^n(J)$ and (5.25), (5.27) hold. Also, let $l, m > 0$ be such that $l + m > 1$. Then, we have the inequality*

$$\int_{t_0}^{a} \left|y_g^{(n)}\left(g^{-(n-1)}(t)\right)\right|^m \left|y\left(g^{-(n-1)}(t)\right)\right|^l dg(t) \leq L\left(\frac{m}{m+l}\right)^{\frac{m}{m+l}} \int_{t_0}^{a} \left|y_g^{(n)}\left(g^{-(n-1)}(t)\right)\right|^{m+l} dg(t),$$

where

$$L = \max\left\{\left(\int_{\frac{t_0+a}{2}}^{a}\left(\int_t^a \left|h_{n-1}\left(s, g^{-(n-1)}(t)\right)\right|^{\frac{l+m}{l+m-1}} dg(s)\right)^{l+m-1} dg(t)\right)^{\frac{l}{l+m}},\right.$$
$$\left.\times\left(\int_{t_0}^{\frac{t_0+a}{2}}\left(\int_{t_0}^t \left|h_{n-1}\left(s, g^{-(n-1)}(t)\right)\right|^{\frac{l+m}{l+m-1}} dg(s)\right)^{l+m-1} dg(t)\right)^{\frac{l}{l+m}}\right\}.$$

Proof. By Theorem 5.39, we have

$$\int_{t_0}^{\frac{a+t_0}{2}} \left|y_g^{(n)}\left(g^{-(n-1)}(t)\right)\right|^m \left|y\left(g^{-(n-1)}(t)\right)\right|^l dg(t)$$
$$\leq \left(\int_{t_0}^{\frac{a+t_0}{2}}\left(\int_{t_0}^t \left|h_{n-1}\left(s, g^{-(n-1)}(t)\right)\right|^{\frac{l+m}{l+m-1}} dg(s)\right)^{l+m-1} dg(t)\right)^{\frac{l}{l+m}}$$
$$\times\left(\frac{m}{m+l}\right)^{\frac{m}{m+l}} \int_{t_0}^{\frac{t_0+a}{2}} \left|y_g^{(n)}\left(g^{-(n-1)}(t)\right)\right|^{m+l} dg(t)$$
$$\leq L\left(\frac{m}{m+l}\right)^{\frac{m}{m+l}} \int_{t_0}^{\frac{t_0+a}{2}} \left|y_g^{(n)}\left(g^{-(n-1)}(t)\right)\right|^{m+l} dg(t).$$

By Theorem 5.40, we have

$$\int_{\frac{t_0+a}{2}}^{a} \left|y_g^{(n)}\left(g^{-(n-1)}(t)\right)\right|^m \left|y\left(g^{-(n-1)}(t)\right)\right|^l dg(t)$$
$$\leq \left(\int_{\frac{a+t_0}{2}}^{a}\left(\int_t^a \left|h_{n-1}\left(s, g^{-(n-1)}(t)\right)\right|^{\frac{l+m}{l+m-1}} dg(s)\right)^{l+m-1} dg(t)\right)^{\frac{l}{l+m}}$$
$$\times\left(\frac{m}{m+l}\right)^{\frac{m}{m+l}} \int_{\frac{t_0+a}{2}}^{a} \left|y_g^{(n)}\left(g^{-(n-1)}(t)\right)\right|^{m+l} dg(t)$$
$$\leq L\left(\frac{m}{m+l}\right)^{\frac{m}{m+l}} \int_{\frac{t_0+a}{2}}^{a} \left|y_g^{(n)}\left(g^{-(n-1)}(t)\right)\right|^{m+l} dg(t).$$

Then,

$$\begin{aligned}
&\int_{t_0}^{a} \left|y_g^{(n)}\left(g^{-(n-1)}(t)\right)\right|^m \left|y\left(g^{-(n-1)}(t)\right)\right|^l dg(t)\\
&= \int_{t_0}^{\frac{t_0+a}{2}} \left|y_g^{(n)}\left(g^{-(n-1)}(t)\right)\right|^m \left|y\left(g^{-(n-1)}(t)\right)\right|^l dg(t)\\
&\quad + \int_{\frac{t_0+a}{2}}^{a} \left|y_g^{(n)}\left(g^{-(n-1)}(t)\right)\right|^m \left|y\left(g^{-(n-1)}(t)\right)\right|^l dg(t)\\
&\le L\left(\frac{m}{m+l}\right)^{\frac{m}{m+l}} \int_{t_0}^{\frac{t_0+a}{2}} \left|y_g^{(n)}\left(g^{-(n-1)}(t)\right)\right|^{m+l} dg(t)\\
&\quad + L\left(\frac{m}{m+l}\right)^{\frac{m}{m+l}} \int_{\frac{t_0+a}{2}}^{a} \left|y_g^{(n)}\left(g^{-(n-1)}(t)\right)\right|^{m+l} dg(t)\\
&= L\Bigg(\left(\frac{m}{m+l}\right)^{\frac{m}{m+l}} \int_{t_0}^{\frac{t_0+a}{2}} \left|y_g^{(n)}\left(g^{-(n-1)}(t)\right)\right|^{m+l} dg(t)\\
&\quad + \left(\frac{m}{m+l}\right)^{\frac{m}{m+l}} \int_{\frac{t_0+a}{2}}^{a} \left|y_g^{(n)}\left(g^{-(n-1)}(t)\right)\right|^{m+l} dg(t)\Bigg)\\
&= L\left(\frac{m}{m+l}\right)^{\frac{m}{m+l}} \int_{t_0}^{a} \left|y_g^{(n)}\left(g^{-(n-1)}(t)\right)\right|^{m+l} dg(t).
\end{aligned}$$

This completes the proof. □

Before deriving our next results, we need the following two auxiliary results.

Lemma 5.1. *Let $f \in \mathscr{C}(J)$. Then,*

$$\begin{aligned}
&\int_{t_0}^{t}\int_{t_0}^{t_{n-1}} \cdots \int_{t_0}^{t_1} f(s)dg(s)dg(t_1)dg(t_2)\ldots dg(t_{n-1})\\
&= \int_{t_0}^{t} h_{n-1}(t,s)f(s)dg(s), \quad t \in J.
\end{aligned}$$

Proof. Let

$$\begin{aligned}
g_1(t) &= \int_{t_0}^{t}\int_{t_0}^{t_{n-1}} \cdots \int_{t_0}^{t_1} f(s)dg(s)dg(t_1)dg(t_2)\ldots dg(t_{n-1}),\\
g_2(t) &= \int_{t_0}^{t} h_{n-1}(t,s)f(s)dg(s), \quad t \in J.
\end{aligned}$$

Then,

$$(g_1)'_g = \int_{t_0}^{t}\int_{t_0}^{t_{n-2}}\cdots\int_{t_0}^{t_1} f(s)dg(s)dg(t_1)\ldots dg(t_{n-2}),$$

$$(g_1)''_g = \int_{t_0}^{t}\int_{t_0}^{t_{n-3}}\cdots\int_{t_0}^{t_1} f(s)dg(s)dg(t_1)\ldots dg(t_{n-3}),$$

$$\vdots$$

$$(g_1)_g^{(n-1)} = \int_{t_0}^{t} f(s)dg(s),$$

$$(g_1)_g^{(n)} = f(t), \quad t \in J,$$

and

$$\begin{aligned}(g_2)'_g &= h_{n-1}(t,t)f(t) + \int_{t_0}^{t} h_{n-2}(t,s)f(s)dg(s)\\ &= \int_{t_0}^{t} h_{n-2}(t,s)f(s)dg(s),\\ (g_2)''_g &= h_{n-2}(t,t)f(t) + \int_{t_0}^{t} h_{n-3}(t,s)f(s)dg(s)\\ &= \int_{t_0}^{t} h_{n-3}(t,s)f(s)dg(s),\\ &\vdots\\ (g_2)_g^{(n-1)} &= h_1(t,t)f(t) + \int_{t_0}^{t} f(s)dg(s)\\ &= \int_{t_0}^{t} f(s)dg(s),\\ (g_2)_g^{(n)} &= f(t), \quad t \in J.\end{aligned}$$

Therefore,

$$(g_1)_g^{(n)} = (g_2)_g^{(n)}, \quad t \in J,$$

or

$$(g_1 - g_2)_g^{(n)} = 0, \quad t \in J.$$

Hence,

$$\begin{aligned}(g_1 - g_2)_g^{(n-1)} &= (g_1)_g^{(n-1)}(t) - (g_2)_g^{(n-1)}(t)\\ &= 0, \quad t \in J,\end{aligned}$$

and then

$$\begin{aligned}(g_1 - g_2)_g^{(n-2)} &= (g_1)_g^{(n-2)}(t) - (g_2)_g^{(n-2)}(t)\\ &= 0, \quad t \in J,\end{aligned}$$

and so on. Thus,

$$\begin{aligned}(g_1 - g_2)(t) &= (g_1 - g_2)(t)\\ &= 0, \quad t \in J.\end{aligned}$$

This completes the proof. □

Lemma 5.2. *Let $f \in \mathscr{C}(J)$. Then,*

$$\begin{aligned}&\int_t^a \int_{t_{n-1}}^a \cdots \int_{t_1}^a f(s)dg(s)dg(t_1)dg(t_2)\ldots dg(t_{n-1})\\ &= \int_t^a h_{n-1}(s,t)f(s)dg(s), \quad t \in J.\end{aligned}$$

Proof. Let

$$\begin{aligned}g_1(t) &= \int_t^a \int_{t_{n-1}}^a \cdots \int_{t_1}^a f(s)dg(s)dg(t_1)dg(t_2)\ldots dg(t_{n-1}),\\ g_2(t) &= \int_t^a h_{n-1}(s,t)f(s)dg(s), \quad t \in J.\end{aligned}$$

Then,

$$\begin{aligned}(g_1)'_g(t) &= \int_t^a \int_{t_{n-2}}^a \cdots \int_{t_1}^a f(s)dg(s)dg(t_1)\ldots dg(t_{n-2}),\\ (g_1)''_g &= \int_t^a \int_{t_{n-3}}^a \cdots \int_{t_1}^a f(s)dg(s)dg(t_1)\ldots dg(t_{n-3}),\\ &\vdots\\ (g_1)_g^{(n-1)} &= \int_t^a f(s)dg(s),\\ (g_1)_g^{(n)} &= -f(t), \quad t \in J,\end{aligned}$$

and

$$\begin{aligned}
(g_2)'_g &= -h_{n-1}(t,t)f(t) + \int_t^a h_{n-2}(s,t)f(s)dg(s) \\
&= \int_t^a h_{n-2}(s,t)f(s)dg(s), \\
(g_2)''_g(t) &= -h_{n-2}(t,t)f(t) + \int_t^a h_{n-3}(s)f(s)dg(s) \\
&= \int_t^a h_{n-3}(t,s)f(s)dg(s), \\
&\vdots \\
(g_2)^{(n-1)}_g &= -h_1(t,t)f(t) + \int_t^a f(s)dg(s) \\
&= \int_t^a f(s)dg(s), \\
(g_2)^{(n)}_g &= -f(t), \quad t \in J.
\end{aligned}$$

Therefore,

$$(g_1)^{(n)}_g(t) = (g_2)^{(n)}_g(t), \quad t \in J,$$

or

$$(g_1 - g_2)^{(n)}_g(t) = 0, \quad t \in J.$$

Hence,

$$\begin{aligned}
(g_1 - g_2)^{(n-1)}_g(t) &= (g_1)^{(n-1)}_g(t) - (g_2)^{(n-1)}_g(t) \\
&= 0, \quad t \in J,
\end{aligned}$$

and then

$$\begin{aligned}
(g_1 - g_2)^{(n-2)}_g(t) &= (g_1)^{(n-2)}_g(t) - (g_2)^{(n-2)}_g(t) \\
&= 0, \quad t \in J,
\end{aligned}$$

and so on. Thus,

$$\begin{aligned}
(g_1 - g_2)(t) &= (g_1 - g_2)(t) \\
&= 0, \quad t \in J.
\end{aligned}$$

This completes the proof. □

Theorem 5.42. *Let $y \in \mathscr{C}_g^n(J)$ and (5.25) hold. Also, let $p \geq 0$ and $q \geq 1$ and $h \in \mathscr{C}(J)$ be a nonnegative nonincreasing function. Then, we have the inequality*

$$\int_{t_0}^{a} h(t)|y(t)|^p|y_g^{(n)}(t)|^q dg(t) \leq \frac{q}{p+q}(g(a)-g(t_0))^{np} \int_{t_0}^{a} h(t)|y_g^{(n)}(t)|^{p+q} dg(t).$$

Proof. Let

$$z(t) = \int_{t_0}^{t}\int_{t_0}^{t_{n-1}} \cdots \int_{t_0}^{t_1} |y_g^{(n)}(s)|d_g(s)dg(t_1)\dots dg(t_{n-1}).$$

By Lemma 5.1, we get

$$z(t) = \int_{t_0}^{t} h_{n-1}(t,s)|y_g^{(n)}(s)|dg(s), \quad t \in J.$$

Then,

$$\begin{aligned}
z_g'(t) &= h_{n-1}(t,t)|y_g^{(n)}(t)| + \int_{t_0}^{t} h_{n-2}(t,s)|y_g^{(n)}(s)|dg(s)\\
&= \int_{t_0}^{t} h_{n-2}(t,s)|y_g^{(n)}(s)|dg(s)\\
&\geq 0,\\
z_g''(t) &= h_{n-2}(t,t)|y_g^{(n)}(t)| + \int_{t_0}^{t} h_{n-3}(t,s)|y_g^{(n)}(s)|dg(s)\\
&= \int_{t_0}^{t} h_{n-3}(t,s)|y_g^{(n)}(s)|dg(s)\\
&\geq 0,\\
&\vdots\\
z_g^{(n)}(t) &= |y_g^{(n)}(t)|\\
&\geq 0, \quad t \in J.
\end{aligned}$$

Moreover, $z(t_0) = z_g'(t_0) = \cdots = z_g^{(n-1)}(t_0) = 0$. Thus, $z, z_g', \dots, z_g^{(n-1)}$ are nondecreasing functions on J. Next,

$$\begin{aligned}
y(t) &= \int_{t_0}^{t} y_g'(s)dg(s)\\
&= \int_{t_0}^{t}\int_{t_0}^{t_{n-1}} y_g''(s)dg(s)dg(t)_{n-1}\\
&\vdots\\
&= \int_{t_0}^{t}\int_{t_0}^{t_{n-1}}\cdots\int_{t_0}^{t_1} y_g^{(n)}(s)dg(s)dg(t_1)\dots dg(t_{n-1}), \quad t\in J.
\end{aligned}$$

Therefore,

$$\begin{aligned}
|y(t)| &= \left|\int_{t_0}^{t}\int_{t_0}^{t_{n-1}}\cdots\int_{t_0}^{t_1} y_g^{(n)}(s)dg(s)dg(t_1)\dots dg(t_{n-1})\right|\\
&\leq \int_{t_0}^{t}\int_{t_0}^{t_{n-1}}\cdots\int_{t_0}^{t_1} |y_g^{(n)}(s)|dg(s)dg(t_1)\dots dg(t_{n-1})\\
&= z(t), \quad t\in J.
\end{aligned}$$

Also,

$$\begin{aligned}
z_g^{(k)}(t) &= \int_{t_0}^{t} z_g^{(k+1)}(s)dg(s)\\
&\leq (g(t)-g(t_0))z_g^{(k+1)}(t), \quad t\in J, \quad k\in\{0,1,\dots,n-1\}.
\end{aligned}$$

Then, applying Theorem 5.16, we find

$$\begin{aligned}
&\int_{t_0}^{a} h(t)|y(t)|^p|y_g^{(n)}(t)|^q dg(t)\\
&\quad= \int_{t_0}^{a} h(t)|y(t)|^p(z_g^{(n)}(t))^q dg(t)\\
&\quad\leq \int_{t_0}^{a} h(t)(z(t))^p(z_g^{(n)}(t))^q dg(t)\\
&\quad\leq \int_{t_0}^{a} h(t)\left((g(t)-g(t_0))z_g'(t)\right)^p\left(z_g^{(n)}(t)\right)^q dg(t)
\end{aligned}$$

$$\le \int_{t_0}^{a} h(t)\left((g(t)-g(t_0))^2 z_g''(t)\right)^p \left(z_g^{(n)}(t)\right)^q dg(t)$$

$$\vdots$$

$$\le \int_{t_0}^{a} h(t)\left((g(t)-g(t_0))^{n-1} z_g^{(n-1)}(t)\right)^p \left(z_g^{(n)}(t)\right)^q dg(t)$$

$$\le \frac{q}{p+q}(g(a)-g(t_0))^p \int_{t_0}^{a} h(t)(g(t)-g(t_0))^{(n-1)p} \left(z_g^{(n)}(t)\right)^{p+q} dg(t)$$

$$\le \frac{q}{p+q}(g(a)-g(t_0))^p (g(a)-g(t_0))^{(n-1)p} \int_{t_0}^{a} h(t)|z_g^{(n)}(t)|^{p+q} dg(t)$$

$$= \frac{q}{p+q}(g(a)-g(t_0))^{np} \int_{t_0}^{a} h(t)|z_g^{(n)}(t)|^{p+q} dg(t)$$

$$= \frac{q}{p+q}(g(a)-g(t_0))^{np} \int_{t_0}^{a} h(t)|y_g^{(n)}(t)|^{p+q} dg(t).$$

This completes the proof. □

Theorem 5.43. *Let $y \in \mathscr{C}_g^n(J)$ and (5.27) hold. Also, let $p \ge 0$ and $q \ge 1$ and $h \in \mathscr{C}(J)$ be a nonnegative nonincreasing function. Then, we have the inequality*

$$\int_{t_0}^{a} h(t)|y(t)|^p |y_g^{(n)}(t)|^q dg(t) \le \frac{q}{p+q}(g(a)-g(t_0))^{np} \int_{t_0}^{a} h(t)|y_g^{(n)}(t)|^{p+q} dg(t).$$

Proof. Let

$$z(t) = \int_{t}^{a} \int_{t_{n-1}}^{a} \cdots \int_{t_1}^{a} |y_g^{(n)}(s)| dg(s) dg(t_1) \ldots dg(t)_{n-1}.$$

By Lemma 5.2, we get

$$z(t) = \int_{t}^{a} h_{n-1}(s,t)|y_g^{(n)}(s)| dg(s), \quad t \in J.$$

Then,

$$\begin{aligned} z_g'(t) &= -h_{n-1}(t,t)|y_g^{(n)}(t)| + \int_{t}^{a} h_{n-2}(s,t)|y_g^{(n)}(s)| dg(s) \\ &= \int_{t}^{a} h_{n-2}(s,t)|y_g^{(n)}(s)| dg(s) \\ &\ge 0, \end{aligned}$$

$$
\begin{aligned}
z_g''(t) &= h_{n-2}(t,t)|y_g^{(n)}(t)| + \int_t^a h_{n-3}(s,t)|y_g^{(n)}(s)|dg(s) \\
&= \int_t^a h_{n-3}(s,t)|y_g^{(n)}(s)|dg(s) \\
&\geq 0, \\
&\vdots \\
z_g^{(n-1)}(t) &= \int_t^a |y_g^{(n)}(s)|dg(s) \\
&\geq 0, \\
z_g^{(n)}(t) &= -|y_g^{(n)}(t)| \\
&\leq 0, \quad t \in J.
\end{aligned}
$$

Moreover, $z(a) = z_g'(a) = \cdots = z_g^{(n-1)}(a) = 0$. Thus, $z, z_g', \ldots, z_g^{(n-1)}$ are nondecreasing functions on J. Next,

$$
\begin{aligned}
y(t) &= -\int_t^a y_g'(s)dg(s) \\
&= \int_t^a \int_{t_{n-1}}^a y_g''(s)dg(s)dg(t_{n-1}) \\
&\vdots \\
&= (-1)^n \int_t^a \int_{t_{n-1}}^a \cdots \int_{t_1}^a y_g^{(n)}(s)dg(s)dg(t_1)\ldots dg(t_{n-1}), \quad t \in J.
\end{aligned}
$$

Therefore,

$$
\begin{aligned}
|y(t)| &= \left|(-1)^n \int_t^a \int_{t_{n-1}}^a \cdots \int_{t_1}^a y_g^{(n)}(s)dg(s)dg(t_1)\ldots dg(t_{n-1})\right| \\
&\leq \int_t^a \int_{t_{n-1}}^a \cdots \int_{t_1}^a |y_g^{(n)}(s)|dg(s)dg(t_1)\ldots dg(t_{n-1}) \\
&= z(t), \quad t \in J.
\end{aligned}
$$

Also,

$$
\begin{aligned}
z_g^{(k)}(t) &= -\int_t^a z_g^{(k+1)}(s)dg(s) \\
&\leq -(g(a) - g(t))z_g^{(k+1)}(t), \quad t \in J, \quad k \in \{0, 1, \ldots, n-1\}.
\end{aligned}
$$

Then, applying Theorem 5.17, we find

$$\begin{aligned}
&\int_{t_0}^{a} h(t)|y(t)|^p|y_g^{(n)}(t)|^q dg(t)\\
&= \int_{t_0}^{a} h(t)|y(t)|^p(z_g^{(n)}(t))^q dg(t)\\
&\le \int_{t_0}^{a} h(t)(z(t))^p(z_g^{(n)}(t))^q dg(t)\\
&\le \int_{t_0}^{a} h(t)\left((g(a)-g(t))z_g'(t)\right)^p\left(z_g^{(n)}(t)\right)^q dg(t)\\
&\le \int_{t_0}^{a} h(t)\left((g(a)-g(t))^2 z_g''(t)\right)^p\left(z_g^{(n)}(t)\right)^q dg(t)\\
&\vdots\\
&\le \int_{t_0}^{a} h(t)\left((g(a)-g(t))^{n-1} z_g^{(n-1)}(t)\right)^p\left(z_g^{(n)}(t)\right)^q dg(t)\\
&\le \frac{q}{p+q}(g(a)-g(t_0))^p h(t)(g(a)-g(t))^{(n-1)p}\left(z_g^{(n)}(t)\right)^{p+q} dg(t)\\
&\le \frac{q}{p+q}(g(a)-g(t_0))^p(g(a)-g(t_0))^{(n-1)p}\int_{t_0}^{a} h(t)|z_g^{(n)}(t)|^{p+q}dg(t)\\
&= \frac{q}{p+q}(g(a)-g(t_0))^{np}\int_{t_0}^{a} h(t)|z_g^{(n)}(t)|^{p+q}dg(t)\\
&= \frac{q}{p+q}(g(a)-g(t_0))^{np}\int_{t_0}^{a} h(t)|y_g^{(n)}(t)|^{p+q}dg(t).
\end{aligned}$$

This completes the proof. □

Theorem 5.44. *Let $y \in \mathscr{C}_g^n(J)$ and (5.25), (5.27) hold. Also, let $p \ge 0$ and $q \ge 1$, and $h \in \mathscr{C}(J)$ be a nonnegative nonincreasing function. Then, we have the following inequality:*

$$\int_{t_0}^{a} h(t)|y(t)|^p|y_g^{(n)}(t)|^q dg(t) \le \frac{q}{p+q}\left(\frac{g(a)-g(t_0)}{2}\right)^{np}\int_{t_0}^{a} h(t)|y_g^{(n)}(t)|^{p+q}dg(t).$$

Proof. By Theorem 5.42, we have

$$\int_{t_0}^{\frac{a+t_0}{2}} h(t)|y(t)|^p|y_g^{(n)}(t)|^q dg(t)$$
$$\leq \frac{q}{p+q}\left(\frac{g(a)-g(t_0)}{2}\right)^{np}\int_{t_0}^{\frac{a+t_0}{2}} h(t)|y_g^{(n)}(t)|^{p+q}dg(t).$$

By Theorem 5.43, we find

$$\int_{\frac{a+t_0}{2}}^{a} h(t)|y(t)|^p|y_g^{(n)}(t)|^q dg(t)$$
$$\leq \frac{q}{p+q}\left(\frac{g(a)-g(t_0)}{2}\right)^{np}\int_{\frac{a+t_0}{2}}^{a} h(t)|y_g^{(n)}(t)|^{p+q}dg(t).$$

Then,

$$\int_{t_0}^{a} h(t)|y(t)|^p|y_g^{(n)}(t)|^q dg(t)$$
$$= \int_{t_0}^{\frac{a+t_0}{2}} h(t)|y(t)|^p|y_g^{(n)}(t)|^q dg(t) + \int_{\frac{a+t_0}{2}}^{a} h(t)|y(t)|^p|y_g^{(n)}(t)|^q dg(t)$$
$$\leq \frac{q}{p+q}\left(\frac{g(a)-g(t_0)}{2}\right)^{np}\int_{t_0}^{\frac{a+t_0}{2}} h(t)|y_g^{(n)}(t)|^{p+q}dg(t)$$
$$+ \frac{q}{p+q}\left(\frac{g(a)-g(t_0)}{2}\right)^{np}\int_{\frac{a+t_0}{2}}^{a} h(t)|y_g^{(n)}(t)|^{p+q}dg(t)$$
$$= \frac{q}{p+q}\left(\frac{g(a)-g(t_0)}{2}\right)^{np}\int_{t_0}^{a} h(t)|y_g^{(n)}(t)|^{p+q}dg(t).$$

This completes the proof. □

Theorem 5.45. *Let* $x, y \in \mathscr{C}_g^n(J)$ *and* (5.25) *hold, and*

$$x(t_0) = x_g'(t_0) = \cdots = x_g^{(n)}(t_0) = 0.$$

Also, let $p \geq 0$ *and* $q \geq 1$, *and* $h \in \mathscr{C}(J)$ *be a nonnegative nonincreasing function. Then, we have the inequality*

$$\int_{t_0}^{a} h(t)\left(|x(t)|^p|y_g^{(n)}(t)|^q + |y(t)|^p|x_g^{(n)}(t)|^q\right)dg(t)$$
$$\leq \frac{2q}{p+q}(g(a)-g(t_0))^{np}\int_{t_0}^{a} h(t)\left(|x_g^{(n)}(t)|^{p+q} + |y_g^{(n)}(t)|^{p+q}\right)dg(t).$$

Proof. Let

$$z(t) = \int_{t_0}^{t}\int_{t_0}^{t_{n-1}}\cdots\int_{t_0}^{t_1} \times \left(|x_g^{(n)}(s)|^{p+q} + |y_g^{(n)}(s)|^{p+q}\right)^{\frac{1}{p+q}} dg(s)dg(t_1)\dots dg(t_{n-1}).$$

By Lemma 5.1, we get

$$z(t) = \int_{t_0}^{t} h_{n-1}(t, g(s))\left(|x_g^{(n)}(s)|^{p+q} + |y_g^{(n)}(s)|^{p+q}\right)^{\frac{1}{p+q}} dg(s), \quad t \in J.$$

Then,

$$\begin{aligned}
z_g'(t) &= h_{n-1}(t,t)\left(|x_g^{(n)}(t)|^{p+q} + |y_g^{(n)}(t)|^{p+q}\right)^{\frac{1}{p+q}} \\
&\quad + \int_{t_0}^{t} h_{n-2}(t, g(s))\left(|x_g^{(n)}(s)|^{p+q} + |y_g^{(n)}(s)|^{p+q}\right)^{\frac{1}{p+q}} dg(s) \\
&= \int_{t_0}^{t} h_{n-2}(t, g(s))\left(|x_g^{(n)}(s)|^{p+q} + |y_g^{(n)}(s)|^{p+q}\right)^{\frac{1}{p+q}} dg(s) \\
&\geq 0, \\
z_g''(t) &= h_{n-2}(t,t)\left(|x_g^{(n)}(t)|^{p+q} + |y_g^{(n)}(t)|^{p+q}\right)^{\frac{1}{p+q}} \\
&\quad + \int_{t_0}^{t} h_{n-3}(t, g(s))\left(|x_g^{(n)}(s)|^{p+q} + |y_g^{(n)}(s)|^{p+q}\right)^{\frac{1}{p+q}} dg(s) \\
&= \int_{t_0}^{t} h_{n-3}(t, g(s))\left(|x_g^{(n)}(s)|^{p+q} + |y_g^{(n)}(s)|^{p+q}\right)^{\frac{1}{p+q}} dg(s) \\
&\geq 0, \\
&\vdots \\
z_g^{(n)}(t) &= \left(|x_g^{(n)}(t)|^{p+q} + |y_g^{(n)}(t)|^{p+q}\right)^{\frac{1}{p+q}} \\
&\geq \max\left\{|x_g^{(n)}(t)|, |y_g^{(n)}(t)|\right\}, \quad t \in J.
\end{aligned}$$

Moreover, $z(t_0) = z_g'(t_0) = \cdots = z_g^{(n-1)}(t_0) = 0$. Thus, $z, z_g', \dots, z_g^{(n-1)}$ are nondecreasing functions on J. Next,

$$\begin{aligned}
y(t) &= \int_{t_0}^{t} y'_g(s)dg(s)\\
&= \int_{t_0}^{t}\int_{t_0}^{t_{n-1}} y''_g(s)dg(s)dg(t)_{n-1}\\
&\vdots\\
&= \int_{t_0}^{t}\int_{t_0}^{t_{n-1}}\cdots\int_{t_0}^{t_1} y_g^{(n)}(s)dg(s)dg(t_1)\dots dg(t_{n-1}), \quad t\in J.
\end{aligned}$$

Therefore,

$$\begin{aligned}
|y(t)| &= \left|\int_{t_0}^{t}\int_{t_0}^{t_{n-1}}\cdots\int_{t_0}^{t_1} y_g^{(n)}(s)dg(s)dg(t_1)\dots dg(t_{n-1})\right|\\
&\le \int_{t_0}^{t}\int_{t_0}^{t_{n-1}}\cdots\int_{t_0}^{t_1} |y_g^{(n)}(s)|dg(s)dg(t_1)\dots dg(t_{n-1})\\
&\le \int_{t_0}^{t}\int_{t_0}^{t_{n-1}}\cdots\int_{t_0}^{t_1} \left(|x_g^{(n)}(s)|^{p+q}+|y_g^{(n)}(s)|^{p+q}\right)^{\frac{1}{p+q}}\\
&\quad\times dg(s)dg(t_1)\dots dg(t_{n-1}) = z(t), \quad t\in J,
\end{aligned}$$

and

$$\begin{aligned}
x(t) &= \int_{t_0}^{t} x'_g(s)dg(s)\\
&= \int_{t_0}^{t}\int_{t_0}^{t_{n-1}} x''_g(s)dg(s)dg(t_{n-1})\\
&\vdots\\
&= \int_{t_0}^{t}\int_{t_0}^{t_{n-1}}\cdots\int_{t_0}^{t_1} x_g^{(n)}(s)dg(s)dg(t_1)\dots dg(t_{n-1}), \quad t\in J.
\end{aligned}$$

Therefore,

$$\begin{aligned}
|x(t)| &= \left|\int_{t_0}^{t}\int_{t_0}^{t_{n-1}}\cdots\int_{t_0}^{t_1} x_g^{(n)}(s)dg(s)dg(t_1)\dots dg(t_{n-1})\right|\\
&\le \int_{t_0}^{t}\int_{t_0}^{t_{n-1}}\cdots\int_{t_0}^{t_1} |x_g^{(n)}(s)|dg(s)dg(t_1)\dots dg(t_{n-1})
\end{aligned}$$

$$
\begin{aligned}
&\le \int_{t_0}^{t}\int_{t_0}^{t_{n-1}}\cdots\int_{t_0}^{t_1}\left(|x_g^{(n)}(s)|^{p+q}+|y_g^{(n)}(s)|^{p+q}\right)^{\frac{1}{p+q}}\\
&\quad\times dg(s)dg(t_1)\dots dg(t_{n-1})\\
&= z(t), \quad t\in J.
\end{aligned}
$$

Also,

$$
\begin{aligned}
z_g^{(k)}(t) &= \int_{t_0}^{t} z_g^{(k+1)}(s)dg(s)\\
&\le (g(t)-g(t_0))z_g^{(k+1)}(t), \quad t\in J, \quad k\in\{0,1,\dots,n-1\}.
\end{aligned}
$$

Then, applying Theorem 5.42, we find

$$
\begin{aligned}
&\int_{t_0}^{a} h(t)\left(|x(t)|^p|y_g^{(n)}(t)|^q+|y(t)|^p|x_g^{(n)}(t)|^q\right)dg(t)\\
&\le \int_{t_0}^{a} h(t)|z(t)|^p\left(|y_g^{(n)}(t)|^q+|x_g^{(n)}(t)|^q\right)dg(t)\\
&\le 2\int_{t_0}^{a} h(t)|z(t)|^p|z_g^{(n)}(t)|^q dg(t)\\
&\le \frac{2q}{p+q}(g(a)-g(t_0))^p(g(a)-g(t_0))^{(n-1)p}\int_{t_0}^{a} h(t)|z_g^{(n)}(t)|^{p+q}dg(t)\\
&= \frac{2q}{p+q}(g(a)-g(t_0))^{np}\int_{t_0}^{a} h(t)|z_g^{(n)}(t)|^{p+q}dg(t)\\
&= \frac{2q}{p+q}(g(a)-g(t_0))^{np}\int_{t_0}^{a} h(t)\left(|x_g^{(n)}(t)|^{p+q}+|y_g^{(n)}(t)|^{p+q}\right)dg(t).
\end{aligned}
$$

This completes the proof. □

Theorem 5.46. *Let $x, y \in \mathscr{C}_g^n(J)$ and (5.27) hold, and $x(a) = x_g'(a) = \cdots x_g^{(n)}(a) = 0$. Also, let $p \ge 0$ and $q \ge 1$, and $h \in \mathscr{C}(J)$ be a nonnegative nonincreasing function. Then, we have the inequality*

$$
\begin{aligned}
&\int_{t_0}^{a} h(t)\left(|y(t)|^p|x_g^{(n)}(t)|^q+|x(t)|^p|y_g^{(n)}(t)|^q\right)dg(t)\\
&\quad\le \frac{2q}{p+q}(g(a)-g(t_0))^{np}\int_{t_0}^{a} h(t)\left(|x_g^{(n)}(t)|+|y_g^{(n)}(t)|^{p+q}\right)dg(t).
\end{aligned}
$$

Proof. Let

$$z(t) = \int_t^a \int_{t_{n-1}}^a \cdots \int_{t_1}^a \left(|x_g^{(n)}(s)|^{p+q} + |y_g^{(n)}(s)|^{p+q}\right)^{\frac{1}{p+q}} \times dg(s)dg(t_1)\ldots dg(t_{n-1}).$$

By Lemma 5.2, we get

$$z(t) = \int_t^a h_{n-1}(s,t)\left(|x_g^{(n)}(s)|^{p+q} + |y_g^{(n)}(s)|^{p+q}\right)^{\frac{1}{p+q}} dg(s), \quad t \in J.$$

Then,

$$\begin{aligned}
z_g'(t) &= -h_{n-1}(t,t)\left(|x_g^{(n)}(t)|^{p+q} + |y_g^{(n)}(t)|^{p+q}\right)^{\frac{1}{p+q}} \\
&\quad + \int_t^a h_{n-2}(s,t)\left(|x_g^{(n)}(s)|^{p+q} + |y_g^{(n)}(s)|^{p+q}\right)^{\frac{1}{p+q}} dg(s) \\
&= \int_t^a h_{n-2}(s,t)\left(|x_g^{(n)}(s)|^{p+q} + |y_g^{(n)}(s)|^{p+q}\right)^{\frac{1}{p+q}} dg(s) \geq 0, \\
z_g''(t) &= h_{n-2}(t,t)\left(|x_g^{(n)}(t)|^{p+q} + |y_g^{(n)}(t)|^{p+q}\right)^{\frac{1}{p+q}} \\
&\quad + \int_t^a h_{n-3}(s,t)\left(|x_g^{(n)}(s)|^{p+q} + |y_g^{(n)}(s)|^{p+q}\right)^{\frac{1}{p+q}} dg(s) \\
&= \int_t^a h_{n-3}(s,t)\left(|x_g^{(n)}(s)|^{p+q} + |y_g^{(n)}(s)|^{p+q}\right)^{\frac{1}{p+q}} dg(s) \geq 0, \\
&\vdots \\
z_g^{(n-1)}(t) &= \int_t^a \left(|x_g^{(n)}(s)|^{p+q} + |y_g^{(n)}(s)|^{p+q}\right)^{\frac{1}{p+q}} dg(s) \geq 0, \\
z_g^{(n)}(t) &= -\left(|x_g^{(n)}(t)|^{p+q} + |y_g^{(n)}(t)|^{p+q}\right)^{\frac{1}{p+q}} \leq 0, \quad t \in J.
\end{aligned}$$

Moreover, $z(a) = z_g'(a) = \cdots = z_g^{(n-1)}(a) = 0$. Thus, $z, z_g', \ldots, z_g^{(n-1)}$ are nondecreasing functions on J. Next,

$$\begin{aligned}
y(t) &= -\int_t^a y_g'(s)dg(s) \\
&= \int_t^a \int_{t_{n-1}}^a y_g''(s)dg(s)dg(t_{n-1})
\end{aligned}$$

$$\vdots$$

$$= (-1)^n \int_t^a \int_{t_{n-1}}^a \cdots \int_{t_1}^a y_g^{(n)}(s)dg(s)dg(t_1)\dots dg(t_{n-1}), \quad t \in J.$$

Therefore,

$$\begin{aligned}
|y(t)| &= \left| (-1)^n \int_t^a \int_{t_{n-1}}^a \cdots \int_{t_1}^a y_g^{(n)}(s)dg(s)dg(t_1)\dots dg(t_{n-1}) \right| \\
&\leq \int_t^a \int_{t_{n-1}}^a \cdots \int_{t_1}^a |y_g^{(n)}(s)|dg(s)dg(t_1)\dots dg(t_{n-1}) \\
&\leq \int_t^a \int_{t_{n-1}}^a \cdots \int_{t_1}^a \left(|x_g^{(n)}(s)|^{p+q} + |y_g^{(n)}(s)|^{p+q} \right)^{\frac{1}{p+q}} \\
&\quad \times dg(s)dg(t_1)\dots dg(t_{n-1}) = z(t), \quad t \in J,
\end{aligned}$$

and

$$\begin{aligned}
x(t) &= -\int_t^a x'_g(s)dg(s) \\
&= \int_t^a \int_{t_{n-1}}^a x''_g(s)dg(s)dg(t_{n-1}) \\
&\vdots \\
&= (-1)^n \int_t^a \int_{t_{n-1}}^a \cdots \int_{t_1}^a x_g^{(n)}(s)dg(s)dg(t_1)\dots dg(t_{n-1}), \quad t \in J.
\end{aligned}$$

Therefore,

$$\begin{aligned}
|x(t)| &= \left| (-1)^n \int_t^a \int_{t_{n-1}}^a \cdots \int_{t_1}^a x_g^{(n)}(s)dg(s)dg(t_1)\dots dg(t_{n-1}) \right| \\
&\leq \int_t^a \int_{t_{n-1}}^a \cdots \int_{t_1}^a |x_g^{(n)}(s)|dg(s)dg(t_1)\dots dg(t_{n-1}) \\
&\leq \int_t^a \int_{t_{n-1}}^a \cdots \int_{t_1}^a \left(|x_g^{(n)}(s)|^{p+q} + |y_g^{(n)}(s)|^{p+q} \right)^{\frac{1}{p+q}} \\
&\quad \times dg(s)dg(t_1)\dots dg(t_{n-1}) = z(t), \quad t \in J,
\end{aligned}$$

Also,

$$\begin{aligned} z_g^{(k)}(t) &= -\int_t^a z_g^{(k+1)}(s)dg(s) \\ &\le -(g(a)-g(t))z_g^{(k+1)}(t), \quad t \in J, \quad k \in \{0,1,\ldots,n-1\}. \end{aligned}$$

Then, applying Theorem 5.43, we find

$$\begin{aligned} &\int_{t_0}^a h(t)\left(|x(t)|^p|y_g^{(n)}(t)|^q + |y(t)|^p|x_g^{(n)}(t)|^q\right)dg(t) \\ &\le \int_{t_0}^a h(t)|z(t)|^p\left(|y_g^{(n)}(t)|^q + |x_g^{(n)}(t)|^q\right)dg(t) \\ &\le 2\int_{t_0}^a h(t)|z(t)|^p|z_g^{(n)}(t)|^q dg(t) \\ &\le \frac{2q}{p+q}(g(a)-g(t_0))^p(g(a)-g(t_0))^{(n-1)p}\int_{t_0}^a h(t)|z_g^{(n)}(t)|^{p+q}dg(t) \\ &= \frac{2q}{p+q}(g(a)-g(t_0))^{np}\int_{t_0}^a h(t)|z_g^{(n)}(t)|^{p+q}dg(t) \\ &= \frac{2q}{p+q}(g(a)-g(t_0))^{np}\int_{t_0}^a h(t)\left(|x_g^{(n)}(t)|^{p+q} + |y_g^{(n)}(t)|^{p+q}\right)dg(t). \end{aligned}$$

This completes the proof. □

Theorem 5.47. *Let* $y \in \mathscr{C}_g^n(J)$ *and* (5.25), (5.27) *hold, and*

$$\begin{aligned} x(t_0) &= x'_g(t_0) = \cdots = x_g^{(n)}(t_0) = 0, \\ x(a) &= x'_g(a) = x''_g(a) = \cdots = x_g^{(n)}(a) = 0. \end{aligned}$$

Also, let $p \ge 0$ *and* $q \ge 1$, *and* $h \in \mathscr{C}(J)$ *be a nonnegative nonincreasing function. Then, we have the inequality*

$$\begin{aligned} &\int_{t_0}^a h(t)\left(|x(t)|^p|y_g^{(n)}(t)|^q + |y(t)|^p|x_g^{(n)}(t)|^q\right)dg(t) \\ &\le \frac{q}{p+q}\left(\frac{g(a)-g(t_0)}{2}\right)^{np}\int_{t_0}^a h(t)\left(|y_g^{(n)}(t)|^{p+q} + |y_g^{(n)}(t)|^{p+q}\right)dg(t). \end{aligned}$$

Proof. By Theorem 5.45, we have

$$\int_{t_0}^{\frac{a+t_0}{2}} h(t)\left(|x(t)|^p|y_g^{(n)}(t)|^q+|y(t)|^p|x_g^{(n)}(t)|^q\right)dg(t)$$
$$\leq \frac{q}{p+q}\left(\frac{g(a)-g(t_0)}{2}\right)^{np}\int_{t_0}^{\frac{a+t_0}{2}} h(t)\left(|x_g^{(n)}(t)|^{p+q}+|y_g^{(n)}(t)|^{p+q}\right)dg(t).$$

By Theorem 5.46, we find

$$\int_{\frac{a+t_0}{2}}^{a} h(t)\left(|x(t)|^p|y_g^{(n)}(t)|^q+|y(t)|^p|x_g^{(n)}(t)|^q\right)dg(t)$$
$$\leq \frac{q}{p+q}\left(\frac{g(a)-g(t_0)}{2}\right)^{np}\int_{\frac{a+t_0}{2}}^{a} h(t)\left(|x_g^{(n)}(t)|^{p+q}+|y_g^{(n)}(t)|^{p+q}\right)dg(t).$$

Then,

$$\int_{t_0}^{a} h(t)\left(|x(t)|^p|y_g^{(n)}(t)|^q+|y(t)|^p|x_g^{(n)}(t)|^q\right)dg(t)$$
$$=\int_{t_0}^{\frac{a+t_0}{2}} h(t)\left(|x(t)|^p|y_g^{(n)}(t)|^q+|y(t)|^p|x_g^{(n)}(t)|^q\right)dg(t)$$
$$+\int_{\frac{a+t_0}{2}}^{a} h(t)\left(|x(t)|^p|y_g^{(n)}(t)|^q+|y(t)|^p|x_g^{(n)}(t)|^q\right)dg(t)$$
$$\leq \frac{2q}{p+q}\left(\frac{g(a)-g(t_0)}{2}\right)^{np}\int_{t_0}^{\frac{a+t_0}{2}} h(t)\left(|x_g^{(n)}(t)|^{p+q}+|y_g^{(n)}(t)|^{p+q}\right)dg(t)$$
$$+\frac{2q}{p+q}\left(\frac{g(a)-g(t_0)}{2}\right)^{np}\int_{\frac{a+t_0}{2}}^{a} h(t)\left(|x_g^{(n)}(t)|^{p+q}+|y_g^{(n)}(t)|^{p+q}\right)dg(t)$$
$$=\frac{2q}{p+q}\left(\frac{g(a)-g(t_0)}{2}\right)^{np}\int_{t_0}^{a} h(t)\left(|x_g^{(n)}(t)|^{p+q}+|y_g^{(n)}(t)|^{p+q}\right)dg(t).$$

This completes the proof. □

5.8 Advanced Practical Problems

Problem 5.5. Let $r \in \mathscr{C}(J)$ be a positive function such that

$$\int_{t_0}^{a} \frac{1}{r(t)}dg(t) < \infty$$

and $y \in \mathscr{C}_g^1(J)$ be such that $y(t_0) = 0$. Then, prove that

$$\int_{t_0}^{a} r(t)|(y(t) + y(t+))y_g'(t)|dg(t)$$
$$\leq \left(\int_{t_0}^{a} \frac{1}{r(t)} dg(t)\right)\left(\int_{t_0}^{a} r(t)q(t)(y_g'(t))^2 dg(t)\right).$$

Problem 5.6. Let $r \in \mathscr{C}(J)$ be a positive function such that

$$\int_{t_0}^{a} \frac{1}{r(t)} dg(t) < \infty$$

and $y \in \mathscr{C}_g^1(J)$ be such that $y(t_0) = 0$ and y_g' does not change its sign on J. Then, prove that

$$\int_{t_0}^{a} r(t)|y(t) + y(t+)||y_g'(t)|dg(t)$$
$$\leq K_1^*(a, t_0) \int_{t_0}^{a} r(t)(y_g'(t))^2 dg(t),$$

where

$$K_1^*(a, t_0) = \sqrt{2}\left(\int_{t_0}^{a} r(t)\left(\int_{t_0}^{t} \frac{1}{r(\tau)} dg(\tau)\right) dg(t)\right)^{\frac{1}{2}} + \sup_{t\in J}((\Delta g(t))).$$

Problem 5.7. Let $r \in \mathscr{C}(J)$ be a positive function such that

$$\int_{t_0}^{a} \frac{1}{r(t)} dg(t) < \infty$$

and $y \in \mathscr{C}_g^1(J)$ be such that $y(a) = 0$ and y_g' does not change its sign on J. Then, prove that

$$\int_{t_0}^{a} r(t)|y(t) + y(t+)||y_g'(t)|dg(t) \leq K_2^*(a, t_0) \int_{t_0}^{a} r(t)(y_g'(t))^2 dg(t),$$

where

$$K_2^*(a, t_0) = \sqrt{2}\left(\int_{t_0}^{a} r(t)\left(\int_{t}^{a} \frac{1}{r(\tau)} dg(\tau)\right) dg(t)\right)^{\frac{1}{2}} + \sup_{t\in J}((\Delta g(t))).$$

Problem 5.8. Let $r \in \mathscr{C}(J)$ be a positive function such that

$$\int_{t_0}^{a} \frac{1}{r(t)} dg(t) < \infty$$

and $y \in \mathscr{C}_g^1(J)$ be such that $y(t_0) = y(a) = 0$ and y_g' does not change its sign on J. Then, prove that, for any $c \in J$, we have

$$\int_{t_0}^{a} r(t)|y(t) + y(t+)||y_g'(t)|dg(t)$$
$$\leq 2\max\{K_1^*(c,t_0), K_2^*(a,c)\} \int_{t_0}^{a} r(t)(y_g'(t))^2 dg(t),$$

where $K_1^*(c,t_0)$ and $K_2^*(a,c)$ are as in Problem 5.6 and Problem 5.7, respectively.

Problem 5.9. Let $r \in \mathscr{C}(J)$ be a positive function such that

$$\int_{t_0}^{a} \frac{1}{r(t)} dg(t) < \infty$$

and $y \in \mathscr{C}_g^1(J)$ be such that $y(t_0) = y(a) = 0$ and y_g' does not change its sign on J. Then, prove that

$$\int_{t_0}^{a} r(t)|y(t) + y(t+)||y_g'(t)|dg(t)$$
$$\leq 2\min_{c\in J}\left(\max\{K_1^*(c,t_0), K_2^*(a,c)\}\right) \int_{t_0}^{a} r(t)(y_g'(t))^2 dg(t),$$

where $K_1^*(c,t_0)$ and $K_2^*(a,c)$ are as in Problem 5.6 and Problem 5.7, respectively.

Problem 5.10. Suppose p and q are positive real numbers such that $p+q > 1$ and $r \in \mathscr{C}(J)$ is a nonnegative function such that

$$\int_{t_0}^{a} \frac{1}{(r(t))^{\frac{1}{p+q-1}}} dg(t) < \infty.$$

If $y \in \mathscr{C}_g^1(J)$ is such that $y(t_0) = 0$ and y_g' does not change its sign on J, then prove that

$$\int_{t_0}^{a} r(t)|y(t)|^p |y_g'(t)|^q dg(t) \leq K_1^*(t_0,a,p,q) \int_{t_0}^{a} r(t)|y_g'(t)|^{p+q} dg(t),$$

where

$$K_1^*(t_0,a,p,q)$$
$$= \left(\frac{q}{p+q}\right)^{\frac{q}{p+q}} \left(\int_{t_0}^{a} r(t)\left(\int_{t_0}^{t} \frac{1}{(r(\tau))^{\frac{1}{p+q-1}}}dg(\tau)\right)^{p+q-1} dg(t)\right)^{\frac{p}{p+q}}.$$

Problem 5.11. Suppose p and q are positive real numbers such that $p+q > 1$ and $r \in \mathscr{C}(J)$ is a nonnegative function such that

$$\int_{t_0}^{a} \frac{1}{(r(t))^{\frac{1}{p+q-1}}}dg(t) < \infty.$$

If $y \in \mathscr{C}_g^1(J)$ is such that $y(a) = 0$ and y'_g does not change its sign on J, then prove that

$$\int_{t_0}^{a} r(t)|y(t)|^p|y'_g(t)|^q dg(t) \le K_2^*(t_0,a,p,q)\int_{t_0}^{a} r(t)|y'_g(t)|^{p+q}dg(t),$$

where

$$K_2^*(t_0,a,p,q)$$
$$= \left(\frac{q}{p+q}\right)^{\frac{q}{p+q}} \left(\int_{t_0}^{a} r(t)\left(\int_{t}^{a} \frac{1}{(r(\tau))^{\frac{1}{p+q-1}}}dg(\tau)\right)^{p+q-1} dg(t)\right)^{\frac{p}{p+q}}.$$

Problem 5.12. Suppose p and q are positive real numbers such that $p+q > 1$ and $r \in \mathscr{C}(J)$ is a nonnegative function such that

$$\int_{t_0}^{a} \frac{1}{(r(t))^{\frac{1}{p+q-1}}}dg(t) < \infty.$$

If $y \in \mathscr{C}_g^1(J)$ is such that $y(t_0) = y(a) = 0$ and y'_g does not change its sign on J, then prove that

$$\int_{t_0}^{a} r(t)|y(t)|^p|y'_g(t)|^q dg(t) \le K^*(t_0,a,p,q)\int_{t_0}^{a} r(t)|y'_g(t)|^{p+q}dg(t),$$

where

$$K^*(t_0,a,p,q) = \max_{c\in J}\left(\max\{K_1^*(t_0,c,p,q), K_2^*(c,a,p,q)\}\right),$$

where $K_1^*(t_0,c,p,q)$ and $K_2^*(c,a,p,q)$ are as in Problem 5.10 and Problem 5.11, respectively.

Problem 5.13. Suppose p and q are positive real numbers such that $p \geq 1$ and $r \in \mathscr{C}(J)$ is a nonnegative function such that

$$\int_{t_0}^{a} \frac{1}{(r(t))^{\frac{1}{p+q-1}}} dg(t) < \infty.$$

Let $y \in \mathscr{C}_g^1(J)$ be such that $y(t_0) = 0$ and y'_g does not change its sign on J. Then, prove the inequality

$$\int_{t_0}^{a} r(t)|y(t) + y(t+)|^p |y'_g(t)|^q dg(t) \leq K_1(t_0, a, p, q) \int_{t_0}^{a} r(t)|y'_g(t)|^{p+q} dg(t),$$

where

$$K_1(t_0, a, p, q) = 2^{2p-1} \left(\frac{q}{p+q}\right)^{\frac{q}{p+q}} \left(\int_{t_0}^{a} (r(t))^{\frac{p^2+pq-q^2}{pq}} \right.$$
$$\left. \times \left(\int_{t_0}^{t} \frac{1}{(r(\tau))^{\frac{1}{p+q-1}}} dg(\tau)\right)^{p+q-1} dg(t)\right)^{\frac{p}{p+q}}$$
$$+ 2^{p-1} \sup_{t\in J} ((\Delta g(t))^p).$$

Problem 5.14. Suppose p and q are positive real numbers such that $p \geq 1$ and $r \in \mathscr{C}(J)$ is a nonnegative function such that

$$\int_{t_0}^{a} \frac{1}{(r(t))^{\frac{1}{p+q-1}}} dg(t) < \infty.$$

Let $y \in \mathscr{C}_g^1(J)$ be such that $y(a) = 0$ and y'_g does not change its sign on J. Then, prove the inequality

$$\int_{t_0}^{a} r(t)|y(t) + y(t+)|^p |y'_g(t)|^q dg(t) \leq K_2(t_0, a, p, q) \int_{t_0}^{a} r(t)|y'_g(t)|^{p+q} dg(t),$$

where

$$K_2(t_0, a, p, q) = 2^{2p-1} \left(\frac{q}{p+q}\right)^{\frac{q}{p+q}}$$
$$\times \left(\int_{t_0}^{a} (r(t))^{\frac{p^2+pq-q^2}{pq}} \left(\int_{t}^{a} \frac{1}{(r(\tau))^{\frac{1}{p+q-1}}} dg(\tau)\right)^{p+q-1}\right)^{\frac{p}{p+q}}$$
$$+ 2^{p-1} \sup_{t\in J} ((\Delta g(t))^p).$$

Problem 5.15. Suppose p and q are positive real numbers such that $p \geq 1$ and $r \in \mathscr{C}(J)$ is a nonnegative function such that

$$\int_{t_0}^{a} \frac{1}{(r(t))^{\frac{1}{p+q-1}}} dg(t) < \infty.$$

Let $y \in \mathscr{C}_g^1(J)$ be such that $y(t_0) = y(a) = 0$ and y_g' does not change its sign on J. Then, prove the inequality

$$\int_{t_0}^{a} r(t)|y(t) + y(t+)|^p |y_g'(t)|^q dg(t) \leq K(t_0, a, p, q) \int_{t_0}^{a} r(t)|y_g'(t)|^{p+q} dg(t),$$

where

$$K(t_0, a, p, q) = \max_{c \in J} (\max \{K_1(t_0, c, p, q), K_2(c, a, p, q)\}),$$

and $K_1(t_0, c, p, q)$ and $K_2(c, a, p, q)$ are as in Problem 5.13 and Problem 5.14, respectively.

Chapter 6

Stieltjes–Lyapunov-Type Inequalities

In this chapter, we investigate some properties of the solutions of second-order linear and half-linear Stieltjes differential equations and second-order Stieltjes differential equations with damping terms. We derive several kinds of Lyapunov-type inequalities in the framework of Stieltjes calculus.

6.1 Second-Order Linear Equations

Consider the Stieltjes–Sturm–Liouville differential equation

$$y_g''(t) + p(t)y(t+) = 0, \quad t \in [a,b], \tag{6.1}$$

and the quadratic functional

$$\mathscr{F}(y) = \int_a^b \left((y_g'(t))^2 - p(t)y(t+)^2\right) dg(t),$$

where $p \in \mathscr{C}([a,b])$ is a given positive function. We start with the following useful result.

Lemma 6.1. *Let $x \in \mathscr{C}_g^2([a,b])$ be a solution of Equation* (6.1). *Then,*

$$\mathscr{F}(y) - \mathscr{F}(x) = \mathscr{F}(y-x) + 2(y-x)(bx_g'(b) - 2(y-x)(a)x_g'(a)$$

for $y \in \mathscr{C}_g^2([a,b])$.

Proof. Since x is a solution to Equation (6.1), we have

$$-p(t)x(t+) = x_g''(t), \quad t \in [a,b].$$

Then, we get the following:

$$
\begin{aligned}
&\mathscr{F}(y) - \mathscr{F}(x) - \mathscr{F}(y-x)\\
&= \int_a^b \Big((y_g'(t))^2 - p(t)(y(t+))^2 - (x_g'(t))^2 + p(t)(x(t+))^2\\
&\quad - (y_g'(t) - x_g'(t))^2 + p(t)(y(t+) - x(t+))^2 \Big) dg(t)\\
&= \int_a^b \Big((y_g'(t))^2 - p(t)(y(t+))^2 - (x_g'(t))^2 + p(t)(x(t+))^2\\
&\quad - (y_g'(t))^2 + 2y_g'(t)x_g'(t) - ((x(t))^2)_g'' + p(t)(y(t+))^2\\
&\quad - 2p(t)y(t+)x(t+) + p(t)(x(t+))^2 \Big) dg(t)\\
&= 2\int_a^b \left(y_g'(t)x_g'(t) - p(t)y(t+)x(t+) + p(t)(x(t+))^2 - (x_g'(t))^2 \right) dg(t)\\
&= 2\int_a^b \left(y_g'(t)x_g'(t) + y(t+)x_g''(t) - x(t+)x_g''(t) - (x_g'(t))^2 \right) dg(t)\\
&= 2\int_a^b \left((yx_g')_g'(t) - (xx_g')_g'(t) \right) dg(t)\\
&= 2\int_a^b \left(yx_g' - xx_g' \right)_g' (t) dg(t)\\
&= 2\int_a^b \left((y-x)x_g' \right)_g' (t) dg(t)\\
&= 2\left((y-x)x_g' \right)(t) \Big|_{t=a}^{t=b}\\
&= 2(y-x)(b)x_g'(b) - 2(y-x)(a)x_g'(a).
\end{aligned}
$$

This completes the proof. □

Theorem 6.1. *For $y \in \mathscr{C}_g^2([a,b])$ and for $a \le r < s \le b$, we have*

$$\int_r^s (y_g'(t))^2 dg(t) \ge \frac{(y(s) - y(r))^2}{s - r}.$$

Proof. Define

$$x(t) = \frac{y(s) - y(r)}{s - r}t + \frac{sy(r) - ry(s)}{s - r}, \quad t \in [a, b].$$

Then,

$$x'_g(t) = \frac{y(s) - y(r)}{s - r},$$
$$x''_g(t) = 0, \quad t \in [a, b],$$

and

$$x(s) = \frac{y(s) - y(r)}{s - r}s + \frac{sy(r) - ry(s)}{s - r} = y(s),$$

and

$$x(r) = \frac{y(s) - y(r)}{s - r}r + \frac{sy(r) - ry(s)}{s - r} = y(r).$$

Thus, x is a solution to Equation (6.1) with $p(t) = 0$, $t \in [a, b]$. Define the functional

$$\mathscr{F}_0(x) = \int_r^s (x'_g(t))^2 dg(t).$$

Then, applying Lemma 6.1, we find

$$\begin{aligned}
\mathscr{F}_0(y) &= \mathscr{F}_0(x) + \mathscr{F}_0(y - x) + (y - x)(s)x'_g(s) - (y - x)(r)x'_g(r) \\
&= \mathscr{F}_0(x) + \mathscr{F}_0(y - x) \\
&\geq \mathscr{F}_0(x) \\
&= \int_r^s \left(\frac{y(s) - y(r)}{s - r}\right)^2 dg(t) \\
&= \frac{(y(s) - y(r))^2}{s - r}.
\end{aligned}$$

This completes the proof. □

Theorem 6.2. *Let $y \in \mathscr{C}_g^2([a,b]$ be a nontrivial solution to Equation* (6.1) *with*

$$y(a) = y(b) = 0. \tag{6.2}$$

Then,

$$\int_a^b p(t)dg(t) > \frac{b-a}{f(d)},$$

where $f(t) = (t-a)(b-t)$, $t \in [a,b]$, and $f(d) = \max\{f(t) \colon t \in [a,b]\}$.

Proof. Note that $f \in \mathscr{C}([a,b])$. Then, there is a point $d \in [a,b]$ such that

$$f(d) = \max_{t\in[a,b]} f(t).$$

Since $y \in \mathscr{C}_g^2([a,b])$ is a nontrivial solution to Equation (6.1) that satisfies (6.2), we get

$$y_g''(t) = -p(t)y(t), \quad t \in [a,b],$$

and

$$\begin{aligned}
\mathscr{F}(y) &= \int_a^b \left((y_g'(t))^2 - p(t)(y(t+))^2\right) dg(t) \\
&= \int_a^b \left((y_g'(t))^2 + y_g''(t)y(t+)\right) dg(t) \\
&= \int_a^b (y_g'(t))^2 dg(t) + \int_a^b y_g''(t)y(t+)dg(t) \\
&= (y_g'(t))y(t)\Big|_{t=a}^{t=b} - \int_a^b y_g''(t)y(t+)dg(t) + \int_a^b y_g''(t)y(t+)dg(t) \\
&= 0,
\end{aligned}$$

whereupon

$$\int_a^b (y_g'(t))^2 dg(t) = \int_a^b p(t)(y(t+))^2 dg(t).$$

Set

$$M = \max\{(y(t))^2 \colon t \in [a,b]\}.$$

Since $y \in \mathscr{C}_g^2([a,b])$, there is a $c \in [a,b]$ such that $M = (y(c))^2$. Then, applying Theorem 6.1, we find

$$\begin{aligned}
M \int_a^b p(t)dg(t) &\geq \int_a^b p(t)(y(t+))^2 dg(t) \\
&= \int_a^b (y_g'(t))^2 dg(t) \\
&= \int_a^c (y_g'(t))^2 dg(t) + \int_c^b (y_g'(t))^2 dg(t) \\
&\geq \frac{(y(c)-y(a))^2}{c-a} + \frac{(y(b)-y(c))^2}{b-c} \\
&= \frac{(y(c))^2}{c-a} + \frac{(y(c))^2}{b-c} \\
&= (y(c))^2 \left(\frac{1}{c-a} + \frac{1}{b-c}\right) \\
&= M \left(\frac{1}{c-a} + \frac{1}{b-c}\right) \\
&= M \frac{b-a}{(c-a)(b-c)} \\
&= N \frac{b-a}{f(c)} \\
&\geq M \frac{b-a}{f(d)}.
\end{aligned}$$

Hence,

$$\int_a^b p(t)dg(t) > \frac{b-a}{f(d)}.$$

This completes the proof. □

Theorem 6.3. *Let y be a nontrivial solution to Equation (6.1) with $y(a) = y_g'(b+) = 0$. Also, let*

$$P(t) = \int_t^{b+} p(s)dg(s), \quad t \in [a, b+].$$

Then,

$$K_P(b+, a) = \left(2 \int_a^{b+} (P(t))^2 (g(t) - g(a))^2 dg(t)\right)^{\frac{1}{2}} \geq 1.$$

Proof. Using integration by parts in the sense of Stieltjes, we have

$$\begin{aligned}\int_a^{b+} y(t+)y_g''(t)dg(t) &= y(t)y_g'(t)\Big|_{t=a}^{t=b+} - \int_a^{b+} (y_g'(t))^2 dg(t)\\ &= -\int_a^{b+} (y_g'(t))^2 dg(t).\end{aligned} \tag{6.3}$$

Now, applying the definition of the function P, we find $P_g'(t) = -p(t)$, $t \in [a, b+]$, and

$$\begin{aligned}\int_a^{b+} p(t)(y(t+))^2 dg(t) &= -\int_a^{b+} P_g'(t)(y(t+))^2 dg(t)\\ &= -P(t)(y(t))^2\Big|_{t=a}^{t=b+} + \int_a^{b+} P(t)((y^2)_g')(t)dg(t)\\ &= \int_a^{b+} P(t)(y^2)_g'(t)dg(t)\\ &= \int_a^{b+} P(t)(y(t)+y(t+))y_g'(t)dg(t).\end{aligned}$$

Now, multiplying (6.1) by $y(t)$, $t \in [a, b+]$, we obtain

$$y(t+)y_g''(t) = -p(t)(y(t+))^2, \quad t \in [a, b],$$

which we integrate in the sense of Stieltjes, from a to $b+$, and find

$$\int_a^{b+} y(t+)y_g''(t)dg(t) - \int_a^{b+} p(t)(y(t+))^2 dg(t).$$

Applying (6.3), we arrive at

$$-\int_a^{b+} (y_g'(t))^2 dg(t) = -\int_a^{b+} p(t)(y(t+))^2 dg(t)$$

or

$$\begin{aligned}\int_a^{b+} (y_g'(t))^2 dg(t) &= \int_a^{b+} p(t)(y(t+))^2 dg(t)\\ &= \int_a^{b+} P(t)(y(t)+y(t+))y_g'(t)dg(t)\\ &\le \int_a^{b+} P(t)|y(t)+y(t+)||y_g'(t)|dg(t).\end{aligned}$$

Next, applying the Stieltjes–Opial-type inequality, Theorem 5.7, we get

$$\int_a^{b+} P(t)|y(t)+y(t+)||y_g'(t)|dg(t) \le K_P(b+,a)\int_a^{b+}(y_g'(t))^2 dg(t),$$

whereupon

$$\int_a^{b+}(y_g'(t))^2 dg(t) \le K_P(b+,a)\int_a^{b+}(y_g'(t))^2 dg(t)$$

and

$$K_P(b+,a) \ge 1.$$

This completes the proof. □

Theorem 6.4. *Let y be a nontrivial solution to Equation* (6.1) *with* $y(a) = y_g'(b+) = 0$. *Also, let*

$$P(t) = \int_a^t p(s)dg(s), \quad t \in [a, b+].$$

Then,

$$L_P(b+,a) = \left(2\int_a^{b+}(P(t))^2(g(b+)-g(t))^2 dg(t)\right)^{\frac{1}{2}} \ge 1.$$

Proof. Integration by parts in the sense of Stieltjes gives us

$$\begin{aligned}\int_a^{b+} y(t+)y_g''(t)dg(t) &= y(t)y_g'(t)\Big|_{t=a}^{t=b+} - \int_a^{b+}(y_g'(t))^2 dg(t)\\ &= -\int_a^{b+}(y_g'(t))^2 dg(t).\end{aligned} \tag{6.4}$$

By the definition of the function P, we obtain $P_g'(t) = p(t)$, $t \in [a, b+]$, and

$$\begin{aligned}\int_a^{b+} p(t)(y(t+))^2 dg(t) &= -\int_a^{b+} P_g'(t)(y(t+))^2 dg(t)\\ &= P(t)(y(t))^2\Big|_{t=a}^{t=b+} - \int_a^{b+} P(t)(y^2)_g'(t)dg(t)\\ &= -\int_a^{b+} P(t)(y^2)_g'(t)dg(t)\\ &= -\int_a^{b+} P(t)(y(t)+y(t+))y_g'(t)dg(t).\end{aligned}$$

Now, multiplying (6.1) by $y(t+)$, $t \in [a, b+]$, we obtain

$$y(t+)y_g''(t) = -p(t)(y(t+))^2, \quad t \in [a, b],$$

which we integrate in the sense of Stieltjes, from a to $b+$, we find

$$\int_a^{b+} y(t+)y_g''(t)dg(t) = -\int_a^{b+} p(t)(y(t+))^2 dg(t).$$

Employing (6.4), we find

$$-\int_a^{b+} (y_g'(t))^2 dg(t) = -\int_a^{b+} p(t)(y(t+))^2 dg(t)$$

or

$$\begin{aligned}\int_a^{b+} (y_g'(t))^2 dg(t) &= \int_a^{b+} p(t)(y(t+))^2 dg(t)\\ &= -\int_a^{b+} P(t)(y(t) + y(t+))y_g'(t)dg(t)\\ &\le \int_a^{b+} P(t)|y(t) + y(t+)||y_g'(t)|dg(t).\end{aligned}$$

Using the Stieltjes–Opial-type inequality, Theorem 5.8, we arrive at

$$\int_a^{b+} P(t)|y(t) + y(t+)||y_g'(t)|dg(t) \le L_P(b+, a)\int_a^{b+} (y_g'(t))^2 dg(t),$$

whereupon

$$\int_a^{b+} (y_g'(t))^2 dg(t) \le L_P(b+, a)\int_a^{b+} (y_g'(t))^2 dg(t)$$

and

$$L_P(b+, a) \ge 1.$$

This completes the proof. □

Next, consider the following second-order Stieltjes differential equation:

$$(ry_g')_g'(t) + q(t)y(t+) = 0, \quad t \in [a, b], \tag{6.5}$$

where $r, q \in \mathscr{C}([a,b])$, r is a positive function on $[a,b]$, and

$$\begin{aligned}&\int_a^b \frac{1}{r(t)} dg(t) < \infty,\\ &\int_a^b |q(t)| dg(t) < \infty.\end{aligned} \tag{6.6}$$

Theorem 6.5. *Let y be a solution to Equation* (6.5) *for which $y(a) = y'_g(b) = 0$, and let* (6.6) *hold. Then,*

$$\sqrt{2}\left(\int_a^b \frac{(Q(t))^2}{r(t)}\left(\int_a^t \frac{1}{r(s)} dg(s)\right) dg(t)\right)^{\frac{1}{2}} + \sup_{t\in[a,b]}\left(\Delta g(t)\frac{|Q(t)|}{r(t)}\right) \geq 1,$$

where

$$Q(t) = \int_t^b q(s) dg(s), \quad t \in [a,b].$$

Proof. Multiplying (6.5) by $y(t+)$, $t \in [a,b]$, we get

$$y(t+)(ry'_g)'_g(t) + (y(t+))^2 q(t) = 0, \quad t \in [a,b],$$

or

$$-q(t)(y(t+))^2 = y(t+)(ry'_g)'_g(t), \quad t \in [a,b].$$

Integrating the last equation in the sense of Stieltjes, from a to b, we find

$$\begin{aligned}-\int_a^b q(t)(y(t+))^2 dg(t) &= \int_a^b y(t+)(ry'_g)'_g(t) dg(t)\\ &= y(t)r(t)y'_g(t)\Big|_{t=a}^{t=b} - \int_a^b r(t)(y'_g(t))^2 dg(t)\\ &= -\int_a^b r(t)(y'_g(t))^2 dg(t),\end{aligned}$$

i.e.,

$$\int_a^b q(t)(y(t+))^2 dg(t) = \int_a^b r(t)(y'_g(t))^2 dg(t). \tag{6.7}$$

On the other hand, we have $Q'_g(t) = -q(t)$, $t \in [a, b]$, and $Q(b) = 0$. Then,

$$\begin{aligned}
\int_a^b q(t)(y(t+))^2 dg(t) &= -\int_a^b Q'_g(t)(y(t+))^2 dg(t) \\
&= -Q(t)y(t)\Big|_{t=a}^{t=b} + \int_a^b Q(t)(y^2)'_g(t) dg(t) \\
&= \int_a^b Q(t)(y^2)'_g(t) dg(t) \\
&= \int_a^b Q(t)(y(t) + y(t+))y'_g(t) dg(t) \\
&\le \int_a^b |Q(t)||y(t) + y(t+)||y'_g(t)| dg(t).
\end{aligned}$$

Hence, with (6.7), we find

$$\int_a^b r(t)(y'_g(t))^2 dg(t) \le \int_a^b |Q(t)||y(t) + y(t+)||y'_g(t)| dg(t). \tag{6.8}$$

Now, applying Theorem 5.12 for the right side of the inequality (6.8), we arrive at

$$\begin{aligned}
\int_a^b r(t)(y'_g(t))^2 dg(t) &\le \sqrt{2}\left(\int_a^b \frac{(Q(t))^2}{r(t)}\left(\int_a^t \frac{1}{r(s)} dg(s)\right) dg(t)\right)^{\frac{1}{2}} \\
&\quad + \sup_{t\in[a,b]}\left(\Delta g(t)\frac{|Q(t)|}{r(t)}\right) \times \int_a^b r(t)(y'_g(t))^2 dg(t),
\end{aligned}$$

whereupon

$$\sqrt{2}\left(\int_a^b \frac{(Q(t))^2}{r(t)}\left(\int_a^t \frac{1}{r(s)} dg(s)\right) dg(t)\right)^{\frac{1}{2}} + \sup_{t\in[a,b]}\left(\Delta g(t)\frac{|Q(t)|}{r(t)}\right) \ge 1.$$

This completes the proof. □

Theorem 6.6. *Let y be a solution to Equation (6.5) for which $y'_g(a) = y(b) = 0$, and let (6.6) hold. Then,*

$$\sqrt{2}\left(\int_a^b \frac{(Q(t))^2}{r(t)}\left(\int_t^b \frac{1}{r(s)} dg(s)\right) dg(t)\right)^{\frac{1}{2}} + \sup_{t\in[a,b]}\left(\Delta g(t)\frac{|Q(t)|}{r(t)}\right) \ge 1,$$

where

$$Q(t) = \int_a^t q(s)dg(s), \quad t \in [a,b].$$

Proof. As in the proof of Theorem 6.5, we find (6.7). Note that $Q'_g(t) = q(t)$, $t \in [a,b]$, and $Q(a) = 0$. Then, we have the following:

$$\begin{aligned}
\int_a^b q(t)(y(t+))^2 dg(t) &= \int_a^b Q'_g(t)(y(t+))^2 dg(t) \\
&= Q(t)y(t)\Big|_{t=a}^{t=b} - \int_a^b Q(t)(y^2)'_g(t)dg(t) \\
&= -\int_a^b Q(t)(y^2)'_g(t)dg(t) \\
&= -\int_a^b Q(t)(y(t)+y(t+))y'_g(t)dg(t) \\
&\leq \int_a^b |Q(t)||y(t)+y(t+)||y'_g(t)|dg(t).
\end{aligned}$$

From this and (6.7), we find (6.8). Employing Theorem 5.14 for the right side of the inequality (6.8), we arrive at

$$\begin{aligned}
\int_a^b r(t)(y'_g(t))^2 dg(t) \leq \sqrt{2}\left(\int_a^b \frac{(Q(t))^2}{r(t)}\left(\int_t^b \frac{1}{r(s)}dg(s)\right)dg(t)\right)^{\frac{1}{2}} \\
+ \sup_{t\in[a,b]}\left(\Delta g(t)\frac{|Q(t)|}{r(t)}\right) \times \int_a^b r(t)(y'_g(t))^2 dg(t),
\end{aligned}$$

whereupon

$$\sqrt{2}\left(\int_a^b \frac{(Q(t))^2}{r(t)}\left(\int_t^b \frac{1}{r(s)}dg(s)\right)dg(t)\right)^{\frac{1}{2}} + \sup_{t\in[a,b]}\left(\Delta g(t)\frac{|Q(t)|}{r(t)}\right) \geq 1.$$

This completes the proof. □

Problem 6.1. Let y be a solution to the following SBVP:

$$\begin{aligned}
y''_g(t) + q(t)y(t+) &= 0, \quad t \in [a,b], \\
y(a) = y'_g(b) &= 0,
\end{aligned}$$

where $q \in \mathscr{C}([a,b])$ is such that

$$\int_a^b |q(t)|dg(t) < \infty.$$

Then, prove that

$$\sqrt{2}\left(\int_a^b (Q(t))^2(g(t)-g(a))dg(t)\right)^{\frac{1}{2}} + \sup_{t\in[a,b]}((\Delta g(t))|Q(t)|) \geq 1,$$

where

$$Q(t) = \int_t^b q(s)dg(s), \quad t \in [a,b].$$

Problem 6.2. Let y be a solution to the following SBVP:

$$y_g''(t) + q(t)y(t+) = 0, \quad t \in [a,b],$$
$$y_g'(a) = y(b) = 0,$$

where $q \in \mathscr{C}([a,b])$ is such that

$$\int_a^b |q(t)|dg(t) < \infty.$$

Then, prove that

$$\sqrt{2}\left(\int_a^b (Q(t))^2(g(b)-g(t))dg(t)\right)^{\frac{1}{2}} + \sup_{t\in[a,b]}(\Delta g(t)|Q(t)|) \geq 1,$$

where

$$Q(t) = \int_a^t q(s)dg(s), \quad t \in [a,b].$$

Theorem 6.7. *Let y be a solution to Equation (6.5) that satisfies the boundary conditions $y(a) = y_g'(b) = 0$ and r, q, Q be as in Theorem 6.5. Then,*

$$\sqrt{2}\left(\sup_{t\in[a,b]}\frac{(Q(t))^2}{r(t)}\int_a^b \frac{1}{r(t)}\left(\int_a^t \frac{1}{r(s)}dg(s)\right)dg(t)\right)^{\frac{1}{2}}$$
$$+ \sup_{t\in[a,b]} \Delta g(t)\frac{|Q(t)|}{r(t)} \geq 1.$$

Proof. As in the proof of Theorem 6.5, we get (6.8). Then,

$$\begin{aligned}\int_a^b r(t)(y_g'(t))^2 dg(t) &\le \int_a^b \frac{|Q(t)|}{r(t)}|y(t)+y(t+)|r(t)|y_g'(t)|dg(t)\\ &\le \sup_{t\in[a,b]}\frac{|Q(t)|}{r(t)}\int_a^b r(t)|y(t)+y(t+)||y_g'(t)|dg(t).\end{aligned}$$

Applying Theorem 5.13, we get

$$\begin{aligned}&\int_a^b r(t)|y(t)+y(t+)||y_g'(t)|dg(t)\\ &\quad\le\left(\sqrt{2}\left(\int_a^b \frac{1}{r(t)}\left(\int_a^t \frac{1}{r(s)}dg(s)\right)dg(t)\right)^{\frac{1}{2}}+\sup_{t\in[a,b]}\Delta g(t)\right)\\ &\qquad\times\int_a^b r(t)(y_g'(t))^2 dg(t).\end{aligned}$$

Therefore,

$$\begin{aligned}&\int_a^b r(t)|y(t)+y(t+)||y_g'(t)|dg(t)\\ &\quad\le\left(\sqrt{2}\left(\sup_{t\in[a,b]}\frac{(Q(t))^2}{r(t)}\int_a^b\frac{1}{r(t)}\left(\int_a^t\frac{1}{r(s)}dg(s)\right)dg(t)\right)^{\frac{1}{2}}\right.\\ &\qquad\left.+\sup_{t\in[a,b]}\Delta g(t)\frac{|Q(t)|}{r(t)}\right)\int_a^b r(t)(y_g'(t))^2dg(t),\end{aligned}$$

whereupon

$$\begin{aligned}&\sqrt{2}\left(\sup_{t\in[a,b]}\frac{(Q(t))^2}{r(t)}\int_a^b\frac{1}{r(t)}\left(\int_a^t\frac{1}{r(s)}dg(s)\right)dg(t)\right)^{\frac{1}{2}}\\ &\quad+\sup_{t\in[a,b]}\Delta g(t)\frac{|Q(t)|}{r(t)}\ge 1.\end{aligned}$$

This completes the proof. □

Theorem 6.8. *Let y be a solution to Equation* (6.5) *that satisfies the boundary conditions $y_g'(a)=y(b)=0$ and r, q, Q be as in Theorem* 6.6.

Then,

$$\sqrt{2}\left(\sup_{t\in[a,b]}\frac{(Q(t))^2}{r(t)}\int_a^b\frac{1}{r(t)}\left(\int_t^b\frac{1}{r(s)}dg(s)\right)dg(t)\right)^{\frac{1}{2}}$$
$$+\sup_{t\in[a,b]}\Delta g(t)\frac{|Q(t)|}{r(t)}\geq 1.$$

Proof. As in the proof of Theorem 6.6, we get

$$\int_a^b r(t)(y_g'(t))^2dg(t)\leq\sup_{t\in[a,b]}\frac{|Q(t)|}{r(t)}\int_a^b r(t)|y(t)+y(t+)||y_g'(t)|dg(t).$$

Applying Theorem 5.14, we get

$$\int_a^b r(t)|y(t)+y(t+)||y_g'(t)|dg(t)$$
$$\leq\left(\sqrt{2}\left(\int_a^b\frac{1}{r(t)}\left(\int_t^b\frac{1}{r(s)}dg(s)\right)dg(t)\right)^{\frac{1}{2}}+\sup_{t\in[a,b]}\Delta g(t)\right)$$
$$\times\int_a^b r(t)(y_g'(t))^2dg(t).$$

Therefore,

$$\int_a^b r(t)|y(t)+y(t+)||y_g'(t)|dg(t)$$
$$\leq\left(\sqrt{2}\left(\sup_{t\in[a,b]}\frac{(Q(t))^2}{r(t)}\int_a^b\frac{1}{r(t)}\left(\int_t^b\frac{1}{r(s)}dg(s)\right)dg(t)\right)^{\frac{1}{2}}\right.$$
$$\left.+\sup_{t\in[a,b]}\Delta g(t)\frac{|Q(t)|}{r(t)}\right)\int_a^b r(t)(y_g'(t))^2dg(t),$$

whereupon

$$\sqrt{2}\left(\sup_{t\in[a,b]}\frac{(Q(t))^2}{r(t)}\int_a^b\frac{1}{r(t)}\left(\int_t^b\frac{1}{r(s)}dg(s)\right)dg(t)\right)^{\frac{1}{2}}$$
$$+\sup_{t\in[a,b]}(\Delta g(t))\frac{|Q(t)|}{r(t)}\geq 1.$$

This completes the proof. □

Problem 6.3. Let y be a solution to the following SBVP:

$$y_g''(t) + q(t)y(t+) = 0, \quad t \in [a, b],$$
$$y(a) = y_g'(b) = 0,$$

where $q \in \mathscr{C}([a, b])$ is such that

$$\int_a^b |q(t)|dg(t) < \infty.$$

Then, prove that

$$\sqrt{2}\left(\sup_{t\in[a,b]} (Q(t))^2 \int_a^b (g(t) - g(a))dg(t)\right)^{\frac{1}{2}} + \sup_{t\in[a,b]} (\Delta g(t)|Q(t)|) \geq 1,$$

where

$$Q(t) = \int_t^b q(s)dg(s), \quad t \in [a, b].$$

Problem 6.4. Let y be a solution to the following SBVP:

$$y_g''(t) + y(t+) = 0, \quad t \in [a, b],$$
$$y_g'(a) = y(b) = 0,$$

where $q \in \mathscr{C}([a, b])$ is such that

$$\int_a^b |q(t)|dg(t) < \infty.$$

Then, prove that

$$\sqrt{2}\left(\sup_{t\in[a,b]} (Q(t))^2 \int_a^b (g(b) - g(t))dg(t)\right)^{\frac{1}{2}} + \sup_{t\in[a,b]} (\Delta g(t)|Q(t)|) \geq 1,$$

where

$$Q(t) = \int_a^t q(s)dg(s), \quad t \in [a, b].$$

Theorem 6.9. *Let y be a solution to Equation (6.5) such that $y(a) = y(b) = 0$. Also, let r, q, and Q be as in Theorem 6.5, and*

$$Q_g'(t) = q(t), \quad t \in [a, b].$$

Then,

$$\sqrt{2}\left(\int_a^b \frac{(Q(t))^2}{r(t)}\left(\int_a^t \frac{1}{r(s)}dg(s)\right)dg(t)\right)^{\frac{1}{2}} + \sup_{t\in[a,b]}\left(\Delta g(t)\frac{|Q(t)|}{r(t)}\right) \geq 1.$$

Proof. As in the proof of Theorem 6.5, using that $y(a) = y(b) = 0$, we get (6.7) and then (6.8). Applying Theorem 5.12 for the right side of the inequality (6.8), we find

$$\int_a^b r(t)(y_g'(t))^2 dg(t) \leq \sqrt{2}\left(\int_a^b \frac{(Q(t))^2}{r(t)}\left(\int_a^t \frac{1}{r(s)}dg(s)\right)dg(t)\right)^{\frac{1}{2}}$$
$$+ \sup_{t\in[a,b]}\left((\Delta g(t))\frac{|Q(t)|}{r(t)}\right) \times \int_a^b r(t)(y_g'(t))^2 dg(t),$$

whereupon

$$\sqrt{2}\left(\int_a^b \frac{(Q(t))^2}{r(t)}\left(\int_a^t \frac{1}{r(s)}dg(s)\right)dg(t)\right)^{\frac{1}{2}} + \sup_{t\in[a,b]}\left(\Delta g(t)\frac{|Q(t)|}{r(t)}\right) \geq 1.$$

This completes the proof. □

Theorem 6.10. *Let y be a solution to Equation* (6.1) *such that $y(a) = y(b) = 0$. Also, let q be as in Theorem* 6.5 *and*

$$Q_g'(t) = q(t), \quad t \in [a,b].$$

Then,

$$\sqrt{2(g(b)-g(a))}\left(\int_a^b (Q(t))^2 dg(t)\right)^{\frac{1}{2}} + \sup_{t\in[a,b]}(\Delta g(t))|Q(t)| \geq 1.$$

Proof. We apply Theorem 6.9 for $r(t) = 1$, $t \in [a,b]$, and we get

$$1 \leq \sqrt{2}\left(\int_a^b (Q(t))^2(g(t)-g(a))dg(t)\right)^{\frac{1}{2}} + \sup_{t\in[a,b]}(\Delta g(t))|Q(t)|$$
$$\leq \sqrt{2}\left(\int_a^b (Q(t))^2(g(b)-g(a))dg(t)\right)^{\frac{1}{2}} + \sup_{t\in[a,b]}(\Delta g(t))|Q(t)|$$
$$= \sqrt{2(g(b)-g(a))}\left(\int_a^b (Q(t))^2 dg(t)\right)^{\frac{1}{2}} + \sup_{t\in[a,b]}(\Delta g(t))|Q(t)|.$$

This completes the proof. □

Problem 6.5. Let $I = [a, b]$, $a < b$, $t \in I$, and $q \in \mathscr{C}([a, b])$ be such that

$$\int_a^b |q(t)| dg(t) < \infty.$$

Also, let y be a nontrivial solution to Equation (6.1) such that $y(a) = y(b) = 0$. Then, prove that

$$\left(\int_a^b \left(\int_a^t q(s) dg(s) \right)^2 dg(t) \right)^{\frac{1}{2}} \geq \frac{1}{\sqrt{2(b-a)}}.$$

6.2 Second-Order Half-Linear Stieltjes Differential Equations

In this section, we consider the second-order half-linear Stieltjes differential equation

$$\left(r(t) \phi \left(y_g' \right) \right)_g' + p(t) \phi(y(t+)) = 0, \quad t \in [a, b], \tag{6.9}$$

where $r, p \in \mathscr{C}([a, b])$ are positive functions, $r(t) \neq 0$, $t \in [a, b]$, and $\phi(x) = |x|^{\gamma - 1} x$, $x \in \mathbb{R}$, for some $\gamma > 0$. Our first result in this section is as follows.

Theorem 6.11. *Let y be a positive solution to Equation (6.9) such that*

$$y(a) = y(b) = 0 \tag{6.10}$$

and $y(t) > 0$, $t \in (a, b)$, and $\max_{t \in [a,b]} y(t) = y(c)$ for some $c \in (a, b)$. Then,

$$\left(\int_a^b (r(t))^{-\frac{1}{\gamma}} dg(t) \right)^{\gamma} \left(\int_a^b p(t) dg(t) \right) > 2^{\gamma + 1}.$$

Proof. Let $M = y(c)$. Since (6.10) holds, we have

$$M = \int_a^c y_g'(t) dg(t) = \left| \int_c^b y_g'(t) dg(t) \right|.$$

Thus,

$$\begin{aligned} 2M &= M + M \\ &= \int_a^c y_g'(t) dg(t) + \left| \int_c^b y_g'(t) dg(t) \right| \end{aligned}$$

$$\leq \int_a^c |y_g'(t)|dg(t) + \int_c^b |y_g'(t)|dg(t)$$
$$= \int_a^b |y_g'(t)|dg(t)$$
$$= \int_a^b (r(t))^{-\frac{1}{\gamma+1}} (r(t))^{\frac{1}{\gamma+1}} |y_g'(t)|dg(t).$$

Therefore,

$$(2M)^{\gamma+1} \leq \left(\int_a^b (r(t))^{-\frac{1}{\gamma+1}} (r(t))^{\frac{1}{\gamma+1}} |y_g'(t)|dg(t) \right)^{\gamma+1}.$$

We apply the Stieltjes–Hölder inequality for

$$f(t) = (r(t))^{-\frac{1}{\gamma+1}} \quad \text{and} \quad g(t) = (r(t))^{\frac{1}{\gamma+1}} |y_g'(t)|, \quad t \in [a, b],$$

and

$$p = \gamma + 1 \quad \text{and} \quad q = \frac{\gamma+1}{\gamma}.$$

We find

$$(2M)^{\gamma+1} \leq \left(\int_a^b \left((r(t))^{-\frac{1}{\gamma+1}} \right)^{\frac{\gamma+1}{\gamma}} dg(t) \right)^{\gamma} \left(\int_a^b r(t)|y_g'(t)|^{\gamma+1} dg(t) \right)$$
$$= \left(\int_a^b (r(t))^{-\frac{1}{\gamma}} dg(t) \right)^{\gamma} \left(\int_a^b r(t)|y_g'(t)|^{\gamma+1} dg(t) \right). \tag{6.11}$$

Using integration by parts in the sense of Stieltjes, we find

$$\int_a^b r(t)|y_g'(t)|^{\gamma+1} dg(t) = \int_a^b y_g'(t) r(t) |y_g'(t)|^{\gamma-1} y_g'(t) dg(t)$$
$$= r(t) y_g'(t) |y_g'(t)|^{\gamma-1} y(t) \Big|_{t=a}^{t=b}$$
$$- \int_a^b \left(y_g'(t) r(t) |y_g'(t)|^{\gamma-1} \right)_g' y(t+) dg(t)$$
$$= - \int_a^b \left(y_g'(t) r(t) |y_g'(t)|^{\gamma-1} \right)_g' y(t+) dg(t).$$

We multiply Equation (6.9) with $y(t+)$, $t \in [a,b]$, and find

$$\begin{aligned}\left(r(t)|y_g'(t)|^{\gamma-1}y_g'(t)\right)_g' y(t+) &= -p(t)\phi(y(t+))y(t+)\\ &= -p(t)(y(t+))^{\gamma-1}y(t+)y(t+)\\ &= -p(t)(y(t+))^{\gamma+1}, \quad t \in [a,b].\end{aligned}$$

Integrating the last equation in the sense of Stieltjes, from a to b, we obtain

$$\begin{aligned}-\int_a^b p(t)(y(t+))^{\gamma+1}dg(t) &= \int_a^b \left(r(t)(y_g'(t))^\gamma\right)_g' y(t+)dg(t)\\ &= r(t)(y_g'(t))^\gamma y(t)\Big|_{t=a}^{t=b}\\ &\quad - \int_a^b r(t)(y_g'(t))^{\gamma+1}dg(t)\\ &= \int_a^b r(t)(y_g'(t))^{\gamma+1}dg(t).\end{aligned}$$

Therefore,

$$\int_a^b r(t)(y_g'(t))^{\gamma+1}dg(t) = \int_a^b r(t)(y_g'(t))^{\gamma+1}dg(t).$$

Hence, with (6.11), we find

$$\begin{aligned}(2M)^{\gamma+1} &\leq \left(\int_a^b (r(t))^{-\frac{1}{\gamma}}dg(t)\right)^\gamma \left(\int_a^b r(t)(y_g'(t))^{\gamma+1}dg(t)\right)\\ &\leq M^{\gamma+1}\left(\int_a^b (r(t))^{-\frac{1}{\gamma}}dg(t)\right)^\gamma \left(\int_a^b p(t)dg(t)\right),\end{aligned}$$

whereupon

$$\left(\int_a^b (r(t))^{-\frac{1}{\gamma}}dg(t)\right)^\gamma \left(\int_a^b p(t)dg(t)\right) \geq 2^{\gamma+1}.$$

This completes the proof. □

Theorem 6.12. *Let r and p satisfy*

$$\int_a^b p(t)dg(t) < \begin{cases} \frac{(r(a))^{\gamma+1}((g(b)-g(c))^{\gamma}+(g(c)-g(a))^{\gamma})}{(r(b))^{\gamma}(g(c)-g(a))^{\gamma}(g(b)-g(c))^{\gamma}} & \text{if } r \text{ is increasing,} \\ \frac{(r(b))^{\gamma+1}((g(b)-g(c))^{\gamma}+(g(c)-g(a))^{\gamma})}{(r(a))^{\gamma}(g(c)-g(a))^{\gamma}(g(b)-g(c))^{\gamma}} & \text{if } r \text{ is decreasing.} \end{cases} \tag{6.12}$$

Then, Equation (6.9) *is disconjugate on* $[a, b]$.

Proof. Assume the contrary, i.e., suppose that Equation (6.9) is not disconjugate. Then, there is a nontrivial solution y of Equation (6.9) for which $y(a) = y(b) = 0$. Using this and integrating by parts in the sense of Stieltjes, from a to b, we find

$$\begin{aligned} \int_a^b r(t)|y_g'(t)|^{\gamma+1}dg(t) &= \int_a^b y_g'(t)r(t)|y_g'(t)|^{\gamma-1}y_g'(t)dg(t) \\ &= r(t)y_g'(t)|y_g'(t)|^{\gamma-1}y(t)\Big|_{t=a}^{t=b} \\ &\quad - \int_a^b \left(y_g'(t)r(t)|y_g'(t)|^{\gamma-1}\right)_g' y(t+)dg(t) \\ &= -\int_a^b \left(y_g'(t)r(t)|y_g'(t)|^{\gamma-1}\right)_g' y(t+)dg(t). \end{aligned}$$

Now, multiplying Equation (6.9) with $y(t)$, $t \in [a, b]$, we find

$$\begin{aligned} \left(r(t)|y_g'(t)|^{\gamma-1}y_g'(t)\right)_g' y(t+) &= -p(t)\phi(y(t+))y(t+) \\ &= -p(t)(y(t+))^{\gamma-1}y(t+)y(t+) \\ &= -p(t)(y(t+))^{\gamma+1}, \quad t \in [a, b]. \end{aligned}$$

Integrating the last equation in the sense of Stieltjes, from a to b, we obtain

$$\begin{aligned} -\int_a^b p(t)(y(t+))^{\gamma+1}dg(t) &= \int_a^b \left(r(t)(y_g'(t))^{\gamma}\right)_g' y(t+)dg(t) \\ &= r(t)(y_g'(t))^{\gamma}y(t)\Big|_{t=a}^{t=b} \\ &\quad - \int_a^b r(t)(y_g'(t))^{\gamma+1}dg(t) \\ &= -\int_a^b r(t)(y_g'(t))^{\gamma+1}dg(t), \end{aligned}$$

from where we arrive at the equality

$$\int_a^b p(t)(y(t+))^{\gamma+1}dg(t) = \int_a^b r(t)(y_g'(t))^{\gamma+1}dg(t).$$

Hence,

$$\begin{aligned} M^{\gamma+1}\int_a^b p(t)dg(t) &\geq \int_a^b p(t)(y(t+))^{\gamma+1}dg(t) \\ &= \int_a^b r(t)(y_g'(t))^{\gamma+1}dg(t) \\ &= \int_a^c r(t)(y_g'(t))^{\gamma+1}dg(t) + \int_c^b r(t)(y_g'(t))^{\gamma+1}dg(t). \end{aligned} \tag{6.13}$$

Since

$$\int_a^c r(t)|y_g'(t)|dg(t) = \int_a^c (r(t))^{\frac{\gamma}{\gamma+1}}\left((r(t))^{\frac{1}{\gamma+1}}|y_g'(t)|\right)dg(t), \tag{6.14}$$

we can apply the Stieltjes–Hölder inequality for

$$f(t) = (r(t))^{\frac{\gamma}{\gamma+1}} \quad \text{and} \quad g(t) = (r(t))^{\frac{1}{\gamma+1}}|y_g'(t)|, \quad t \in [a,b],$$

and

$$p = \gamma + 1 \quad \text{and} \quad q = \frac{\gamma+1}{\gamma},$$

and find

$$\begin{aligned} &\int_a^c (r(t))^{\frac{\gamma}{\gamma+1}}\left((r(t))^{\frac{1}{\gamma+1}}|y_g'(t)|\right)dg(t) \\ &\leq \left(\int_a^c \left((r(t))^{\frac{\gamma}{\gamma+1}}\right)^{\frac{\gamma+1}{\gamma}} dg(t)\right)^{\frac{\gamma}{\gamma+1}} \left(\int_a^c \left((r(t))^{\frac{1}{\gamma+1}}|y_g'(t)|\right)^{\gamma+1} dg(t)\right)^{\frac{1}{\gamma+1}} \\ &= \left(\int_a^c r(t)dg(t)\right)^{\frac{\gamma}{\gamma+1}} \left(\int_a^c r(t)|y_g'(t)|^{\gamma+1}dg(t)\right)^{\frac{1}{\gamma+1}}. \end{aligned}$$

Therefore, employing (6.14), we find

$$\int_a^c r(t)|y_g'(t)|dg(t) \leq \left(\int_a^c r(t)dg(t)\right)^{\frac{\gamma}{\gamma+1}} \left(\int_a^c r(t)|y_g'(t)|^{\gamma+1}dg(t)\right)^{\frac{1}{\gamma+1}},$$

and then

$$\left(\int_a^c r(t)|y_g'(t)|dg(t)\right)^{\gamma+1} \leq \left(\int_a^c r(t)dg(t)\right)^{\gamma} \left(\int_a^c r(t)|y_g'(t)|^{\gamma+1}dg(t)\right),$$

whereupon

$$\int_a^c r(t)|y_g'(t)|^{\gamma+1}dg(t) \geq \frac{\left(\int_a^c r(t)|y_g'(t)|dg(t)\right)^{\gamma+1}}{\left(\int_a^c r(t)dg(t)\right)^{\gamma}}.$$

As above,

$$\int_c^b r(t)|y_g'(t)|^{\gamma+1}dg(t) \geq \frac{\left(\int_c^b r(t)|y_g'(t)|dg(t)\right)^{\gamma+1}}{\left(\int_c^b r(t)dg(t)\right)^{\gamma}}.$$

From the above two inequalities and the inequality (6.13), we get

$$\begin{aligned} M^{\gamma+1}\int_a^b p(t)dg(t) &\geq \frac{\left(\int_a^c r(t)|y_g'(t)|dg(t)\right)^{\gamma+1}}{\left(\int_a^c r(t)dg(t)\right)^{\gamma}} + \frac{\left(\int_c^b r(t)|y_g'(t)|dg(t)\right)^{\gamma+1}}{\left(\int_c^b r(t)dg(t)\right)^{\gamma}} \\
&\geq \begin{cases} \frac{\left(r(a)\int_a^c |y_g'(t)|dg(t)\right)^{\gamma+1}}{\left(\int_a^c r(t)dg(t)\right)^{\gamma}} + \frac{\left(r(a)\int_c^b |y_g'(t)|dg(t)\right)^{\gamma+1}}{\left(\int_c^b r(t)dg(t)\right)^{\gamma}} & \text{if } r \text{ is increasing} \\ \frac{\left(r(b)\int_a^c |y_g'(t)|dg(t)\right)^{\gamma+1}}{\left(\int_a^c r(t)dg(t)\right)^{\gamma}} + \frac{\left(r(b)\int_c^b |y_g'(t)|dg(t)\right)^{\gamma+1}}{\left(\int_c^b r(t)dg(t)\right)^{\gamma}} & \text{if } r \text{ is decreasing} \end{cases} \\
&\geq \begin{cases} \frac{(r(a))^{\gamma+1}M^{\gamma+1}}{(r(b))^{\gamma}(g(c)-g(a))^{\gamma}} + \frac{(r(a))^{\gamma+1}M^{\gamma+1}}{(r(b))^{\gamma}(g(b)-g(c))^{\gamma}} & \text{if } r \text{ is increasing} \\ \frac{(r(b))^{\gamma+1}M^{\gamma+1}}{(r(a))^{\gamma}(g(c)-g(a))^{\gamma}} + \frac{(r(b))^{\gamma+1}M^{\gamma+1}}{(r(a))^{\gamma}(g(b)-g(c))^{\gamma}} & \text{if } r \text{ is increasing} \end{cases} \\
&= \begin{cases} \frac{M^{\gamma+1}(r(a))^{\gamma+1}((g(b)-g(c))^{\gamma}+(g(c)-g(a))^{\gamma})}{(r(b))^{\gamma}(g(c)-g(a))^{\gamma}(g(b)-g(c))^{\gamma}} & \text{if } r \text{ is increasing} \\ \frac{M^{\gamma+1}(r(b))^{\gamma+1}((g(b)-g(c))^{\gamma}+(g(c)-g(a))^{\gamma})}{(r(a))^{\gamma}(g(c)-g(a))^{\gamma}(g(b)-g(c))^{\gamma}} & \text{if } r \text{ is decreasing,} \end{cases} \end{aligned}$$

whereupon

$$\int_a^b p(t)dg(t) \geq \begin{cases} \frac{(r(a))^{\gamma+1}((g(b)-g(c))^{\gamma}+(g(c)-g(a))^{\gamma})}{(r(b))^{\gamma}(g(c)-g(a))^{\gamma}(g(b)-g(c))^{\gamma}} & \text{if } r \text{ is increasing,} \\ \frac{(r(b))^{\gamma+1}((g(b)-g(c))^{\gamma}+(g(c)-g(a))^{\gamma})}{(r(a))^{\gamma}(g(c)-g(a))^{\gamma}(g(b)-g(c))^{\gamma}} & \text{if } r \text{ is decreasing.} \end{cases}$$

This is a contradiction with (6.12), and this completes the proof. □

Theorem 6.13. *Let r and p satisfy*

$$\int_a^b p(t)dg(t) < \begin{cases} \dfrac{(r(a))^{\gamma+1}2^{\gamma+1}}{(r(b))^{\gamma}(g(b)-g(a))^{\gamma}} & \text{if } r \text{ is increasing,} \\[2ex] \dfrac{(r(b))^{\gamma+1}2^{\gamma+1}}{(r(a))^{\gamma}(g(b)-g(a))^{\gamma}} & \text{if } r \text{ is decreasing.} \end{cases} \tag{6.15}$$

Then, Equation (6.9) *is disconjugate.*

Proof. Let $x_1 = g(c) - g(a)$ and $x_2 = g(b) - g(c)$. Then, applying the inequality

$$x_1^{\gamma} + x_2^{\gamma} \geq 2^{\gamma+1}\frac{x_1^{\gamma}x_2^{\gamma}}{(x_1+x_2)^{\gamma}},$$

we get

$$(g(b)-g(c))^{\gamma} + (g(c)-g(a))^{\gamma} > 2^{\gamma+1}\frac{(g(b)-g(c))^{\gamma}(g(c)-g(a))^{\gamma}}{(g(b)-g(a))^{\gamma}}$$

and

$$\frac{(g(b)-g(c))^{\gamma} + (g(c)-g(a))^{\gamma}}{(g(b)-g(c))^{\gamma}(g(c)-g(a))^{\gamma}} \geq \frac{2^{\gamma+1}}{(g(b)-g(a))^{\gamma}}.$$

Hence, with (6.15), we find

$$\begin{aligned} \int_a^b p(t)dg(t) &< \begin{cases} \frac{(r(a))^{\gamma+1}2^{\gamma+1}}{(r(b))^{\gamma}(g(b)-g(a))^{\gamma}} & \text{if } r \text{ is increasing,} \\ \frac{(r(b))^{\gamma+1}2^{\gamma+1}}{(r(a))^{\gamma}(g(b)-g(a))^{\gamma}} & \text{if } r \text{ is decreasing} \end{cases} \\ &\leq \begin{cases} \frac{(r(a))^{\gamma+1}((g(b)-g(c))^{\gamma}+(g(c)-g(a))^{\gamma})}{(r(b))^{\gamma}(g(c)-g(a))^{\gamma}(g(b)-g(c))^{\gamma}} & \text{if } r \text{ is increasing,} \\ \frac{(r(b))^{\gamma+1}((g(b)-g(c))^{\gamma}+(g(c)-g(a))^{\gamma})}{(r(a))^{\gamma}(g(c)-g(a))^{\gamma}(g(b)-g(c))^{\gamma}} & \text{if } r \text{ is decreasing.} \end{cases} \end{aligned}$$

Now, applying Theorem 6.12, we conclude that (6.9) is disconjugate. This completes the proof. □

Now, we consider the second-order Stieltjes differential equation

$$\left(r(t)\left(y_g'(t)\right)^{\gamma}\right)_g' + q(t)(y(t+))^{\gamma} = 0, \quad t \in [a,b], \tag{6.16}$$

where $\gamma \in (0,1]$, $r, q \in \mathscr{C}([a,b])$, $r > 0$ on $[a,b]$. Define

$$M(b) = \sup_{t\in[a,b]} (\Delta g(t))^{\gamma} \frac{|Q(t)|}{r(t)}, \quad \text{where} \quad Q(t) = \int_t^b q(s)dg(s),$$

$$M(a) = \sup_{t\in[a,b]} (\Delta g(t))^{\gamma} \frac{|Q(t)|}{r(t)}, \quad \text{where} \quad Q(t) = \int_a^t q(s)dg(s).$$

Theorem 6.14. *Let y be a solution to Equation* (6.16) *such that $y(a) = y_g'(b) = 0$. Then,*

$$2^{\gamma}(1+\gamma)^{\frac{\gamma}{\gamma+1}} \left(\int_a^b \frac{|Q(t)|^{\gamma+1}}{(r(t))^{\frac{1}{\gamma}}} \left(\int_a^t \frac{1}{(r(s))^{\frac{1}{\gamma+2}}} dg(s) \right)^{\gamma} dg(t) \right)^{\frac{1}{1+\gamma}}$$
$$+ 2^{\gamma-1} M(b) \geq 1,$$

where

$$Q(t) = \int_t^b q(s)dg(s), \quad t \in [a,b].$$

Proof. Without any restrictions, suppose that $y_g'(t) > 0$, $t \in [a,b]$. Multiplying Equation (6.16) with $y(t+)$, $t \in [a,b]$, we get

$$-q(t)(y(t+))^{\gamma+1} = \left(r(t)(y_g'(t))^{\gamma}\right)_g' y(t+), \quad t \in [a,b].$$

Integrating the last equation in the sense of Stieltjes, from a to b, we find

$$\begin{aligned} -\int_a^b q(t)(y(t+))^{\gamma+1} dg(t) &= \int_a^b \left(r(t)(y_g'(t))^{\gamma}\right)_g' y(t+)dg(t) \\ &= r(t)(y_g'(t))^{\gamma} y(t)\Big|_{t=a}^{t=b} - \int_a^b r(t)(y_g'(t))^{\gamma+1} dg(t) \\ &= -\int_a^b r(t)(y_g'(t))^{\gamma+1} dg(t), \end{aligned}$$

whereupon

$$\int_a^b r(t)(y_g'(t))^{\gamma+1} dg(t) = \int_a^b q(t)(y(t+))^{\gamma+1} dg(t).$$

Since

$$Q(t) = \int_t^b q(s)dg(s), \quad t \in [a,b],$$

we have $Q'_g(t) = -q(t)$, $t \in [a, b]$, and $Q(b) = 0$. Therefore,

$$\begin{aligned}\int_a^b r(t)(y'_g(t))^{\gamma+1}dg(t) &= -\int_a^b Q'_g(t)(y(t+))^{\gamma+1}dg(t)\\ &= -Q(t)(y(t))^{\gamma+1}\Big|_{t=a}^{t=b} + \int_a^b Q(t)\left(y^{\gamma+1}\right)'_g(t)dg(t)\\ &= \int_a^b Q(t)\left(y^{\gamma+1}\right)'_g(t)dg(t).\end{aligned}$$

By applying the chain rule, Theorem 1.42, we get

$$\begin{aligned}\left|\left(y^{\gamma+1}\right)'_g(t)\right| &= \left|(\gamma+1)\left(\int_a^1 (hy(t)+(1-h)y(t))^{\gamma}dh\right)y'_g(t)\right|\\ &\le (\gamma+1)y'_g(t)\left(\int_0^1 |hy(t)+(1-h)y(t)|^{\gamma}dg(t)\right)\\ &\le (\gamma+1)y'_g(t)\left(\int_a^1 |hy(t)|^{\gamma}dh + \int_0^1 |(1-h)y(t)|^{\gamma}dh\right)\\ &\le (\gamma+1)y'_g(t)\left(|y(t)|^{\gamma}+|y(t)|^{\gamma}\right)\\ &\le 2^{1-\gamma}(1+\gamma)y'_g(t)|y(t)+y(t+)|^{\gamma}, \quad t\in[a,b].\end{aligned}$$

Therefore,

$$\begin{aligned}\int_a^b r(t)(y'_g(t))^{\gamma+1}dg(t) &\le \int_a^b |Q(t)||(y^{\gamma+1})'_g(t)|dg(t)\\ &\le 2^{1-\gamma}(1+\gamma)\int_a^b |Q(t)|y'_g(t)|y(t)+y(t+)|^{\gamma}dg(t).\end{aligned}$$

Now, applying Theorem 5.26 for $s(t) = |Q(t)|$, $t \in [a, b]$, $p = \gamma$, $q = 1$, we find

$$\begin{aligned}&\int_a^b |Q(t)|y'_g(t)|y(t)+y(t+)|^{\gamma}dg(t)\\ &\quad\le \left(2^{2\gamma-1}\left(\frac{1}{\gamma+1}\right)^{\frac{1}{\gamma+1}}\left(\int_a^b \frac{|Q(t)|^{\gamma+1}}{(r(t))^{\frac{1}{\gamma}}}\left(\int_a^t \frac{1}{(r(s))^{\frac{1}{\gamma+2}}}dg(s)\right)^{\gamma}dg(t)\right)^{\frac{1}{\gamma+1}}\right.\\ &\qquad\left.+\,2^{\gamma-1}\sup_{t\in[a,b]}\left((\Delta g(t))^{\gamma}\frac{|Q(t)|}{r(t)}\right)\right)\times\int_a^b r(t)(y'_g(t))^{\gamma+1}dg(t).\end{aligned}$$

Therefore,

$$\begin{aligned}&\int_a^b r(t)(y_g'(t))^{\gamma+1}dg(t)\\ &\quad\leq (1+\gamma)2^{1-\gamma}\left(2^{2\gamma-1}\left(\frac{1}{\gamma+1}\right)^{\frac{1}{\gamma+1}}\right.\\ &\qquad\times\left(\int_a^b \frac{|Q(t)|^{\gamma+1}}{(r(t))^{\frac{1}{\gamma}}}\left(\int_a^t \frac{1}{(r(s))^{\frac{1}{\gamma+2}}}dg(s)\right)^{\gamma}dg(t)\right)^{\frac{1}{\gamma+1}}\\ &\qquad\left.+2^{\gamma-1}\sup_{t\in[a,b]}\left((\Delta g(t))^{\gamma}\frac{|Q(t)|}{r(t)}\right)\right)\times\int_a^b r(t)(y_g'(t))^{\gamma+1}dg(t).\end{aligned}$$

Hence,

$$\begin{aligned}&(1+\gamma)2^{1-\gamma}\left(2^{2\gamma-1}\left(\frac{1}{\gamma+1}\right)^{\frac{1}{\gamma+1}}\right.\\ &\quad\times\left(\int_a^b \frac{|Q(t)|^{\gamma+1}}{(r(t))^{\frac{1}{\gamma}}}\left(\int_a^t \frac{1}{(r(s))^{\frac{1}{\gamma+2}}}dg(s)\right)^{\gamma}dg(t)\right)^{\frac{1}{\gamma+1}}\\ &\quad\left.+2^{\gamma-1}\sup_{t\in[a,b]}\left((\Delta g(t))^{\gamma}\frac{|Q(t)|}{r(t)}\right)\right)\geq 1\end{aligned}$$

or

$$\begin{aligned}&2^{\gamma}(1+\gamma)^{\frac{\gamma}{\gamma+1}}\left(\int_a^b \frac{|Q(t)|^{\gamma+1}}{(r(t))^{\frac{1}{\gamma}}}\left(\int_a^t \frac{1}{(r(s))^{\frac{1}{\gamma+2}}}dg(s)\right)^{\gamma}dg(t)\right)^{\frac{1}{\gamma+1}}\\ &\quad+2^{\gamma-1}M(b)\geq 1.\end{aligned}$$

This completes the proof. □

Theorem 6.15. *Let y be a solution to Equation* (6.16) *such that $y_g'(a) = y(b) = 0$. Then,*

$$\begin{aligned}&2^{\gamma}(1+\gamma)^{\frac{\gamma}{\gamma+1}}\left(\int_a^b \frac{|Q(t)|^{\gamma+1}}{(r(t))^{\frac{1}{\gamma}}}\left(\int_t^b \frac{1}{(r(s))^{\frac{1}{\gamma+2}}}dg(s)\right)^{\gamma}dg(t)\right)^{\frac{1}{1+\gamma}}\\ &\quad+2^{\gamma-1}M(a)\geq 1,\end{aligned}$$

where

$$Q(t)=\int_a^t q(s)dg(s),\quad t\in[a,b].$$

Proof. Without any restrictions, suppose that $y'_g(t) > 0$, $t \in [a, b]$. Multiplying Equation (6.16) with $y(t+)$, $t \in [a, b]$, we get

$$-q(t)(y(t+))^{\gamma+1} = \left(r(t)(y'_g(t))^{\gamma}\right)'_g y(t+), \quad t \in [a, b].$$

Integrating the last equation in the sense of Stieltjes, from a to b, we find

$$\begin{aligned}
-\int_a^b q(t)(y(t+))^{\gamma+1}dg(t) &= \int_a^b \left(r(t)(y'_g(t))^{\gamma}\right)'_g y(t+)dg(t) \\
&= r(t)(y'_g(t))^{\gamma}y(t)\Big|_{t=a}^{t=b} - \int_a^b r(t)(y'_g(t))^{\gamma+1}dg(t) \\
&= -\int_a^b r(t)(y'_g(t))^{\gamma+1}dg(t),
\end{aligned}$$

whereupon

$$\int_a^b r(t)(y'_g(t))^{\gamma+1}dg(t) = \int_a^b q(t)(y(t+))^{\gamma+1}dg(t).$$

Since

$$Q(t) = \int_a^t q(s)dg(s), \quad t \in [a, b],$$

we have $Q'_g(t) = q(t)$, $t \in [a, b]$, and $Q(a) = 0$. Therefore,

$$\begin{aligned}
\int_a^b r(t)(y'_g(t))^{\gamma+1}dg(t) &= \int_a^b Q'_g(t)(y(t+))^{\gamma+1}dg(t) \\
&= Q(t)(y(t))^{\gamma+1}\Big|_{t=a}^{t=b} - \int_a^b Q(t)\left(y^{\gamma+1}\right)'_g(t)dg(t) \\
&= -\int_a^b Q(t)\left(y^{\gamma+1}\right)'_g(t)dg(t) \\
&\leq \int_a^b |Q(t)| \left|(y^{\gamma+1})'_g\right|(t)dg(t).
\end{aligned}$$

By applying the chain rule, Theorem 1.42, we get

$$\begin{aligned}\left|\left(y^{\gamma+1}\right)_g'(t)\right| &= \left|(\gamma+1)\left(\int_a^1 (hy(t)+(1-h)y(t))^\gamma dh\right) y_g'(t)\right|\\ &\le (\gamma+1)y_g'(t)\left(\int_0^1 |hy(t)+(1-h)y(t)|^\gamma dh\right)\\ &\le (\gamma+1)y_g'(t)\left(\int_a^1 |hy(t)|^\gamma dh + \int_0^1 |(1-h)y(t)|^\gamma dh\right)\\ &\le (\gamma+1)y_g'(t)\left(|y(t)|^\gamma + |y(t)|^\gamma\right)\\ &\le 2^{1-\gamma}(1+\gamma)y_g'(t)|y(t)+y(t+)|^\gamma, \quad t\in[a,b].\end{aligned}$$

Therefore,

$$\begin{aligned}\int_a^b r(t)(y_g'(t))^{\gamma+1}dg(t) &\le \int_a^b |Q(t)||(y^{\gamma+1})_g'(t)|dg(t)\\ &\le 2^{1-\gamma}(1+\gamma)\int_a^b |Q(t)|y_g'(t)|y(t)+y(t+)|^\gamma dg(t).\end{aligned}$$

Now, we apply Theorem 5.28 for $s(t)=|Q(t)|$, $t\in[a,b]$, $p=\gamma$, and $q=1$, and find

$$\begin{aligned}&\int_a^b |Q(t)|y_g'(t)|y(t)+y(t+)|^\gamma dg(t)\\ &\le \left(2^{2\gamma-1}\left(\frac{1}{\gamma+1}\right)^{\frac{1}{\gamma+1}}\left(\int_a^b \frac{|Q(t)|^{\gamma+1}}{(r(t))^{\frac{1}{\gamma}}}\left(\int_t^b \frac{1}{(r(s))^{\frac{1}{\gamma+2}}}dg(s)\right)^\gamma dg(t)\right)^{\frac{1}{\gamma+1}}\right.\\ &\quad\left.+2^{\gamma-1}\sup_{t\in[a,b]}\left((\Delta g(t))^\gamma\frac{|Q(t)|}{r(t)}\right)\right)\times\int_a^b r(t)(y_g'(t))^{\gamma+1}dg(t).\end{aligned}$$

Therefore,

$$\begin{aligned}&\int_a^b r(t)(y_g'(t))^{\gamma+1}dg(t)\\ &\quad\le (1+\gamma)2^{1-\gamma}\left(2^{2\gamma-1}\left(\frac{1}{\gamma+1}\right)^{\frac{1}{\gamma+1}}\right.\end{aligned}$$

$$\times \left(\int_a^b \frac{|Q(t)|^{\gamma+1}}{(r(t))^{\frac{1}{\gamma}}} \left(\int_t^b \frac{1}{(r(s))^{\frac{1}{\gamma+2}}} dg(s) \right)^{\gamma} dg(t) \right)^{\frac{1}{\gamma+1}}$$

$$+ 2^{\gamma-1} \sup_{t\in[a,b]} \left((\Delta g(t))^{\gamma} \frac{|Q(t)|}{r(t)} \right) \Bigg) \times \int_a^b r(t)(y_g'(t))^{\gamma+1} dg(t).$$

Hence,

$$(1+\gamma)2^{1-\gamma} \Bigg(2^{2\gamma-1} \left(\frac{1}{\gamma+1} \right)^{\frac{1}{\gamma+1}}$$

$$\times \left(\int_a^b \frac{|Q(t)|^{\gamma+1}}{(r(t))^{\frac{1}{\gamma}}} \left(\int_t^b \frac{1}{(r(s))^{\frac{1}{\gamma+2}}} dg(s) \right)^{\gamma} dg(t) \right)^{\frac{1}{\gamma+1}}$$

$$+ 2^{\gamma-1} \sup_{t\in[a,b]} \left((\Delta g(t))^{\gamma} \frac{|Q(t)|}{r(t)} \right) \Bigg) \geq 1$$

or

$$2^{\gamma}(1+\gamma)^{\frac{\gamma}{\gamma+1}} \left(\int_a^b \frac{|Q(t)|^{\gamma+1}}{(r(t))^{\frac{1}{\gamma}}} \left(\int_t^b \frac{1}{(r(s))^{\frac{1}{\gamma+2}}} dg(s) \right)^{\gamma} dg(t) \right)^{\frac{1}{\gamma+1}}$$

$$+ 2^{\gamma-1} M(a) \geq 1.$$

This completes the proof. □

Problem 6.6. Let y be a solution to the second-order Stieltjes differential Equation (6.16) for which $y(a) = y_g'(b) = 0$. Then, prove that

$$2^{\gamma}(1+\gamma)^{\frac{\gamma}{\gamma+1}} \left(\int_a^b |Q(t)|^{\gamma+1} (t-a)^{\gamma} dg(t) \right)^{\frac{1}{1+\gamma}} + 2^{\gamma-1} M(b) \geq 1,$$

where

$$Q(t) = \int_t^b q(s) dg(s), \quad t \in [a,b].$$

Problem 6.7. Let y be a solution to the second-order Stieltjes differential equation (6.16) for which $y_g'(a) = y(b) = 0$. Then, prove that

$$2^{\gamma}(1+\gamma)^{\frac{\gamma}{\gamma+1}} \left(\int_a^b |Q(t)|^{\gamma+1} (t-a)^{\gamma} dg(t) \right)^{\frac{1}{1+\gamma}} + 2^{\gamma-1} M(a) \geq 1,$$

where

$$Q(t) = \int_a^t q(s)dg(s), \quad t \in [a,b].$$

6.3 Second-Order Stieltjes Differential Equations with Damping Arguments

In this section, we investigate the second-order Stieltjes differential equation

$$\left(r(t)(y_g'(t))^\gamma\right)_g' + p(t)(y_g'(t))^\gamma + q(t)(y(t+))^\gamma = 0, \quad t \in [a,b], \qquad (6.17)$$

where $\gamma \geq 1$ is a quotient of odd positive integers, $r, p, q \in \mathscr{C}([a,b])$, $r(t) > 0$, $t \in [a,b]$, and

$$(\Delta g(t))|p(t)| \leq \frac{1}{c} r(t), \quad t \in [a,b],$$

for some constant $c \geq 1$. Define

$$\Lambda(b) = \max_{t\in[a,b]} (\Delta g(t))^\gamma \frac{|Q(t)|}{r(t)}, \quad \text{where} \quad Q(t) = \int_t^b q(s)dg(s),$$

$$\Lambda(a) = \max_{t\in[a,b]} (\Delta g(t))^\gamma \frac{|Q(t)|}{r(t)}, \quad \text{where} \quad Q(t) = \int_a^t q(s)dg(s),$$

$$R_a(t) = \int_a^t \frac{1}{(r(s))^{\frac{1}{\gamma}}} dg(s),$$

$$R_b(t) = \int_a^t \frac{1}{(r(s))^{\frac{1}{\gamma}}} dg(s).$$

Theorem 6.16. *Let y be a nontrivial solution of Equation* (6.17) *that does not change sign in* (a,b) *and* $y(a) = y_g'(b) = 0$. *Then,*

$$2^{2\gamma-1}\left(\frac{1}{\gamma+1}\right)^{\frac{1}{\gamma+1}} \left(\int_a^b \frac{|Q(t)|^{\gamma+1}}{(r(t))^\gamma}(R_a(t))^\gamma dg(t)\right)^{\frac{1}{\gamma+1}} + \Lambda(b)$$

$$+\left(\frac{\gamma}{\gamma+1}\right)^{\gamma+1} \left(\int_a^b \frac{|p(t)|^{\gamma+1}}{(r(t))^\gamma}(R_a(t))^\gamma dg(t)\right)^{\frac{1}{\gamma+1}} \geq 1 - \frac{1}{c},$$

where

$$Q(t) = \int_t^b q(s)dg(s), \quad t \in [a,b].$$

Proof. Without any restrictions, assume that $y'_g(t) > 0$, $t \in (a, b)$. Multiplying Equation (6.17) with $y(t+)$, $t \in [a, b]$, we find

$$-q(t)(y(t+))^{\gamma+1} = \left(r(t)(y'_g(t))^{\gamma}\right)'_g y(t+) + p(t)(y'_g(t))^{\gamma} y(t+), \quad t \in [a, b].$$

Integrating the last equation in the sense of Stieltjes, from a to b, we obtain

$$\begin{aligned}
&-\int_a^b q(t)(y(t+))^{\gamma+1} dg(t)\\
&\quad = \int_a^b \left(r(t)(y'_g(t))^{\gamma}\right)'_g y(t+) dg(t)\\
&\qquad + \int_a^b p(t)(y'_g(t))^{\gamma} y(t+) dg(t)\\
&\quad = r(t)(y'_g(t))^{\gamma} y(t)\Big|_{t=a}^{t=b} - \int_a^b r(t)(y'_g(t))^{\gamma+1} dg(t)\\
&\qquad + \int_a^b p(t)(y'_g(t))^{\gamma} y(t+) dg(t)\\
&\quad = -\int_a^b r(t)(y'_g(t))^{\gamma+1} dg(t) + \int_a^b p(t)(y'_g(t))^{\gamma} y(t+) dg(t).
\end{aligned}$$

Since

$$Q(t) = \int_t^b q(s) dg(s), \quad t \in [a, b],$$

we have $Q'_g(t) = -q(t)$, $t \in [a, b]$, and $Q(b) = 0$. Therefore,

$$\begin{aligned}
&\int_a^b r(t)(y'_g(t))^{\gamma+1} dg(t)\\
&\quad = \int_a^b q(t)(y(t+))^{\gamma+1} dg(t) + \int_a^b p(t)(y'_g(t))^{\gamma} y(t+) dg(t)\\
&\quad = \int_a^b p(t)(y'_g(t))^{\gamma} y(t+) dg(t) - \int_a^b Q'_g(t)(y(t+))^{\gamma+1} dg(t)\\
&\quad = \int_a^b p(t)(y'_g(t))^{\gamma} y(t+) dg(t) - Q(t)(y(t))^{\gamma+1}\Big|_{t=a}^{t=b}\\
&\qquad + \int_a^b Q(t)(y^{\gamma+1})'_g(t) dg(t)\\
&\quad = \int_a^b p(t)(y'_g(t))^{\gamma} y(t+) dg(t) + \int_a^b Q(t)(y^{\gamma+1})'_g(t) dg(t),
\end{aligned}$$

where we have used that $y(a)=0$ and $Q(b)=0$. Now, applying the chain rule, Theorem 1.42, we find

$$\begin{aligned}
(y^{\gamma+1})'_g(t) &= (\gamma+1)\left(\int_0^1 (hy(t)+(1-h)y(t))^\gamma dh\right) y'_g(t)\\
&\le 2^{\gamma-1}(\gamma+1)y'_g(t)\left(\int_0^1 (hy(t))^\gamma dh + \int_0^1 (1-h)(y(t))^\gamma dh\right)\\
&\le 2^{\gamma-1}(\gamma+1)y'_g(t)(|y(t)|^\gamma + |y(t+)|^\gamma)\\
&\le 2^{\gamma-1}(\gamma+1)y'_g(t)|y(t)+y(t+)|^\gamma, \quad t\in[a,b].
\end{aligned}$$

Therefore,

$$\begin{aligned}
&\int_a^b r(t)(y'_g(t))^{\gamma+1}dg(t)\\
&\quad\le \int_a^b |p(t)|(y'_g(t))^\gamma |y(t+)|dg(t)\\
&\quad\quad + 2^{\gamma-1}(\gamma+1)\int_a^b |Q(t)||y(t)+y(t+)|^\gamma (y'_g(t))dg(t).
\end{aligned}$$

Now, for the integral

$$\int_a^b |Q(t)||y(t)+y(t+)|^\gamma (y'_g(t))dg(t),$$

we apply Theorem 5.26 with $s(t)=|Q(t)|$, $p=\gamma$, and $q=1$, and find

$$\begin{aligned}
&\int_a^b |Q(t)||y(t)+y(t+)|^\gamma y'_g(t)dg(t)\\
&\le \left(2^{2\gamma-1}\left(\frac{1}{\gamma+1}\right)^{\frac{1}{\gamma+1}}\left(\int_a^b \frac{|Q(t)|^{\gamma+1}}{(r(t))^{\frac{1}{\gamma}}}\left(\int_a^t \frac{1}{(r(s))^{\frac{1}{\gamma}}}dg(s)\right)^\gamma dg(t)\right)^{\frac{1}{\gamma+1}}\right.\\
&\quad\left. + \sup_{t\in[a,b]}\left((\Delta g(t))\frac{|Q(t)|}{r(t)}\right)\right)\int_a^b r(t)(y'_g(t))^{\gamma+1}dg(t)\\
&\le \left(2^{2\gamma-1}\left(\frac{1}{\gamma+1}\right)^{\frac{1}{\gamma+1}}\left(\int_a^b \frac{|Q(t)|^{\gamma+1}}{(r(t))^\gamma}(R_a(t))^\gamma dg(t)\right)^{\frac{1}{\gamma+1}} + \Lambda(b)\right)\\
&\quad\times\int_a^b r(t)(y'_g(t))^{\gamma+1}dg(t).
\end{aligned}$$

Next,

$$\begin{aligned}\int_a^b &|p(t)|(y_g'(t))^\gamma|y(t+)|dg(t)\\ &= \int_a^b |p(t)|(y_g'(t))^\gamma|y(t)+(\Delta g(t))y_g'(t)|dg(t)\\ &\leq \int_a^b |p(t)|(y_g'(t))^\gamma|y(t)|dg(t)+\int_a^b |p(t)|(\Delta g(t))(y_g'(t))^{\gamma+1}dg(t).\end{aligned}$$

For the integral

$$\int_a^b |p(t)|(y_g'(t))^\gamma|y(t)|dg(t),$$

we apply Theorem 5.22 with $s(t)=|p(t)|$, $p=1$, and $q=\gamma$, and get

$$\begin{aligned}\int_a^b &|p(t)|(y_g'(t))^\gamma|y(t)|dg(t)\\ &\leq \left(\left(\frac{\gamma}{\gamma+1}\right)^{\frac{\gamma}{\gamma+1}}\left(\int_a^b \frac{|p(t)|^{\gamma+1}}{(r(t))^\gamma}\left(\int_a^t \frac{1}{(r(s))^{\frac{1}{\gamma}}}dg(s)\right)^\gamma dg(t)\right)^{\frac{1}{\gamma+1}}\right)\\ &\quad\times\int_a^b r(t)(y_g'(t))^{\gamma+1}dg(t)\\ &= \left(\left(\frac{\gamma}{\gamma+1}\right)^{\frac{\gamma}{\gamma+1}}\left(\int_a^b \frac{|p(t)|^{\gamma+1}}{(r(t))^\gamma}(R_a(t))^\gamma dg(t)\right)^{\frac{1}{\gamma+1}}\right)\\ &\quad\times\int_a^b r(t)(y_g'(t))^{\gamma+1}dg(t).\end{aligned}$$

Now, using the fact that

$$|p(t)|(\Delta g(t))\leq \frac{1}{c}r(t),\quad t\in[a,b],$$

we arrive at

$$\int_a^b |p(t)|(\Delta g(t))(y_g'(t))^{\gamma+1}dg(t)\leq \frac{1}{c}\int_a^b r(t)(y_g'(t))^{\gamma+1}dg(t).$$

Therefore,

$$\begin{aligned}
&\int_a^b |p(t)|(y_g'(t))^\gamma |y(t+)| dg(t)\\
&\quad \le \left(\left(\frac{\gamma}{\gamma+1}\right)^{\frac{\gamma}{\gamma+1}} \left(\int_a^b \frac{|p(t)|^{\gamma+1}}{(r(t))^\gamma}(R_a(t))^\gamma dg(t)\right)^{\frac{1}{\gamma+1}}\right)\\
&\qquad \times \int_a^b r(t)(y_g'(t))^{\gamma+1} dg(t) + \frac{1}{c}\int_a^b r(t)(y_g'(t))^{\gamma+1} dg(t)
\end{aligned}$$

and

$$\begin{aligned}
&\int_a^b r(t)(y_g'(t))^{\gamma+1} dg(t)\\
&\quad \le \left(\left(\frac{\gamma}{\gamma+1}\right)^{\frac{\gamma}{\gamma+1}} \left(\int_a^b \frac{|p(t)|^{\gamma+1}}{(r(t))^\gamma}(R_a(t))^\gamma dg(t)\right)^{\frac{1}{\gamma+1}}\right)\\
&\quad \times \int_a^b r(t)(y_g'(t))^{\gamma+1} dg(t) + \frac{1}{c}\int_a^b r(t)(y_g'(t))^{\gamma+1} dg(t)\\
&\quad + \left(2^{2\gamma-1}\left(\frac{1}{\gamma+1}\right)^{\frac{1}{\gamma+1}} \left(\int_a^b \frac{|Q(t)|^{\gamma+1}}{(r(t))^\gamma}(R_a(t))^\gamma dg(t)\right)^{\frac{1}{\gamma+1}} + \Lambda(b)\right)\\
&\quad \times \int_a^b r(t)(y_g'(t))^{\gamma+1} dg(t),
\end{aligned}$$

whereupon

$$\begin{aligned}
&\left(1-\frac{1}{c}\right)\int_a^b r(t)(y_g'(t))^{\gamma+1} dg(t)\\
&\quad \le \left(\left(\left(\frac{\gamma}{\gamma+1}\right)^{\frac{\gamma}{\gamma+1}} \left(\int_a^b \frac{|p(t)|^{\gamma+1}}{(r(t))^\gamma}(R_a(t))^\gamma dg(t)\right)^{\frac{1}{\gamma+1}}\right)\right.\\
&\quad \left. + \left(2^{2\gamma-1}\left(\frac{1}{\gamma+1}\right)^{\frac{1}{\gamma+1}} \left(\int_a^b \frac{|Q(t)|^{\gamma+1}}{(r(t))^\gamma}(R_a(t))^\gamma dg(t)\right)^{\frac{1}{\gamma+1}} + \Lambda(b)\right)\right)\\
&\quad \times \int_a^b r(t)(y_g'(t))^{\gamma+1} dg(t).
\end{aligned}$$

By the last inequality, we find

$$\left(\left(\frac{\gamma}{\gamma+1}\right)^{\frac{\gamma}{\gamma+1}}\left(\int_a^b \frac{|p(t)|^{\gamma+1}}{(r(t))^{\gamma}}(R_a(t))^{\gamma}dg(t)\right)^{\frac{1}{\gamma+1}}\right)$$

$$+\left(2^{2\gamma-1}\left(\frac{1}{\gamma+1}\right)^{\frac{1}{\gamma+1}}\left(\int_a^b \frac{|Q(t)|^{\gamma+1}}{(r(t))^{\gamma}}(R_a(t))^{\gamma}dg(t)\right)^{\frac{1}{\gamma+1}}+\Lambda(b)\right)$$

$$\geq 1-\frac{1}{c}.$$

This completes the proof. □

Problem 6.8. Let y be a nontrivial solution of Equation (6.17) that does not change sign in (a,b) and $y(a) = y'_g(b) = 0$. Then, prove that

$$2^{2\gamma-1}\left(\frac{1}{\gamma+1}\right)^{\frac{1}{\gamma+1}}\left(\int_a^b |Q(t)|^{\gamma+1}(t-a)^{\gamma}dg(t)\right)^{\frac{1}{\gamma+1}} + \max_{t\in[a,b]}(\Delta g(t))|Q(t)|$$

$$+\left(\frac{\gamma}{\gamma+1}\right)^{\gamma+1}\left(\int_a^b |p(t)|^{\gamma+1}(t-a)^{\gamma}dg(t)\right)^{\frac{1}{\gamma+1}} \geq 1-\frac{1}{c},$$

where

$$Q(t) = \int_t^b q(s)dg(s), \quad t\in[a,b].$$

Theorem 6.17. *Let y be a nontrivial solution of Equation* (6.17) *that does not change sign in (a,b) and $y'_g(a) = y(b) = 0$. Then,*

$$2^{2\gamma-1}\left(\frac{1}{\gamma+1}\right)^{\frac{1}{\gamma+1}}\left(\int_a^b \frac{|Q(t)|^{\gamma+1}}{(r(t))^{\gamma}}(R_b(t))^{\gamma}dg(t)\right)^{\frac{1}{\gamma+1}} + \Lambda(a)$$

$$+\left(\frac{\gamma}{\gamma+1}\right)^{\gamma+1}\left(\int_a^b \frac{|p(t)|^{\gamma+1}}{(r(t))^{\gamma}}(R_b(t))^{\gamma}dg(t)\right)^{\frac{1}{\gamma+1}} \geq 1-\frac{1}{c},$$

where

$$Q(t) = \int_a^t q(s)dg(s), \quad t\in[a,b].$$

Proof. Without any restrictions, assume that $y'_g(t) > 0$, $t \in (a, b)$. As in the proof of Theorem 6.16, we arrive at

$$
\begin{aligned}
-\int_a^b q(t)(y(t+))^{\gamma+1}dg(t) &= \int_a^b \left(r(t)(y'_g(t))^{\gamma}\right)'_g y(t+)dg(t) \\
&= -\int_a^b r(t)(y'_g(t))^{\gamma+1}dg(t) + \int_a^b p(t)(y'_g(t))^{\gamma}y(t+)dg(t).
\end{aligned}
$$

Since

$$
Q(t) = \int_a^t q(s)dg(s), \quad t \in [a, b],
$$

we have $Q'_g(t) = q(t)$, $t \in [a, b]$, and $Q(a) = 0$. Therefore,

$$
\begin{aligned}
&\int_a^b r(t)(y'_g(t))^{\gamma+1}dg(t) \\
&\quad = \int_a^b q(t)(y(t+))^{\gamma+1}dg(t) + \int_a^b p(t)(y'_g(t))^{\gamma}y(t+)dg(t) \\
&\quad = \int_a^b p(t)(y'_g(t))^{\gamma}y(t+)dg(t) - \int_a^b Q'_g(t)(y(t+))^{\gamma+1}dg(t) \\
&\quad = \int_a^b p(t)(y'_g(t))^{\gamma}y(t+)dg(t) - Q(t)(y(t))^{\gamma+1}\Big|_{t=a}^{t=b} \\
&\qquad + \int_a^b Q(t)(y^{\gamma+1})'_g(t)dg(t) \\
&\quad = \int_a^b p(t)(y'_g(t))^{\gamma}y(t+)dg(t) + \int_a^b Q(t)(y^{\gamma+1})'_g(t)dg(t),
\end{aligned}
$$

where we have used the fact that $y(b) = 0$ and $Q(a) = 0$. As in the proof of Theorem 6.16, we find

$$
\begin{aligned}
&\int_a^b r(t)(y'_g(t))^{\gamma+1}dg(t) \\
&\quad \le \int_a^b |p(t)|(y'_g(t))^{\gamma}|y(t+)|dg(t) \\
&\qquad + 2^{\gamma-1}(\gamma+1)\int_a^b |Q(t)||y(t) + y(t+)|^{\gamma}(y'_g(t))dg(t).
\end{aligned}
$$

Now, for the integral

$$\int_a^b |Q(t)||y(t)+y(t+)|^\gamma (y_g'(t))dg(t),$$

we apply Theorem 5.28 with $s(t)=|Q(t)|$, $p=\gamma$, $q=1$, and find

$$\begin{aligned}
&\int_a^b |Q(t)||y(t)+y(t+)|^\gamma y_g'(t)dg(t)\\
&\quad\le \left(2^{2\gamma-1}\left(\frac{1}{\gamma+1}\right)^{\frac{1}{\gamma+1}}\left(\int_a^b \frac{|Q(t)|^{\gamma+1}}{(r(t))^{\frac{1}{\gamma}}}\left(\int_t^b \frac{1}{(r(s))^{\frac{1}{\gamma}}}dg(s)\right)^\gamma dg(t)\right)^{\frac{1}{\gamma+1}}\right.\\
&\qquad\left. + \sup_{t\in[a,b]}\left((\Delta g(t))\frac{|Q(t)|}{r(t)}\right)\right)\int_a^b r(t)(y_g'(t))^{\gamma+1}dg(t)\\
&\quad\le \left(2^{2\gamma-1}\left(\frac{1}{\gamma+1}\right)^{\frac{1}{\gamma+1}}\left(\int_a^b \frac{|Q(t)|^{\gamma+1}}{(r(t))^{\gamma}}(R_b(t))^\gamma dg(t)\right)^{\frac{1}{\gamma+1}}+\Lambda(b)\right)\\
&\qquad\times\int_a^b r(t)(y_g'(t))^{\gamma+1}dg(t).
\end{aligned}$$

Moreover, we have

$$\begin{aligned}
&\int_a^b |p(t)|(y_g'(t))^\gamma |y(t+)|dg(t)\\
&\quad\le \int_a^b |p(t)|(y_g'(t))^\gamma |y(t)|dg(t)+\int_a^b |p(t)|(\Delta g(t))(y_g'(t))^{\gamma+1}dg(t).
\end{aligned}$$

For the integral

$$\int_a^b |p(t)|(y_g'(t))^\gamma |y(t)|dg(t),$$

we apply Theorem 5.23 with $s(t)=|p(t)|$, $p=1$, $q=\gamma$, and get

$$\begin{aligned}
&\int_a^b |p(t)|(y_g'(t))^\gamma |y(t)|dg(t)\\
&\quad\le \left(\left(\frac{\gamma}{\gamma+1}\right)^{\frac{\gamma}{\gamma+1}}\left(\int_a^b \frac{|p(t)|^{\gamma+1}}{(r(t))^{\gamma}}\left(\int_t^b \frac{1}{(r(s))^{\frac{1}{\gamma}}}dg(s)\right)^\gamma dg(t)\right)^{\frac{1}{\gamma+1}}\right)
\end{aligned}$$

$$\times \int_a^b r(t)(y_g'(t))^{\gamma+1}dg(t)$$

$$= \left(\left(\frac{\gamma}{\gamma+1}\right)^{\frac{\gamma}{\gamma+1}} \left(\int_a^b \frac{|p(t)|^{\gamma+1}}{(r(t))^\gamma} (R_b(t))^\gamma dg(t) \right)^{\frac{1}{\gamma+1}} \right)$$

$$\times \int_a^b r(t)(y_g'(t))^{\gamma+1}dg(t).$$

Now, using the fact that

$$|p(t)|(\Delta g(t)) \le \frac{1}{c} r(t), \quad t \in [a,b],$$

we find

$$\int_a^b |p(t)|(\Delta g(t))(y_g'(t))^{\gamma+1}dg(t) \le \frac{1}{c}\int_a^b r(t)(y_g'(t))^{\gamma+1}dg(t).$$

Therefore,

$$\int_a^b |p(t)|(y_g'(t))^\gamma |y(t+)|dg(t)$$

$$\le \left(\left(\frac{\gamma}{\gamma+1}\right)^{\frac{\gamma}{\gamma+1}} \left(\int_a^b \frac{|p(t)|^{\gamma+1}}{(r(t))^\gamma} (R_b(t))^\gamma dg(t) \right)^{\frac{1}{\gamma+1}} \right)$$

$$\times \int_a^b r(t)(y_g'(t))^{\gamma+1}dg(t) + \frac{1}{c}\int_a^b r(t)(y_g'(t))^{\gamma+1}dg(t)$$

and

$$\int_a^b r(t)(y_g'(t))^{\gamma+1}dg(t)$$

$$\le \left(\left(\frac{\gamma}{\gamma+1}\right)^{\frac{\gamma}{\gamma+1}} \left(\int_a^b \frac{|p(t)|^{\gamma+1}}{(r(t))^\gamma} (R_b(t))^\gamma dg(t) \right)^{\frac{1}{\gamma+1}} \right)$$

$$\times \int_a^b r(t)(y_g'(t))^{\gamma+1}dg(t) + \frac{1}{c}\int_a^b r(t)(y_g'(t))^{\gamma+1}dg(t)$$

$$+ \left(2^{2\gamma-1} \left(\frac{1}{\gamma+1}\right)^{\frac{1}{\gamma+1}} \left(\int_a^b \frac{|Q(t)|^{\gamma+1}}{(r(t))^\gamma} (R_b(t))^\gamma dg(t) \right)^{\frac{1}{\gamma+1}} + \Lambda(a) \right)$$

$$\times \int_a^b r(t)(y_g'(t))^{\gamma+1}dg(t),$$

whereupon

$$\left(1-\frac{1}{c}\right)\int_a^b r(t)(y_g'(t))^{\gamma+1}dg(t)$$
$$\leq\left(\left(\left(\frac{\gamma}{\gamma+1}\right)^{\frac{\gamma}{\gamma+1}}\left(\int_a^b\frac{|p(t)|^{\gamma+1}}{(r(t))^{\gamma}}(R_b(t))^{\gamma}dg(t)\right)^{\frac{1}{\gamma+1}}\right)\right.$$
$$\left.+\left(2^{2\gamma-1}\left(\frac{1}{\gamma+1}\right)^{\frac{1}{\gamma+1}}\left(\int_a^b\frac{|Q(t)|^{\gamma+1}}{(r(t))^{\gamma}}(R_b(t))^{\gamma}dg(t)\right)^{\frac{1}{\gamma+1}}+\Lambda(a)\right)\right)$$
$$\times\int_a^b r(t)(y_g'(t))^{\gamma+1}dg(t).$$

By the last inequality, we obtain

$$\left(\left(\frac{\gamma}{\gamma+1}\right)^{\frac{\gamma}{\gamma+1}}\left(\int_a^b\frac{|p(t)|^{\gamma+1}}{(r(t))^{\gamma}}(R_b(t))^{\gamma}dg(t)\right)^{\frac{1}{\gamma+1}}\right)$$
$$+\left(2^{2\gamma-1}\left(\frac{1}{\gamma+1}\right)^{\frac{1}{\gamma+1}}\left(\int_a^b\frac{|Q(t)|^{\gamma+1}}{(r(t))^{\gamma}}(R_b(t))^{\gamma}dg(t)\right)^{\frac{1}{\gamma+1}}+\Lambda(a)\right)$$
$$\geq 1-\frac{1}{c}.$$

This completes the proof. □

Problem 6.9. Let y be a nontrivial solution of Equation (6.17) that does not change sign in (a,b) and $y_g'(a)=y(b)=0$. Then, prove that

$$2^{2\gamma-1}\left(\frac{1}{\gamma+1}\right)^{\frac{1}{\gamma+1}}\left(\int_a^b|Q(t)|^{\gamma+1}(g(b)-g(t))^{\gamma}dg(t)\right)^{\frac{1}{\gamma+1}}$$
$$+\max_{t\in[a,b]}(\Delta g(t))|Q(t)|+\left(\frac{\gamma}{\gamma+1}\right)^{\gamma+1}$$
$$\times\left(\int_a^b|p(t)|^{\gamma+1}(g(b)-g(t))^{\gamma}dg(t)\right)^{\frac{1}{\gamma+1}}\geq 1-\frac{1}{c},$$

where

$$Q(t)=\int_a^t q(s)dg(s),\quad t\in[a,b].$$

In the following, assume that $Q'_g(t) = q(t)$, $t \in [a,b]$, and there is a $h \in (a,b)$ such that $R(h) = R_a(h) = R_b(h)$. Also, let

$$\Lambda = \max_{t\in[a,b]} (\Delta g(t))^{\gamma} \frac{|Q(t)|}{r(t)}.$$

Theorem 6.18. *Let y be a nontrivial solution of Equation (6.17) for which $y(a) = y(b) = 0$. Then,*

$$2^{3\gamma-2}\left(\frac{1}{\gamma+1}\right)^{\frac{1}{\gamma+1}}\left(\int_a^b \frac{|Q(t)|^{\gamma+1}}{(r(t))^{\gamma}}(R(h))^{\gamma}dg(t)\right)^{\frac{1}{\gamma+1}} + 2^{2\gamma-2}\Lambda$$
$$+\left(\frac{\gamma}{\gamma+1}\right)^{\gamma+1}\left(\int_a^b \frac{|p(t)|^{\gamma+1}}{(r(t))^{\gamma}}(R(h))^{\gamma}dg(t)\right)^{\frac{1}{\gamma+1}} \geq 1-\frac{1}{c}.$$

Proof. As in the proof of Theorem 6.16, we find

$$\int_a^b r(t)|y'_g(t)|^{\gamma+1}dg(t)$$
$$\leq \int_a^b |p(t)||y'_g(t)|^{\gamma}|y(t+)|dg(t) + \int_a^b |Q(t)||(y^{\gamma+1})'_g(t)|dg(t)$$

and

$$\int_a^b |Q(t)||(y^{\gamma+1})'_g(t)|dg(t) \leq 2^{\gamma-1}(\gamma+1)\int_a^b |Q(t)||y'_g(t)||y(t)+y(t+)|^p dg(t).$$

For the integral

$$\int_a^b |Q(t)||y'_g(t)||y(t)+y(t+)|^{\gamma}dg(t),$$

applying Theorem 5.26 with $s(t) = |Q(t)|$, $p = \gamma$, $q = 1$, we find

$$\int_a^b |Q(t)||(y^{\gamma+1})'_g(t)|dg(t)$$
$$\leq \left(2^{3\gamma-2}\left(\frac{1}{\gamma+1}\right)^{\frac{1}{\gamma+1}}\left(\int_a^b \frac{|Q(t)|^{\gamma+1}}{(r(t))^{\gamma}}(R(h))^{\gamma}dg(t)\right)^{\frac{1}{\gamma+1}} + 2^{2\gamma-2}\Lambda\right)$$
$$\times \int_a^b r(t)|y'_g(t)|^{\gamma+1}dg(t).$$

As in the proof of Theorem 6.16, we get

$$\begin{aligned}
&\int_a^b |p(t)|(y_g'(t))^\gamma |y(t+)|dg(t) \\
&\leq \left(\left(\frac{\gamma}{\gamma+1}\right)^{\frac{\gamma}{\gamma+1}} \left(\int_a^b \frac{|p(t)|^{\gamma+1}}{(r(t))^\gamma}(R(h))^\gamma dg(t)\right)^{\frac{1}{\gamma+1}}\right) \\
&\quad \times \int_a^b r(t)(y_g'(t))^{\gamma+1}dg(t) + \frac{1}{c}\int_a^b r(t)(y_g'(t))^{\gamma+1}dg(t).
\end{aligned}$$

Consequently,

$$\begin{aligned}
&\int_a^b r(t)|y_g'(t)|^{\gamma+1}dg(t) \\
&\leq \left(2^{3\gamma-2}\left(\frac{1}{\gamma+1}\right)^{\frac{1}{\gamma+1}} \left(\int_a^b \frac{|Q(t)|^{\gamma+1}}{(r(t))^\gamma}(R(h))^\gamma dg(t)\right)^{\frac{1}{\gamma+1}} + 2^{2\gamma-2}\Lambda\right) \\
&\quad \times \int_a^b r(t)|y_g'(t)|^{\gamma+1}dg(t) \\
&\quad + \left(\left(\frac{\gamma}{\gamma+1}\right)^{\frac{\gamma}{\gamma+1}} \left(\int_a^b \frac{|p(t)|^{\gamma+1}}{(r(t))^\gamma}(R(h))^\gamma dg(t)\right)^{\frac{1}{\gamma+1}}\right) \\
&\quad \times \int_a^b r(t)(y_g'(t))^{\gamma+1}dg(t) + \frac{1}{c}\int_a^b r(t)(y_g'(t))^{\gamma+1}dg(t),
\end{aligned}$$

whereupon

$$\begin{aligned}
&\left(1-\frac{1}{c}\right)\int_a^b r(t)|y_g'(t)|^{\gamma+1}dg(t) \\
&\leq \left(\left(2^{3\gamma-2}\left(\frac{1}{\gamma+1}\right)^{\frac{1}{\gamma+1}} \left(\int_a^b \frac{|Q(t)|^{\gamma+1}}{(r(t))^\gamma}(R(h))^\gamma dg(t)\right)^{\frac{1}{\gamma+1}} + 2^{2\gamma-2}\Lambda\right)\right. \\
&\quad \left. + \left(\left(\frac{\gamma}{\gamma+1}\right)^{\frac{\gamma}{\gamma+1}} \left(\int_a^b \frac{|p(t)|^{\gamma+1}}{(r(t))^\gamma}(R(h))^\gamma dg(t)\right)^{\frac{1}{\gamma+1}}\right)\right) \\
&\quad \times \int_a^b r(t)|y_g'(t)|^{\gamma+1}dg(t)
\end{aligned}$$

and

$$2^{3\gamma-2}\left(\frac{1}{\gamma+1}\right)^{\frac{1}{\gamma+1}}\left(\int_a^b \frac{|Q(t)|^{\gamma+1}}{(r(t))^{\gamma}}(R(h))^{\gamma}dg(t)\right)^{\frac{1}{\gamma+1}} + 2^{2\gamma-2}\Lambda$$

$$+\left(\frac{\gamma}{\gamma+1}\right)^{\gamma+1}\left(\int_a^b \frac{|p(t)|^{\gamma+1}}{(r(t))^{\gamma}}(R(h))^{\gamma}dg(t)\right)^{\frac{1}{\gamma+1}} \geq 1-\frac{1}{c}.$$

This completes the proof. □

Problem 6.10. Let y be a nontrivial solution of the Stieltjes differential equation

$$\left((y_g'(t))^{\gamma}\right)_g' + p(t)(y_g'(t))^{\gamma} + q(t)(y(t+))^{\gamma} = 0, \quad t \in [a,b],$$

for which $y(a) = y(b) = 0$. Then, prove that

$$\left(\frac{g(b)-g(a)}{2}\right)^{\frac{\gamma}{\gamma+1}} 2^{3\gamma-2}\left(\frac{1}{\gamma+1}\right)^{\frac{1}{\gamma+1}}\left(\int_a^b |Q(t)|^{\gamma+1}dg(t)\right)^{\frac{1}{\gamma+1}}$$

$$+\, 2^{2\gamma-2}\max_{t\in[a,b]}\left((\Delta g(t))^{\gamma}|Q(t)|\right)$$

$$+\left(\frac{g(b)-g(a)}{2}\right)^{\frac{\gamma}{\gamma+1}}\left(\frac{\gamma}{\gamma+1}\right)^{\gamma+1}\left(\int_a^b |p(t)|^{\gamma+1}dg(t)\right)^{\frac{1}{\gamma+1}} \geq 1-\frac{1}{c}.$$

6.4 Advanced Practical Problems

Problem 6.11. Let $I = (-\sqrt{2}, \sqrt{2})$, $g(t) = \frac{1}{2}t^3$, $t \in I$, and $r \in \mathscr{C}_g^1((-\sqrt{2}, \sqrt{2}))$ be a positive function such that

$$\int_{-\sqrt{\frac{3}{2}}}^{\sqrt{\frac{3}{2}}} \frac{1}{r(t)}dg(t) < \infty.$$

Let y be a solution to the following SBVP:

$$(ry_g')_g'(t) + \left(\frac{1}{4}t^6\frac{1}{2}t^4 - \frac{1}{2}t^3 + t^2 - t + 1\right)y\left(\frac{1}{2}t^3\right) = 0, \quad t \in \left[-\sqrt{\frac{3}{2}}, \sqrt{\frac{3}{2}}\right],$$

$$y\left(-\sqrt{\frac{3}{2}}\right) = y_g'\left(\sqrt{\frac{3}{2}}\right) = 0.$$

Then, prove that

$$\sqrt{2}\left(\int_{-\sqrt{\frac{3}{2}}}^{\sqrt{\frac{3}{2}}} \frac{(t^3-t^2+t)^2}{r(t)}\left(\int_{-\sqrt{\frac{3}{2}}}^{t} \frac{1}{r(s)}dg(s)\right)dg(t)\right)^{\frac{1}{2}} + \sup_{t\in\left[-\sqrt{\frac{3}{2}},\sqrt{\frac{3}{2}}\right]}\left(\left(\frac{1}{2}t^3-t\right)\frac{|t^3-t^2+t|}{r(t)}\right) \geq 1.$$

Problem 6.12. Let $I=\mathbb{R}$, $g(t)=\frac{1}{2}t+2$, $t\in I$, and $r\in\mathscr{C}_g^1((-1,1))$ be a positive function such that

$$\int_{-1}^{1}\frac{1}{r(t)}dg(t)<\infty.$$

Let y be a solution to the following SBVP:

$$(ry_g')_g'(t)+\left(\frac{3}{2}t+1\right)y\left(\frac{1}{2}t+2\right)=0, \quad t\in\left[-\sqrt{\frac{3}{2}},\sqrt{\frac{3}{2}}\right],$$

$$y(-1)=y_g'(1)=0.$$

Then, prove that

$$\sqrt{2}\left(\int_{-1}^{1}\frac{(t^2-t+1)^2}{r(t)}\left(\int_{-1}^{t}\frac{1}{r(s)}dg(s)\right)dg(t)\right)^{\frac{1}{2}} + \sup_{t\in[-1,1]}\left(\left(2-\frac{1}{2}t\right)\frac{t^2-t+1}{r(t)}\right)\geq 1.$$

Problem 6.13. Let y be a nontrivial solution of the Stieltjes differential equation

$$\left((y_g'(t))^\gamma\right)_g'+p(t)(y_g'(t))^\gamma+t(y(t+))^\gamma=0, \quad t\in[a,b],$$

for which $y(a)=y(b)=0$. Then, prove that

$$(g(b)-g(a))\left(\frac{1}{2}\right)^{\frac{\gamma}{\gamma+1}}2^{3\gamma-2}\left(\frac{1}{\gamma+1}\right)^{\frac{1}{\gamma+1}}+2^{2\gamma-2}\max_{t\in[a,b]}\left((\Delta g(t))^\gamma\right) + \left(\frac{g(b)-g(a)}{2}\right)^{\frac{\gamma}{\gamma+1}}\left(\frac{\gamma}{\gamma+1}\right)^{\gamma+1}\left(\int_a^b|p(t)|^{\gamma+1}dg(t)\right)^{\frac{1}{\gamma+1}}\geq 1-\frac{1}{c}.$$

Bibliography

Albés, Ignacio Marquéz. Notes on the linear equation with Stieltjes derivatives. *Electronic Journal of Qualitative Theory of Differential Equations* 2021.42 (2021): 1–18.

Albés, Ignacio Marquéz, and Monteiro, Giselle A. Notes on the existence and uniqueness of solutions of Stieltjes differential equations. *Mathematische Nachrichten* 294.4 (2021): 794–814.

Albés, Ignacio Márquez, and Slavík, Antonín. The logistic equation in the context of Stieltjes differential and integral equations. *Electronic Journal of Qualitative Theory of Differential Equations* 10 (2023): 1–35. https://doi.org/10.14232/ejqtde.2023.1.10.

Albés, Ignacio Márquez, and Slavík, Antonín. Generalized chain rules and applications to Stieltjes differential and integral equations. *Results in Mathematics* 79, 78 (2024). https://doi.org/10.1007/s00025-023-02104-0.

Albés, Ignacio Márquez, Slavík, Antonín, and Tvrdý, Milan. Duality for Stieltjes differential and integral equations. *Journal of Mathematical Analysis and Applications* 519.1 (2023): 126789.

Albés, Ignacio Márquez, and Tojo, F., Adrián, F. Existence and uniqueness of solution for Stieltjes differential equations with several derivators. *Mediterranean Journal of Mathematics* 18 (2021): 1–31.

Bohner, Martin, and Peterson, Allan. *Dynamic Equations on Time Scales: An Introduction with Applications*, Springer Science & Business Media, 2001.

Chen, Yu, ORegan, Donal, and Wang, JinRong. Existence and stability of solutions for linear and nonlinear stieltjes differential equations. *Quaestiones Mathematicae* 43.11 (2020): 1613–1638.

Dhage, B. C. Fixed point theorems in ordered Banach algebras and applications, *PanAmerican Mathematical Journal* 9.4 (1999): 83–102.

Dhage, B. C. A fixed point theorem in Banach algebras with applications to functional integral equations. *Kyungpook Mathematical Journal* 44.1 (2004): 145–155.

Fernández, Francisco J., Albés, Ignacio Márquez, and Tojo, F. Adrián F. Consequences of the product rule in Stieltjes differentiability. arXiv preprint, arXiv:2205.10090 (2022a).

Fernández, Francisco J., Albés, Ignacio Márquez, and Tojo, F. Adrián F. The Wronskian and the variation of parameters method in the theory of linear Stieltjes differential equations of second order. arXiv preprint, arXiv:2206.10855 (2022b).

Fernández, Francisco J., Albés, Ignacio Márquez, and Tojo, F. Adrián F. On first and second order linear Stieltjes differential equations. *Journal of Mathematical Analysis and Applications* 511.1 (2022c): 126010.

Fernández, Francisco J., Albés, Ignacio Marquéz, and Tojo, Fernández. On second-order linear Stieltjes differential equations with non-constant coefficients. *Open Mathematics*, 22.1 (2024): 20240018. https://doi.org/10.1515/math-2024-0018.

Frigon, Marléne, and López Pouso, Rodrigo. Theory and applications of first-order systems of Stieltjes differential equations. *Advances in Nonlinear Analysis* 6.1 (2017): 13–36.

Frigon, Marléne, and Tojo, F. Adrián, F. Stieltjes differential systems with nonmonotonic derivators. *Boundary Value Problems* 2020.1 (2020): 1–24.

Gallegos, Claudio A., Albés, Ignacio Marquéz, and Slavík, Antonín. A general form of Gronwall inequality with Stieltjes integrals. *Journal of Mathematical Analysis and Applications* 541.1 (2025): 128674.

Georgiev, Svetlin G., and Tikare, Sanket. *Stieltjes Differential Calculus with Applications*, Vol. 12, *Trends in Abstract and Applied Analysis*, World Scientific Publishing, 2024.

Georgiev, Svetlin G., and Tikare, Sanket. *Generalized Quantum Calculus with Applications*, Elsevier, 2025.

Kim, Yung-Jin. Stieltjes derivatives and its applications to integral inequalities of Stieltjes type. *The Pure and Applied Mathematics* 18.1 (2011): 63–78.

Kloeden, P. E., and Panadiwal, J. Taylor expansions for continuous Stieltjes differential equations. *Bulletin of the Australian Mathematical Society* 48.2 (1993): 325–336.

López Pouso, Rodrigo and Albés, Ignacio Marquéz. Systems of Stieltjes differential equations with several derivators. *Mediterranean Journal of Mathematics* 16 (2019a): 1–17.

López Pouso, Rodrigo and Albés, Ignacio Marquéz. Resolution methods for mathematical models based on differential equations with Stieltjes derivatives. *Electronic Journal of Qualitative Theory of Differential Equations* 2019.72 (2019b): 1–15.

López Pouso, Rodrigo and Albés, Ignacio Marquéz. Existence of extremal solutions for discontinuous Stieltjes differential equations. *Journal of Inequalities and Applications* 2020.1 (2020): 1–21.

López Pouso, Rodrigo, Albés, Ignacio Marquéz, and Monteiro, Giselle. Extremal solutions of systems of measure differential equations and applications in the study of Stieltjes differential problems. *Electronic Journal of Qualitative Theory of Differential Equations* 2018.38 (2018): 1–24.

López Pouso, Rodrigo, Albés, Ignacio Marquéz, and Rodríguez-López, Jorge. Solvability of non-semicontinuous systems of Stieltjes differential inclusions and equations. *Advances in Difference Equations* 2020.1 (2020): 1–14.

Maia, Lamiae, Khattabi, Noha El, and Frigon, Marléne. Prolongation of solutions and Lyapunov stability for Stieltjes dynamical systems. arXiv preprint, arXiv:2409.03408 (2024).

Marraffa, Valeria, and Satco, Bianca. Stieltjes differential inclusions with periodic boundary conditions without upper semicontinuity. *Mathematics* 10.1 (2021): 55.

Marraffa, Valeria, and Satco, Bianca. Relaxation theorem for Stieltjes differential inclusions on infinite intervals. *Mediterranean Journal of Mathematics* 20. 331 (2023a). https://doi.org/10.1007/s00009-023-02515-x.

Marraffa, Valeria, and Satco, Bianca. Relaxation result for differential inclusions with Stieltjes derivative. *Journal of Mathematical Analysis and Applications*, 528.2 (2023b) 127533.

Marraffa, Valeria, and Satco, Bianca. Stieltjes differential problems with general boundary value conditions. Existence and bounds of solutions. *Czechoslovak Mathematical Journal* (2024). https://doi.org/10.21136/CMJ.2024.0125-23.

Marraffa, Valeria, and Satco, Bianca. The method of lower and upper solutions for second order periodic Stieltjes differential equations. *Journal of Fixed Point Theory and Applications* 27.9 (2025). https://doi.org/10.1007/s11784-025-01162-x.

Monteiro, Giselle, and Satco, Bianca. Distributional, differential and integral problems: equivalence and existence results. *Electronic Journal of Qualitative Theory of Differential Equations* 2017.7 (2017): 1–26.

Monteiro, Giselle Antunes, and Satco, Bianca. Extremal solutions for measure differential inclusions via Stieltjes derivatives. *Advances in Difference Equations* 2019.1 (2019): 1–18.

Monteiro, Giselle Antunes, Slavíkm Antonín, and Tvrdý, Milan. *Kurzweil–Stieltjes Integral: Theory and Applications*, 2019.

Pouso, Rodrigo López, and Albés, Ignacio Márquez. General existence principles for Stieltjes differential equations with applications to mathematical biology. *Journal of Differential Equations* 264.8 (2018): 5388–5407.

Pouso, Rodrigo López, and Rodríguez, Adrián. A new unification of continuous, discrete and impulsive calculus through Stieltjes derivatives. *Real Analysis Exchange* 40.2 (2015): 319–354.

Satco, Bianca. A Filippov-type lemma for Stieltjes differential inclusions with supremum. *Mathematics* 12.22 (2024): 3605. https://doi.org/doi:10.3390/math12223605.

Satco, Bianca, and Smyrlis, George. Applications of Stieltjes derivatives to periodic boundary value inclusions. *Mathematics* 8.12 (2020a): 2142.

Satco, Bianca, and Smyrlis, George. Periodic boundary value problems involving Stieltjes derivatives. *Journal of Fixed Point Theory and Applications* 22.4 (2020b): 94.

Satco, Bianca, and Smyrlis, George. Viability and Filippov-type lemma for Stieltjes differential inclusions. *Set-Valued and Variational Analysis* 31.31 (2023). https://doi.org/10.1007/s11228-023-00690-3.

Satco, Bianca, and Smyrlis, George. State-dependent sweeping processes with stieltjes derivative. *Applied Mathematics and Optimization* 90.24 (2024). https://doi.org/10.1007/s00245-024-10169-8.

Schwabik, Stefan. *Generalized Ordinary Differential Equations*, Vol. 5, World Scientific Publishing, 1992.

Slavík, Antonín. Dynamic equations on time scales and generalized ordinary differential equations. *Journal of Mathematical Analysis and Applications* 385.1 (2012): 534–550.

Index

www.ingramcontent.com/pod-product-compliance
Ingram Content Group UK Ltd.
Pitfield, Milton Keynes, MK11 3LW, UK
UKHW061932100926
8556IPUK00007B/885

9 789819 82329